Printreading
for Installing and Troubleshooting Electrical Systems

AMERICAN TECHNICAL PUBLISHERS, INC.
HOMEWOOD, ILLINOIS 60430-4600

Glen A. Mazur
William J. Weindorf

Printreading for Installing and Troubleshooting Electrical Systems contains procedures commonly practiced in industry and the trade. Specific procedures vary with each task and must be performed by a qualified person. For maximum safety, always refer to specific manufacturer recommendations, insurance regulations, specific job site and plant procedures, applicable federal, state, and local regulations, and any authority having jurisdiction. The material contained is intended to be an educational resource for the user. American Technical Publishers, Inc. assumes no responsibility or liability in connection with this material or its use by any individual or organization.

American Technical Publishers, Inc., Editorial Staff

Editor in Chief:
    Jonathan F. Gosse
Vice President—Production:
    Peter A. Zurlis
Art Manager:
    James M. Clarke
Technical Editor:
    Russell G. Burris
Copy Editor:
    Catherine A. Mini
Cover Design:
    Nicole S. Polak
Illustration/Layout:
    Thomas E. Zabinski
    Eric T. Comiza
    Nicole S. Polak
Multimedia Coordinator:
    Carl R. Hansen
CD-ROM Development:
    Robert E. Stickley
    Gretje Dahl

Adobe, Acrobat, and Reader are either registered trademarks or trademarks of Adobe Systems Incorporated in the United States and/or other countries. Quick Quiz and Quick Quizzes are registered trademarks of American Technical Publishers, Inc. Intel and Pentium are registered trademarks of Intel Corporation or its subsidiaries in the United States and other countries. Microsoft, Windows, Windows NT, Windows XP, and Internet Explorer are either registered trademarks or trademarks of Microsoft Corporation in the United States and/or other countries. Netscape is a registered trademark of Netscape Communications Corporation in the United States and other countries. MasterFormat is a trademark of the Construction Specifications Institute.

© 2009 by American Technical Publishers, Inc.
All rights reserved

1 2 3 4 5 6 7 8 9 – 09 – 9 8 7 6 5 4 3 2

Printed in the United States of America

    ISBN 978-0-8269-2050-8

 This book is printed on 30% recycled paper.

Acknowledgments

The authors and publisher are grateful to the following companies and organizations for providing photographs, prints, and technical assistance.

ASI Robicon
Atlas Technologies, Inc.
Baldor Electric Company
Benson Systems, Inc.
Carrier Corporation
Continental Hydraulics
Cooper Wiring Devices
CSI MasterFormat™
DoALL Company
Eaton Corporation
Flexicon Corporation
FLIR Systems
Fluke Corporation
Fluke Networks
The Gates Rubber Company
General Electric Company
Graco-Trabon
Grand Pointe Homes
Home Arcades
Honeywell
Integrus Architecture
International Rectifier

The Numina Group
Pass & Seymour
PLC Multipoint
Prescolite
Reed Manufacturing Co.
Rockwell Automation, Allen-Bradley Company, Inc.
Rockwell Automation/Reliance Electric
Saftronics, Inc.
Siemens
Snell Infrared
Sonny's Enterprises, Inc.
Southern Forest Products Association
Spokane Intercollegiate Research and Technology Institute
Square D Company
Staedtler, Inc.
Steel Tube Institute
Unico, Inc.
United States Gypsum Company
U.S. Department of Agriculture – Forest Service
Victaulic Company of America
Wendy's International, Inc.
Zircon Corporation

Contents

1. Printreading Fundamentals — 1

Prints · Lines · Abbreviations · Schedules · Print Divisions · Title Blocks · Revision Information · Print Conventions · Notes · Section View and Detail Drawing Symbols · Building Column Numbers and Letters · Print Scales · Architect's Scale · Specifications · CSI MasterFormat™

Examples	27
Review Questions	33
Activities	35
Trade Competency Test	39

2. Residential and Commercial Electrical Symbols — 41

Standards Organizations · Trade Associations · Technical Societies · United States Government Departments · National and International Standards Organizations · Private Organizations · Residential and Commercial Electrical Prints · Symbols and Components · Lighting Symbols · Switch Symbols · Receptacle Symbols · Power Symbols · Signal Symbols · Plot Plan Symbols · Private Property Symbols · Public Property Symbols · Aboveground and Underground Distribution and Lighting Symbols

Examples	63
Review Questions	69
Activities	71
Trade Competency Test	75

3. Industrial Electrical and Electronic Symbols — 77

Industrial Equipment · Power Sources · Direct Current and Alternating Current · Disconnects and Overcurrent Protection Devices · Contacts · Control Switches · Relays and Timers · Contactors and Motor Starters · Resistors · Capacitors · Diodes · Thyristors · Transistors · Digital Logic Gates · Coils · Solenoids · Transformers · Motors · Lights, Alarms, and Meters · General Wiring

Examples	103
Review Questions	107
Activities	109
Trade Competency Test	113

4. Electrical Drawings and Plans — 115

Prints · Drawings · Pictorial Drawings · Orthographic Drawings · Application Drawings · Location Drawings · Detail Drawings · Assembly Drawings · Instructional Drawings · Elevation Drawings · Sectional Drawings · Plans · Plot (Site) Plans · Floor Plans · Foundation Plans · Structural Plans · Utility Plans

Examples	137
Review Questions	143
Activities	145
Trade Competency Test	149

Electrical and Electronic Systems — 151

One-Line Diagrams · Ladder (Line) Diagrams · PLC Programming Diagrams · Wiring Diagrams · Schematic Diagrams · Interconnecting Diagrams · Operational Diagrams · Block Diagrams · Function-Block Diagrams

Examples	165
Review Questions	173
Activities	175
Trade Competency Test	179

Facility Construction and Maintenance Systems — 181

Responsibilities of Construction Personnel · Architects · Engineers · Contractors · Tradesworkers · Building Inspectors · Overview of Construction Process · Site Preparation · Building Core · Electrical Construction · Mechanical Construction · Construction Documentation · Responsibilities of Maintenance Personnel · Overview of Maintenance Process · Preventive Maintenance · Predictive Maintenance · Maintenance Documentation · Rules and Regulations

Examples	199
Review Questions	205
Activities	207
Trade Competency Test	211

Residential and Commercial Power and Lighting Systems — 213

Site Plans · Power Prints · Power Floor Plans · Single-Line Diagrams · Lighting Prints · Lighting Floor Plans · Light Fixture Schedules · Electrical Details · Electrical Elevation Drawings · Mounting and Installation Details · Schedules · Diagrams

Examples	239
Review Questions	247
Activities	249
Trade Competency Test	253

Residential and Commercial VDV Systems — 255

CSI MasterFormat™—Division 27 · VDV System Prints · VDV Symbols, Abbreviations and Legends · VDV Riser Diagrams · VDV Floor Plans · VDV Detail Drawings

Review Questions	271
Activities	273
Trade Competency Test	277

Fire Alarm, Life Safety, and Security Systems — 279

CSI MasterFormat™—Division 28 · Fire Alarm and Life Safety Systems · Fire Alarm and Life Safety Abbreviations and Symbols · Fire Alarm and Life Safety Riser Diagrams · Fire Alarm and Life Safety Floor Plans · Fire Alarm and Life Safety Detail Drawings · Security Systems · Security System Abbreviations and Symbols · Security System Floor Plans · Security System Detail Drawings

Review Questions	295
Activities	297
Trade Competency Test	301

10 HVAC Systems · 303

CSI MasterFormat™—Division 25 · HVAC Control Systems · HVAC Control System Abbreviations and Symbols · HVAC System Prints · HVAC Wiring Diagrams · HVAC Detail Drawings · Sequence of Operation

Review Questions — 317
Activities — 319
Trade Competency Test — 323

11 Industrial Control Systems · 325

Power and Control Circuits · Water Tower Application · Basic Rules of Ladder (Line) Diagrams · One Load Per Line · Load (Component) Connections · Control Device Connections · Control Circuit Numbering Systems · Line Reference Numbers · Numerical Cross-References · Wire Reference Numbers · Manufacturer Terminal Numbers · Cross-Referencing Mechanically Connected Contacts · Control Circuit Logic Functions · AND Circuit Logic · OR Circuit Logic · NOT Circuit Logic · NOR Circuit Logic · NAND Circuit Logic · Combination Circuit Logic

Review Questions — 345
Activities — 347
Trade Competency Test — 351

12 Industrial Power Systems · 353

Power Distribution · Types of Power Distribution · 120/240 V, 1ϕ, 3-Wire Service · Receptacle and Plug Configurations · Grounding · NEC® Phase Arrangement and High-Phase Markings · 120/208 V, 3ϕ, 4-Wire Service · 277/480 V, 3ϕ, 4-Wire Service · 120/240 V, 3ϕ, 4-Wire Service · Power Distribution System Conductor Color-Coding · Busways · Industrial Power Circuit Application

Review Questions — 377
Activities — 379
Trade Competency Test — 383

13 Industrial Equipment · 385

Circuit Wiring · Wiring Methods · Component Layout and Location · Direct Hardwiring · Wiring Variations · Hardwired Reversing Circuit · Hardwiring Using Terminal Strips · Dual Compressor Application · Pump Application · Electric Motor Drive Control Variations · PLC Wiring

Review Questions — 407
Activities — 409
Trade Competency Test — 413

14 Fluid Power Systems · 415

Fluid Power Symbols · Fluid Power · Hydraulic Systems · Pneumatic Systems · Pumps · Fluid Conditioners · Actuators · Directional Control Valves · Normally Closed and Normally Open Directional Valves · Flow Control Valves and Check Valves · Pressure Control Valves · Miscellaneous Fluid Power Devices and Components

Review Questions — 445
Activities — 447
Trade Competency Test — 451

15 Process and Instrumentation Systems   453

Piping and Instrumentation Diagrams · Lines · P&ID Print Process Lines · Instrument Lines · Instruments · Instrument Type and Location Symbols · Instrument and Control Element Identification · Control Element Symbols · Manual Valves · Control Valves · Process Equipment Symbols · Process Vessels and Tanks

Review Questions _____ 467
Activities _____ 469
Trade Competency Test _____ 473

A Appendix   475

G Glossary   519

I Index   527

CD-ROM Contents

- Using this Interactive CD-ROM
- Quick Quizzes®
- Illustrated Glossary
- Resource Library
- Prints
- CSI MasterFormat™
- Virtual Motor Control Enclosure
- Flash Cards
- ATPeResources.com

Introduction

Printreading for Installing and Troubleshooting Electrical Systems presents foundational printreading skills needed to install and troubleshoot electrical systems. Each chapter includes text and multiple print examples on residential, commercial, and/or industrial topics.

With an emphasis on print symbols and print elements, 180 electrical prints along with supplemental information are used to cover electrical printreading throughout the text. Each chapter also includes photographs, electrical and printreading tech tips, and graphic icons that designate files or prints located on the textbook CD-ROM.

The chapters offer a comprehensive learning experience and include an introduction, numerous illustrations, two-scenario based Examples that include graphic answers, Review Questions that include true-false, multiple choice, and completion style questions, two scenario-based Activities complete with prints, and Trade Competency Tests where three different prints are used to answer questions.

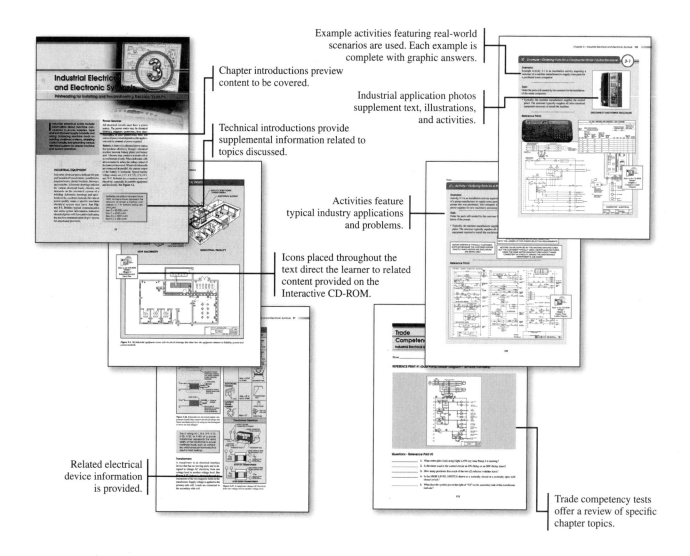

Chapter introductions preview content to be covered.

Example activities featuring real-world scenarios are used. Each example is complete with graphic answers.

Technical introductions provide supplemental information related to topics discussed.

Industrial application photos supplement text, illustrations, and activities.

Activities feature typical industry applications and problems.

Icons placed throughout the text direct the learner to related content provided on the Interactive CD-ROM.

Related electrical device information is provided.

Trade competency tests offer a review of specific chapter topics.

Interactive CD-ROM Contents

The included Interactive CD-ROM enhances content in the textbook with the following features:

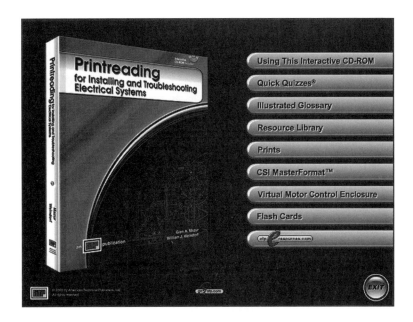

- Quick Quizzes® that reinforce fundamental concepts, with 10 question per chapter
- An Illustrated Glossary that provides a helpful reference to key terms
- A Resource Library that includes a complete set of prints for a commercial building, car wash prints, and assorted other prints
- In the print section over 100 prints from the book that are navigable, with some in color
- The CSI MasterFormat™, which includes multiple MasterFormat listings
- A Virtual Motor Control Enclosure with two activities that test a learner's understanding of prints
- Flash Cards that provide a review of terms and definitions, electrical symbols, and non-electrical symbols
- ATPeResources.com, which provides a comprehensive array of instructional resources

To obtain information about related training products, visit the American Tech web site at www.go2atp.com

The Publisher

Printreading Fundamentals

Printreading for Installing and Troubleshooting Electrical Systems

Every day, various types of prints are used to perform work. Prints are used to construct buildings, modify process control circuits, and troubleshoot systems in commercial and industrial facilities. The type of print used is dependent on the type of work being performed.

Conventions and standards have been developed to ensure consistency among prints and to enhance the communication between the people involved in a project. These conventions and standards create a framework for how information is displayed on a print, how prints are organized, and how information is interpreted. In addition to prints, specifications are used to define the materials and procedures required for a project.

PRINTS

Prints are reproductions of original drawings created by an architect or engineer. The original drawings are made using traditional drafting equipment or computer-aided design (CAD) software. **See Figure 1-1.** Traditional drafting involves the use of pencils, triangles, compasses, scales, and T-squares to produce drawings by hand. CAD involves the use of a computer (typically a PC), CAD software, and a printer or plotter to produce original computer-generated prints.

Today, almost all original drawings are produced by architects or engineers using CAD systems. CAD systems produce drawings that have several advantages over traditional drawings. The advantages of using CAD are as follows:

- accuracy and consistency of drawings
- ease of making changes to drawings
- ability to create drawings in many colors
- ability to store, send, and move drawings electronically
- ability to create drawings from other drawings through layering

Many types of drawings are used to create a set of prints. Sets of prints can include site plans (plot plans), floor plans, elevations, sectional views, wiring diagrams, details, and schedules. **See Figure 1-2.** Different types of prints are used to display the wide variety of information required to complete a project, such as a plot plan being used to determine what conduits will be installed in the ground, or a valve schedule being used to determine what type of actuator should be installed on a globe valve.

2 Printreading for Installing and Troubleshooting Electrical Systems

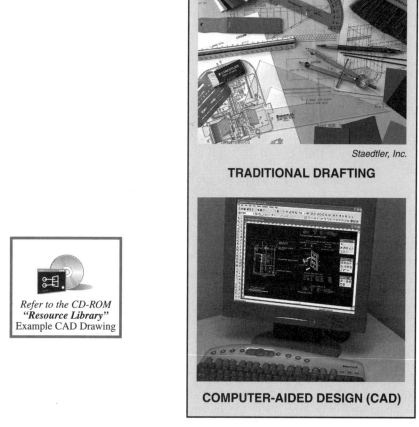

Figure 1-1. Original drawings are drawn with traditional drafting tools or by using CAD software.

Refer to the CD-ROM "Resource Library" Example CAD Drawing

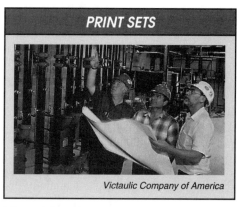

Figure 1-2. Architects and engineers use a variety of print types to provide the required information for a project.

In addition to different types of prints, prints come in a variety of sizes. Typically, a full-sized set of drawings (34″ × 44″) is used for construction purposes. Small sets of drawings (8½″ × 11″ or 11″ × 17″) are used for reference purposes, small parts, or small assemblies. A letter in the title block of a print identifies the size of the print as follows:

- A = 8½″ × 11″ sheet
- B = 11″ × 17″ sheet
- C = 17″ × 22″ sheet
- D = 22″ × 34″ sheet
- E = 34″ × 44″ sheet
- E+ = 34″ × 44″ roll

LINES

A *line* is a straight mark that begins at a starting point and stops at an endpoint. A variety of line types are used to depict objects and items on prints. The type of line used depends on the type of object or item to be shown, the location of the object or item, and the type of dimensioning system being used. The basic types of lines are object, hidden, dimension, extension, leader, center, cutting-plane, section, and break lines. **See Appendix.** Line types and when to use a specific type of line are defined by drafting standards.

An *object line* is a line that indicates the visible shape of an object. Object lines are solid and are found without breaks. Object lines are typically the most common lines shown on a print. **See Figure 1-3.**

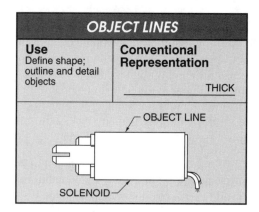

Figure 1-3. Object lines are solid lines that indicate the shape of an object.

Size E engineering sheets (34" × 44") fit neatly into an 8½" × 11" binder when folded four times.

A *hidden line* is a line that represents the shape of an object that cannot be seen. Hidden lines are drawn as dashed lines and are used wherever there is an edge that is not in view. **See Figure 1-4.**

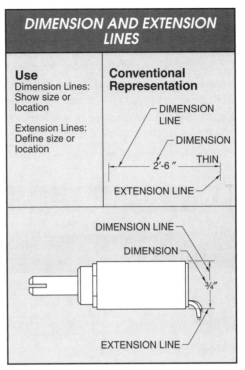

Figure 1-5. Dimension lines are lines with a break for written dimensions that indicate size or location. Extension lines are solid lines that project from an object to terminate dimension lines.

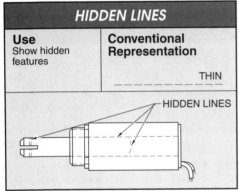

Figure 1-4. Hidden lines are dashed lines that show the hidden features of an object.

A *dimension line* is a line that is used with a written dimension to indicate size or location. Dimension lines are thinner than object lines. A dimension line typically has a gap for the placement of a dimension and has arrows at the ends. An *extension line* is a line that extends from the surface features of an object and is used to terminate dimension lines. Extension lines are projected from the surface and extend beyond the dimension lines. Extension lines do not touch object lines. **See Figure 1-5.**

A *leader line* is a line with a bent knee that connects a written description such as a dimension, note, or specification with a specific feature of a drawn object. A leader line has an arrow at the end that contacts the edge of the object. Leader lines are drawn at any angle required to make the connection between the written description and the object. **See Figure 1-6.**

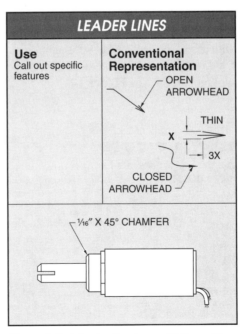

Figure 1-6. A leader line is a solid line with a bent knee that connects dimensions, notes, and specifications to a specific feature of an object.

A *centerline* is a line that locates the center of an object. Centerlines are thin dark lines broken into long and short dashes. Centerlines are used to locate the centers of windows, doors, electrical enclosures, and to indicate the symmetry of an object. **See Figure 1-7.** Dimension lines, centerlines, and extension lines are used together to create a dimensioning system.

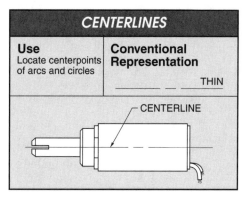

Figure 1-7. Centerlines are lines that are broken into long and short dashes to indicate the center of an object or to indicate that an object is round or cylindrical in shape.

A *cutting-plane line* is a line that indicates the path through which an object will be cut so that its internal features can be seen. A cutting-plane line is a thick line with arrows on each end at 90° that indicate the direction in which the resulting section will be viewed. **See Figure 1-8.**

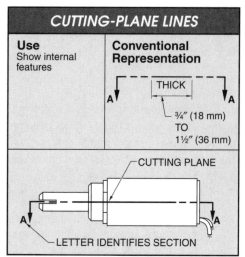

Figure 1-8. Cutting-plane lines are lines that indicate the path through which an object will be cut in order for its internal features to be seen.

A *section line* is a line that identifies the materials cut by a cutting-plane line in a section view. Section lines are typically drawn at an angle. This is different from object lines, which border the area. **See Figure 1-9.** Sections can also consist of hatch patterns and gradient fills.

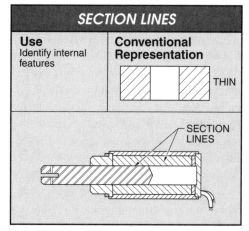

Figure 1-9. Section lines are lines with a variety of appearances that are used to identify the interior of an object.

The American Society of Mechanical Engineers has a standard (ASME Y14.3) covering sectional view drawings.

Baldor Electric Company

Cutting planes are used to cut away part of an object. In this application the 90° cutting plane only cuts the stator of the motor, not the rotor.

A *break line* is a line used to indicate internal features or to avoid showing continuous features of long or large objects. Break lines are drawn as a straight line with a zigzag in the middle or, if drawn freehand, as a jagged line. Break lines are used to eliminate a piece of an object where the whole length does not need to be shown. **See Figure 1-10.**

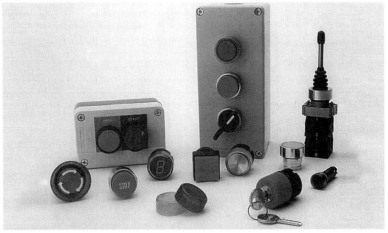

Square D Company

Abbreviations can vary dramatically such as for a stop switch, which can have any of the following abbreviations: SW, S, SS, EMGSW, STPSW, or SPST.

Abbreviations are used in a wide variety of prints, such as electrical and electronic drawings, building elevations, detail plans, and power floor plans. Some abbreviations are universal, such as electrical/electronic abbreviations, and have the same meaning no matter what type of print the abbreviation is found on. Other abbreviations have meanings that are specific to a certain type of print.

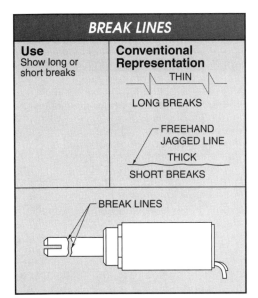

Figure 1-10. Break lines are either jagged lines or straight lines with a zigzag used to indicate that a piece or section of a long object has been removed.

ABBREVIATIONS

An *abbreviation* is a letter or group of letters that represents a term or phrase. Abbreviations allow information to be placed on a print without cluttering the print. Letters that are used for abbreviations are usually capitalized. Abbreviations that form a word are followed by a period to avoid confusion. For example, "ARM." is the abbreviation for armature when using electrical/electronic abbreviations, and "PAN." is the abbreviation for pantry when using floor plan abbreviations. **See Figure 1-11.**

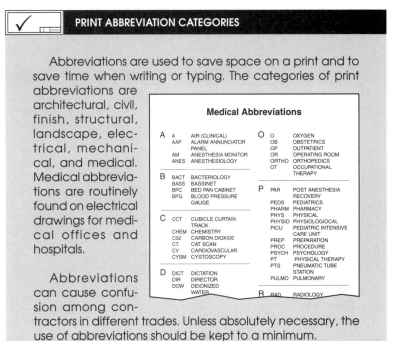

Abbreviations are used to save space on a print and to save time when writing or typing. The categories of print abbreviations are architectural, civil, finish, structural, landscape, electrical, mechanical, and medical. Medical abbreviations are routinely found on electrical drawings for medical offices and hospitals.

Abbreviations can cause confusion among contractors in different trades. Unless absolutely necessary, the use of abbreviations should be kept to a minimum.

ABBREVIATIONS

ELECTRICAL/ELECTRONIC ABBREVIATIONS

ABBR	TERM	ABBR	TERM	ABBR	TERM
A	ammeter; ampere; anode; armature	FU	fuse	PNP	positive-negative-positive
AC	alternating current	FWD	forward	POS	positive
AC/DC	alternating current; direct current	G	gate; giga; green; conductance	POT.	potentiometer
A/D	analog to digital	GEN	generator	P-P	peak-to-peak
AF	audio frequency	GRD	ground	PRI	primary switch
AFC	automatic frequency control	GY	gray	PS	pressure switch
Ag	silver	H	henry; high side of transformer; magnetic flux	PSI	pounds per square inch
ALM	alarm	HF	high frequency	PUT	pull-up torque
AM	ammeter; amplitude modulation	HP	horsepower	Q	transistor
AM/FM	amplitude modulation/frequency modulation	Hz	hertz	R	radius; red; resistance; reverse
ARM.	armature	I	current	RAM	random-access memory

PRINT ABBREVIATIONS

ELEVATION ABBREVIATIONS

ABBREVIATION	TERM	ABBREVIATION	TERM
AC	air conditioner	MET J	metal jalousie
ALUM	aluminum	MGS	metal gravel stop
ANT	antenna	OB	obscure
AWN	awning	OBSC GL or OGL	obscure glass
BC	bookcase	OPG or OPNG	opening
BD	board	OVHG	overhang
BEV	beveled	P	pitch
BK SH	book shelves	PK	peak
BRK	brick	PL GL	plate glass
BRS	brass	PLAS	plaster

SUPPLY ABBREVIATIONS

ABBREVIATION	TERM
AS	air supply
IA	instrument air
PA	plant air
ES	electrical supply
GS	gas supply
HS	hydraulic supply
NS	nitrogen supply
SS	steam supply
WS	water supply

FLOOR PLAN ABBREVIATIONS

ABBREVIATION	TERM	ABBREVIATION	TERM
ACS	access	LINO or LINOL	linoleum
ACSP or AP	access panel	LKT	lookout
ADD or ADH	adhesive	LNG	lining
AT.	asphalt tile	LR	living room
B	bathroom	MC	medicine cabinet
BC	between centers	OP	operator
BCL	broom closet	OVHD DR	overhead door
BLK	block	P	porch
BLKG	blocking	PAN.	pantry
BPL or BRG PL	bearing plate	PASS.	passage

SECTION AND DETAIL ABBREVIATIONS

ABBREVIATION	TERM	ABBREVIATION	TERM
ACT or AT	acoustical tile	FNSH	finish
ACST	acoustic	FBCK	firebrick
CAB.	cabinet	FP	fireplace
CSG	casing	FRWK	framework
CM	center-matched	FR	frame
CER	ceramic	HBD	hardboard
CHM	chimney	JB or JMB	jamb
CO	cleanout	JT	joint
COMB.	combination	LAM	laminate
CTR	counter	MIR	mirror

Figure 1-11. Abbreviations allow for greater amounts of information to be placed on a print.

In some cases, an abbreviation can have different meanings depending on which type of print the abbreviation is used on. For example, "BC" is the abbreviation for "bookcase" on an elevation drawing and "between centers" on a floor plan. Typically, the context of a drawing clarifies the meaning of the abbreviation. When two abbreviations appear for the same word, the first abbreviation and any information pertaining to it should be used. For example, "FIN." and "FNSH" are both abbreviations for finish. **See Appendix.**

The use of abbreviations is standard among architects and engineers, and a list of these abbreviations is typically provided with a set of prints. When a large set of prints includes abbreviations, the abbreviations appear on a legend. The legend sheet is typically found at the beginning of a print set. In addition to abbreviations, the legend sheet may contain symbols, equipment identification information, and drawing conventions used throughout the set of prints. **See Figure 1-12.**

> When abbreviations must be used on an electrical print because of space limitations, use uppercase lettering without periods, use the same abbreviation for singular or plural, and do not use spaces within an abbreviation.

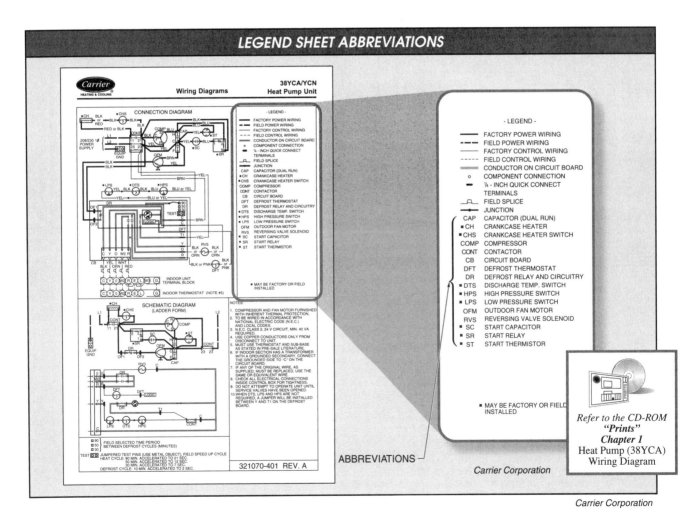

Figure 1-12. A legend sheet is found at the beginning of a set of prints and identifies the abbreviations used throughout the prints.

SCHEDULES

Small amounts of information can be displayed (noted) on a print without causing the print to become cluttered. However, when a large amount of detailed information is required, a schedule is used. A *schedule* is a chart used to conserve space and display information in a concise and organized format. For example, because it is not practical to display in-depth light fixture information on a lighting floor plan, a lighting or fixture schedule is used to identify the various lighting fixtures used on a project. **See Figure 1-13.**

Schedules resemble spreadsheets with information being displayed in columns and rows. A wide variety of schedules are found on architectural, electrical, and mechanical prints. Types of electrical schedules include fixtures, feeders, main switchboard, branch circuit panels, and transformers. Depending upon the amount of information and room, multiple schedules, such as lighting fixtures and switch schedules, are grouped together on a single electrical print. The number and type of schedules increases with the complexity of the project.

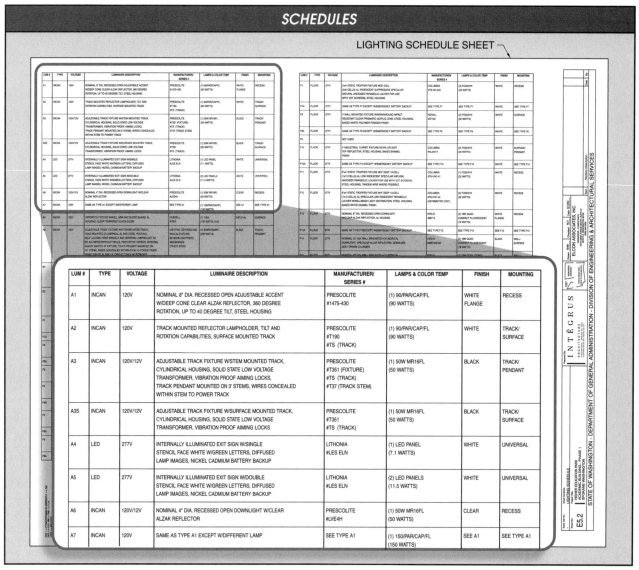

Integrus Architecture

Figure 1-13. *Schedules that are found on electrical, architectural, and mechanical prints are typically set in chart form to conserve space and to display information in an organized manner.*

PRINT DIVISIONS

Prints are separated into different divisions to allow for quick and easy access to information. Print divisions are denoted by a capital letter. For example, architectural prints are denoted by a capital "A". In a similar manner, mechanical prints are denoted by an "M", structural prints by an "S", civil prints by a "C", and electrical prints by an "E". The number that comes after the capital letter denotes the sheet number. The print division and number are always found in the title block of a print. **See Figure 1-14.**

Print numbering begins with the number 1 within each division. For example, architectural prints may run from page A1 to A65. Print divisions are subdivided when several pages apply to the same elements. For example, a set of electrical prints can begin with E1, but have pages subdivided as E1.1, E1.2, and E1.3.

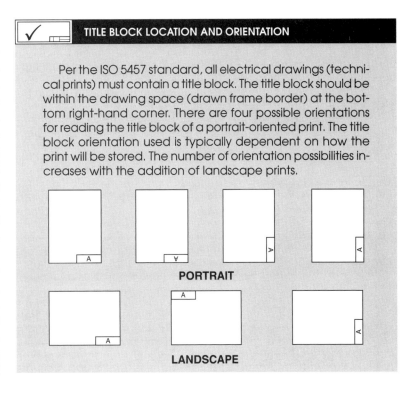

Per the ISO 5457 standard, all electrical drawings (technical prints) must contain a title block. The title block should be within the drawing space (drawn frame border) at the bottom right-hand corner. There are four possible orientations for reading the title block of a portrait-oriented print. The title block orientation used is typically dependent on how the print will be stored. The number of orientation possibilities increases with the addition of landscape prints.

TITLE BLOCKS

A *title block* is the area of a print that contains important information about the contents of the print. **See Figure 1-15.** The title block is located along the right side of a print or to the right and bottom. Information in the title block must be clearly understood before accessing information from a print. Title block information typically includes the following:

- subject or sheet contents
- project title and location
- architect's and/or engineer's names with office locations
- state stamp identifying the architect or engineer as a licensed professional
- print division and print number
- date the print was drawn and the initials of the person who drew the print
- date the print was checked and the initials of the person who checked the print
- drawing scale
- other job-related information, such as the project job number
- any revision information

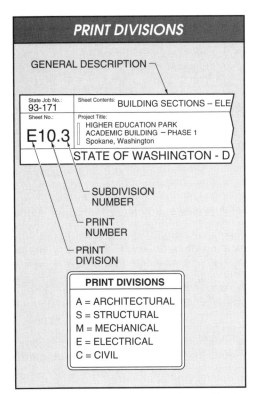

Figure 1-14. Print divisions are denoted in the title blocks by capital letters, such as "E" for electrical.

10 Printreading for Installing and Troubleshooting Electrical Systems

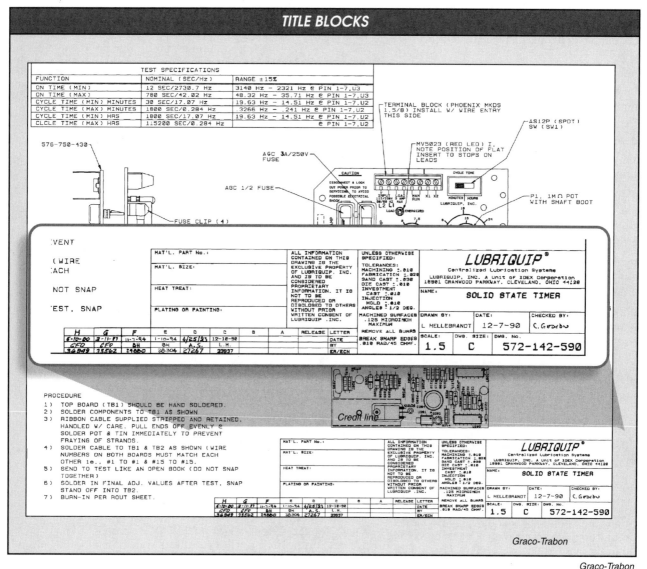

Figure 1-15. Title blocks contain important information and details that must be clearly understood before using information from a print.

Revision Information

One of the most important pieces of information in the title block is the revision information. Architects and engineers typically make several revisions to a set of prints during the course of a project. The revisions are labeled sequentially as 1, 2 and 3, or A, B, and C. A minor revision may be labeled with a decimal, such as revision 1.1. The revision number or letter, a brief description of the revision, the date, and the initials of the person making the revision are all indicated in the revision block. **See Figure 1-16.**

A revision is identified in a drawing on a print by a revision symbol. The symbol is a number or letter inside a triangle, circle, or square. In some cases, a cloud is drawn around the revision symbol to indicate that a change was made. As new prints are issued, they must be checked for any new revisions. Typically, revision information is found next to the title block or in the upper right-hand corner of a print. Construction delays, added costs, and penalties can result if the most recent set of prints is not used.

Chapter 1—Printreading Fundamentals **11**

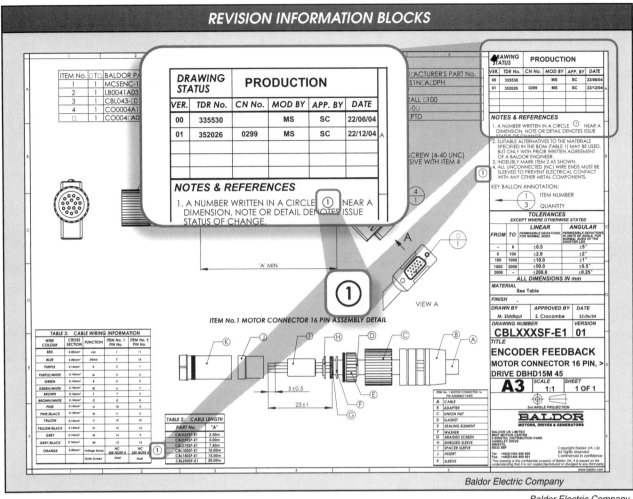

Figure 1-16. Print revisions require a revision number or letter, a brief description of the revision, the date of the revision, and the initials of the person making the revision to be indicated in the revision block of a print, with the revision itself being identified on the print by a symbol or a cloud.

PRINT CONVENTIONS

A *print convention* is an agreed-upon method of displaying information on prints. Most print conventions were developed from years of practice. Conventions save space and prevent clutter on prints, provide for consistency between prints, and enhance communication between all parties using the prints.

Print conventions typically govern the display of notes, detail symbols, section symbols, and column numbers and/or letters. Conventions often come from standards developed by a standards organization. Minor variations in print conventions may exist, depending on the personal preference of the architect or engineer.

The American Standards Institute (ANSI) is a national organization that helps identify industrial and public needs for national standards. These standards are commonly produced by technical societies, trade associations, and governmental agencies and are copublished with ANSI. While drawings for many prints are developed based on ANSI standards, other drawings may not always reflect all current standards.

Notes

A *note* is a sentence or two that provides drawing information that does not fit within the space of the drawing. The two types of notes are general (construction) notes and sheet notes. Notes are numbered sequentially by type.

A *general note* is a note that applies to the entire print that the note appears on. Although general notes are identified by a number, they are not represented by a corresponding note symbol on the print. General notes cover broad or general topics, such as coordination between trades, contact information, and company- or contractor-specific procedures. **See Figure 1-17.**

A *sheet note* is a note that applies to a specific item in the drawing that the note appears with. Sheet notes provide information about specific items, such as receptacle part numbers, motor overload numbers, and junction box information. Sheet notes are identified by a number and have a corresponding symbol on the print. The symbol (a sheet note number within a circle) identifies the item addressed by the sheet note. A leader line may be used to connect the symbol with the item, or the symbol may be located adjacent to the item. **See Figure 1-18.**

> On many occasions, electrical notes for items such as dishwashers and garbage disposals show up on architectural drawings but not on electrical drawings.

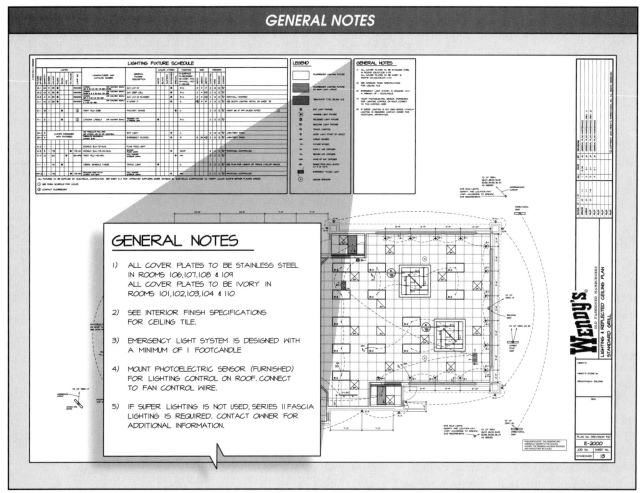

Figure 1-17. General notes cover broad topics and apply to the entire print on which they appear.

Wendy's International, Inc.

Chapter 1—Printreading Fundamentals 13

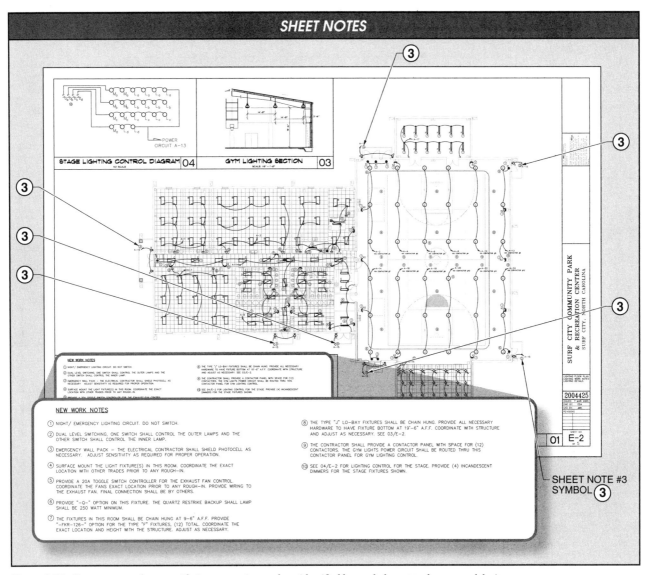

Figure 1-18. Sheet notes apply to specific items on prints and are identified by symbols next to the note and the item.

Section View and Detail Drawing Symbols

Section views and detail drawings are enlarged drawings of an item or feature on a print, such as underground conduit installations, conduit support methods, or contact design. Section views and detail drawings provide additional information about an object. Section views and detail drawings cannot always be included on the print from which they originate because of a lack of space.

Section symbols and detail symbols on the print identify the item or feature that the section view or detail drawing is taken from. Section views and detail drawings have detail header symbols that correspond to the section view and detail drawing symbols found in the drawing space. The detail header symbols used for section views and detail drawings are similar. **See Figure 1-19.** The section view or detail drawing header symbol may have the bottom of the circle divided into two parts. The left side denotes the sheet where the section view or detail drawing originated and the right side denotes where the view or drawing is located.

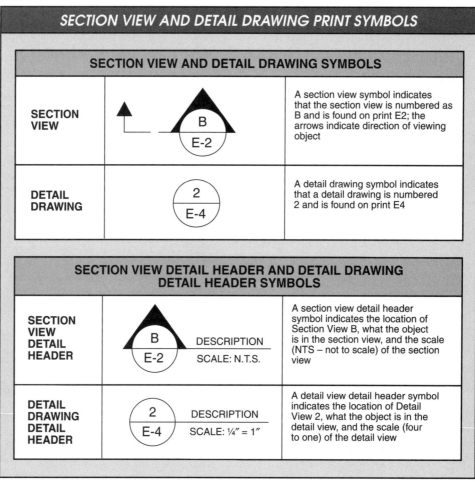

Figure 1-19. Section views and detail drawings have detail header symbols that correspond to the section view and detail drawing symbols found in the original drawing space.

The symbol for a section view consists of a cutting-plane line drawn through the selected item and a circle divided in half by a horizontal line. **See Figure 1-20.** An arrow on the cutting-plane line and/or arrow on the header circle indicates the direction in which the section is viewed. The upper half of the header circle contains the reference number or letter of the section view. The bottom half of the header circle contains the sheet the section view is found on.

> Section and detail header symbols are found as major section symbols, partial section symbols, detail symbols, and interior elevation symbols.

CAD drawings can be used to create realistic renderings of partially sectioned electrical equipment.

Chapter 1—Printreading Fundamentals **15**

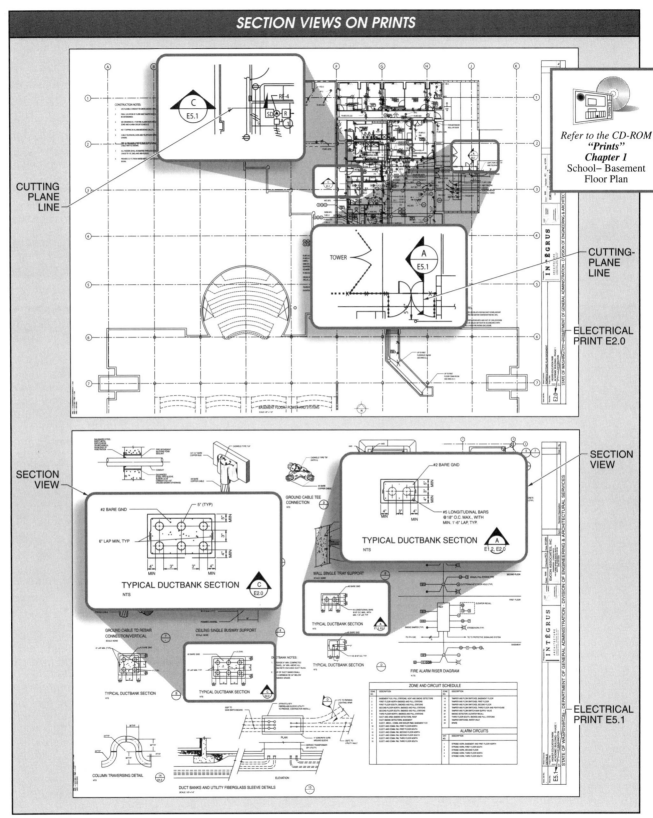

Figure 1-20. *Section views can be drawn on the same print as the original or on a different print, but the section view symbol and section view detail header must be used.*

Integrus Architecture

16 Printreading for Installing and Troubleshooting Electrical Systems

The symbol for a detail drawing consists of a circle divided in half by a horizontal line. **See Figure 1-21.** The upper half of the circle contains the reference number or letter of the detail drawing. The bottom half of the circle contains the sheet the detail drawing is found on. A leader line is drawn from the symbol to the item or feature that appears in the detail drawing.

The symbol for a detail header consists of a circle divided in half by a horizontal line that extends beyond the circle. **See Figure 1-22.** The upper half of the circle contains the reference number or letter for the section view or detail drawing. Typically, the bottom half of the circle contains the sheet the section view or detail drawing is found on. Alternately, the bottom half of the circle may contain the sheet from which the section view or detail drawing originated. A description of the section view or detail drawing appears above the horizontal line that extends beyond the circle. The scale of the section or detail appears below the same horizontal line. Some section views and detail drawings are not drawn to a scale and will have the NTS (not to scale) abbreviation.

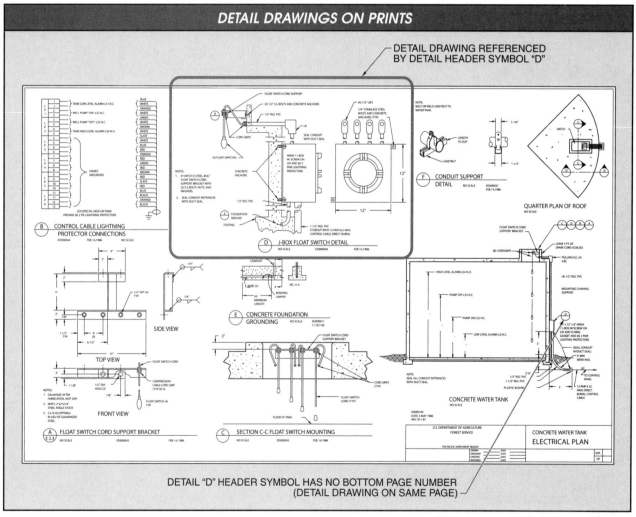

U.S. Department of Agriculture – Forest Service

Figure 1-21. *Detail drawings can be drawn on the same print as the original or on a different print, but the detail drawing symbol and detail drawing detail header must be used.*

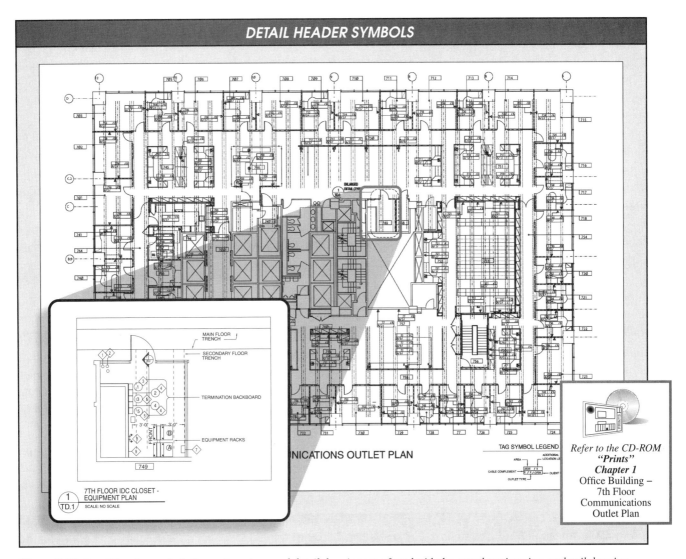

Figure 1-22. Detail header symbols for section views and detail drawings are found with the actual section view or detail drawing.

Building Column Numbers and Letters

Commercial and industrial buildings are constructed using columns of reinforced concrete and structural steel. On a print, horizontal and vertical lines are drawn through the centers of the columns to form a center-to-center grid system. Typically, horizontal lines are identified by letters and vertical lines by numbers, but horizontal lines can be identified by numbers and vertical lines by letters.

The grid system provides dimension information, identifies columns, and serves as a point of reference. The grid system is used throughout a set of prints, whether the prints are architectural prints, electrical prints, or mechanical prints. **See Figure 1-23.**

Structural and architectural prints use the grid system to provide dimension information for placement of equipment, spacing between columns, and the distance between columns and major architectural features, such as walls. All dimensions are from the center of a column. Electrical and mechanical prints use the grid system as a point of reference. Column numbers and letters are frequently used in "request for information" (RFI) documents to identify a location. For example, an electrical room may be located at the intersection of line A and line 7 (A7).

18 Printreading for Installing and Troubleshooting Electrical Systems

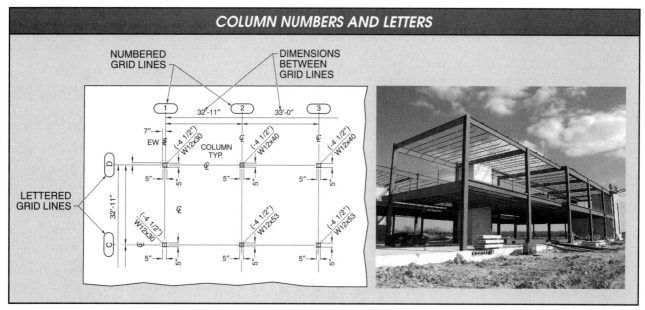

Figure 1-23. A column grid system that includes numbers and letters is used in commercial buildings and industrial facilities to identify specific locations.

PRINT SCALES

Prints are drawn at a reduced size in order to fit an entire project on standard-sized sheets. Frequently, prints must be drawn to a specific scale. The actual dimensions of the objects on prints drawn to scale are reduced proportionally in order to maintain the correct relationship between the objects. **See Figure 1-24.** Some prints are not drawn to scale. Prints that are not drawn to scale include dimensions where necessary. These prints display the notation "Not To Scale" or the abbreviation "NTS."

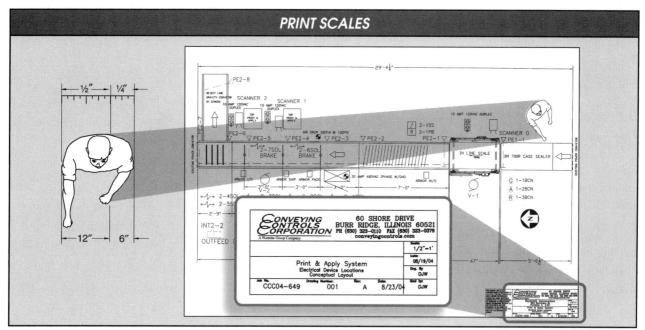

The Numina Group

Figure 1-24. Print scales allow large objects to fit on a standard-sized sheet.

Different scales are used for various types of prints. The scale of a print is found in the title block. Architectural prints use scales ranging in size from 1/32″ = 1′ to 3″ = 1′. A common scale for commercial and industrial building plans is 1/8″ = 1′. An architect's scale is used for architectural prints and building plans.

Detail drawings use large scales to provide an accurate representation of the features of an item. Typical scales for detail drawings are 1/2″ = 1′ or 3/4″ = 1′. Large site plans use scales of one inch equaling 10′, 20′, 30′, or more. An engineer's scale is typically used for site plans. **See Figure 1-25.**

Architect's Scale

An architect's scale is used to measure the size of a drawn object by using a specific scale. The architect's scale is triangular in shape with six faces. Five faces have two scales each and one face is divided into inches and fractional parts of an inch. Scales that are multiples of one another are shown on the same edge to conserve space. For example, the 1/8″ scale and the 1/4″ scale are located on the same edge, with the 1/8″ scale read from left to right and the 1/4″ scale from right to left.

SIX INCH STEEL RULE

The standard 6″ steel rule is a highly accurate linear measuring device graduated in inches and fractions of an inch. The fractional divisions (half, quarter, eighth, sixteenth, thirty-second, and sixty-fourth of an inch) are indicated by division marks of different lengths.

Using a Steel Rule

To take a measurement with the highest accuracy, align the graduation mark with the edge of the workpiece. Do not butt the end of the rule up against shoulders or surfaces of the workpiece to take a measurement. The end of rules may be worn, changing the measurement. Measuring from a midscale graduation might initially be more challenging, but fewer errors will be made.

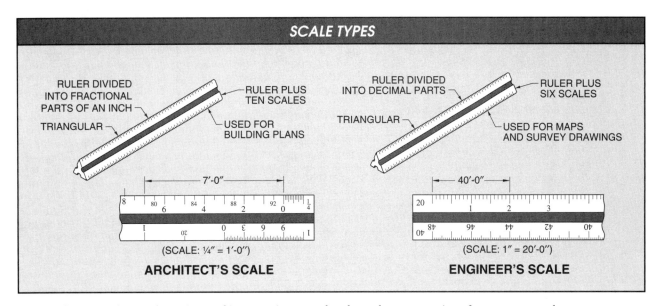

Figure 1-25. Depending on the project, architects, engineers, and tradesworkers use a variety of measurement scales.

Because most measurements taken are not an exact foot measurement, such as 2′, 3′, or 4′, a scale is also provided on each face to measure inches. Inches are read between the 0 designation and the scale designation near the end. **See Figure 1-26.**

An architect's scale is used to measure the distance between two lines as follows:
1. Place the 0 of the appropriate scale on one end of the line and read the largest increment falling on the other end of the line.
2. When the measurement is not an exact foot multiple, move the scale to the smaller whole number and add the inches to the whole foot measurement.

An engineer's scale allows precise measurements in a decimal format (10 scale = 0.1″ per division or 1″ = 10′). The scales are 10, 20, 30, 40, 50, and 60.

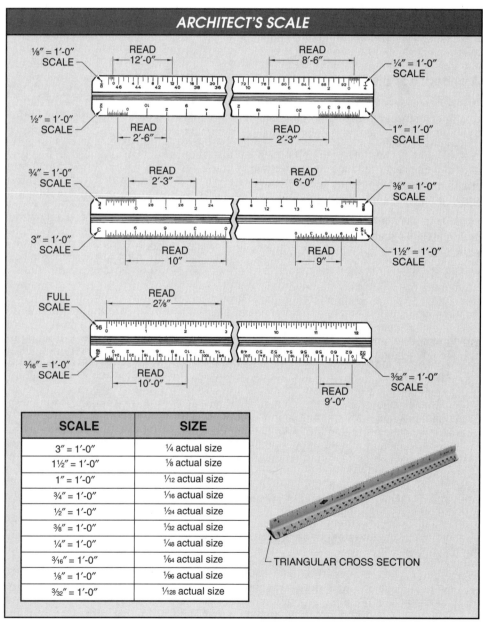

Figure 1-26. An architect's scale measures the size of a drawn object by using a specific scale.

SPECIFICATIONS

A *specification* is additional information that is included with a set of prints. Specifications provide details that cannot be shown on a print. Specifications along with the various drawings describe the entire building, facility process, or project. Specifications contain information related to legal issues, building materials, equipment placement, installation, construction, and quality. **See Figure 1-27.**

Architects and engineers develop specifications based on the requirements of the owner, building, and/or facility process. Specifications list the codes, ordinances, and company policies that must be followed for the project. An example might be, "Perform all electrical work in full accordance with the National Electrical Code (NEC®)."

Municipal building departments use a project's specifications and prints to verify that the proposed project complies with local building codes and zoning ordinances. Contractors use the specifications and prints to accurately bid on a project and then construct the project to meet the requirements of the owner.

Specifications are intended to supplement print drawings, and the specifications must agree with the set of prints. When a conflict exists between what the specifications state and what the prints indicate, the architect or engineer must be contacted for clarification.

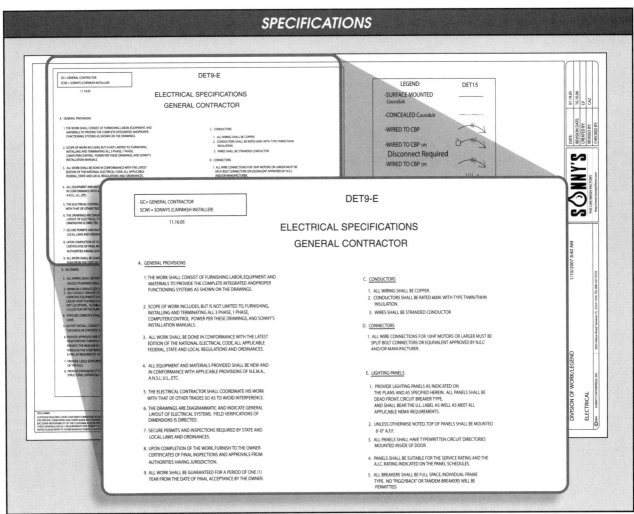

Sonny's Enterprises, Inc.

Figure 1-27. *Specifications provide information about a drawing that cannot be shown on a print.*

A large set of specifications contains many sections. If a conflict exists between different sections of the specifications, the architect or engineer must be contacted. At times a contractor may want to deviate from the specifications, such as substituting a different brand or model of electric motor drive, but before deviating from the specifications, the contractor must obtain permission from the engineer. Failure to follow specifications can result in nonoperational systems, monetary penalties, and legal action.

Residential, commercial, and industrial projects require a set of specifications in addition to a set of prints. The size, format, and complexity of the specifications vary with the project. The specifications for a small residential project may consist of a page or two of requirements attached to the prints. The specifications for a large commercial project may consist of a hundred pages or more of detailed requirements bound together in a book-type format. Specifications for an industrial project will typically have thousands of pages of detailed requirements.

CSI MasterFormat™

The *Construction Specifications Institute (CSI)* is an organization that develops standardized construction specifications. The CSI, in cooperation with the American Institute of Architects (AIA), the Associated General Contractors of America (AGC), the Associated Specialty Contractors (ASC), and other industry groups, has developed the *CSI MasterFormat™ for Construction Specifications,* which is a uniform system for construction specifications, data filing, and cost accounting. CSI continually promotes the CSI MasterFormat™ and updates it periodically.

The MasterFormat is a master list of numbers and titles for organizing information about construction requirements, products, and activities into a standard sequence. The MasterFormat consists of front end documents and 50 divisions. **See Figure 1-28.** The front end documents contain information on bidding requirements, contracting requirements, and conditions related to the construction project. The 50 divisions of the body are numbered and are designed to give complete written information about individual construction requirements for building and material needs.

Division 26—Electrical. Division 26 contains electrical specifications that provide wiring, equipment, and finish information for electrical systems. Division 26 information includes descriptions of electrical site work, raceways and conduits, panelboards, and lighting systems. The specifications also contain quality-assurance requirements and acceptable manufacturer names and product numbers for each section. **See Figure 1-29.**

Refer to the CD-ROM "CSI MasterFormat™" Division Descriptions

CSI MASTERFORMAT™ 2004

- The CSI MasterFormat™ is a list for organizing construction bidding, contract requirements, and building operations by using standardized numbers and titles.
- The MasterFormat does not create trade jurisdictions, design disciplines, or product classifications.
- MasterFormat has a proven 40-yr history.
- Revisions are required to allow for new technologies and flexibility of future projects.
- The levels of organization for the MasterFormat are Groups, Subgroups, and Divisions.
- Common work is assigned numbers and titles.

Scope	Levels	MasterFormat™2004
Division	Level 1	**26** 30 00
Broad scope	Level 2	26 **30** 00
Medium scope	Level 3	26 30 **00**
Narrow scope	Level 4	26 30 00.**13**
User-defined	Level 5	26 30 00.13.**66ATP**

- Numbers in reserved divisions should not be defined by user.
- Time and costs for projects are less when the CSI MasterFormat is adopted early in a project.

The current version of the CSI MasterFormat™ was issued in 2004 and has 50 divisions. Some architects and engineers still use the previous 1995 version. The previous version of the CSI MasterFormat consisted of 16 divisions.

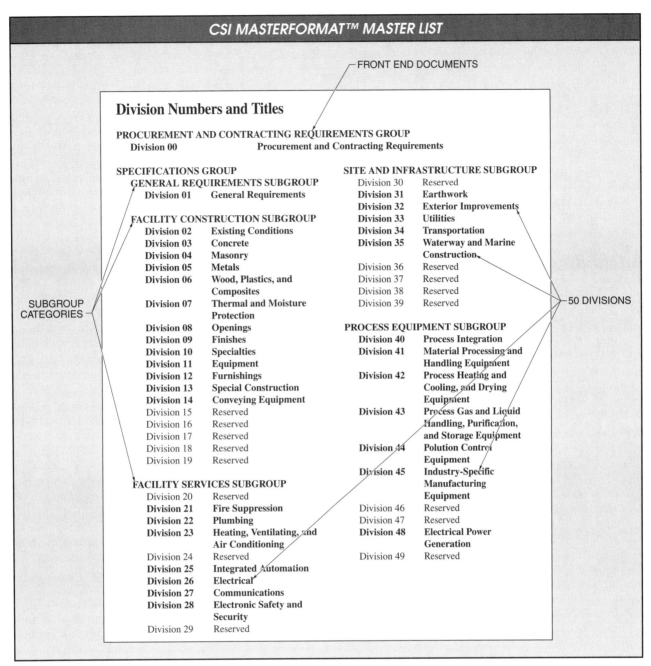

Figure 1-28. The CSI MasterFormat™ contains front end documents, five subgroups, and fifty divisions that are used to organize construction information.

Section Numbering System. Every division is divided into sections. Each section has a six-digit reference number. For example, the reference number for "Facility Electrical Power Generating and Storing Equipment" is 26 30 00. Additional items related to electrical power generation and storage have reference numbers starting with 26 31 00 and ending with 26 36 00. **See Figure 1-30.** The Division 26 (Electrical) number 26 32 13.16 represents four levels. The division number, 26, represents level one. The level two number is 32, the level three number is 13, and when additional levels of clarification are needed, the fourth level number is preceded by a point, in this case, .16. The CSI MasterFormat™ also includes a keyword index of requirements, products, and activities.

Refer to the CD-ROM "CSI MasterFormat™" Numbers & Titles

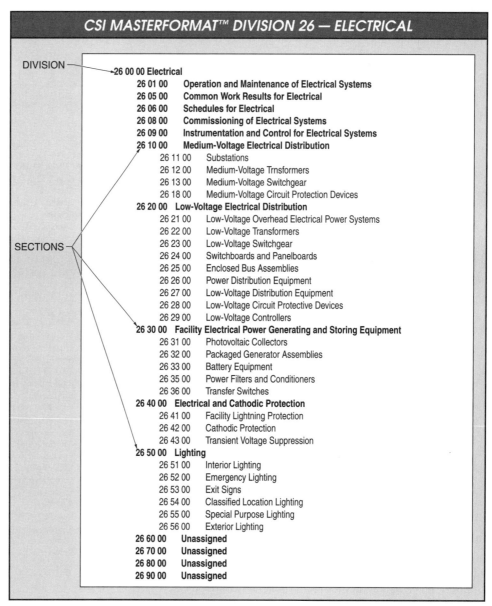

Figure 1-29. Division 26—Electrical is divided into 10 sections that cover electrical site work, distribution systems, power generation, electrical equipment protection, and lighting.

Zircon Corporation

The electrical division of the CSI MasterFormat™ is consulted for quality assurance information on conduit, panelboards, and lighting system site work.

The CSI has created a product data guide (the GreenFormat) that tracks the green properties of building materials and components used during construction.

Not all sections of the CSI MasterFormat™ appear in every set of print specifications. Only the sections that are applicable to the construction project appear in the specifications.

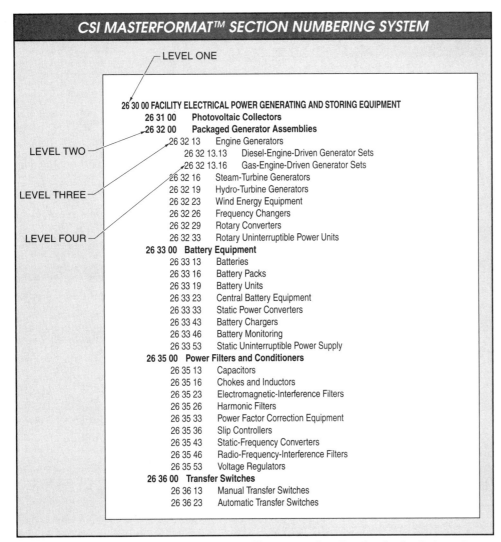
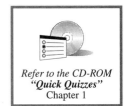

Figure 1-30. *Each section of the CSI MasterFormat™ is numbered to allow four levels of information.*

☑ Example—Installing PLCs

1-1

Scenario:
Example 1-1 is an installation activity requiring that four (4) programmble logic controllers (PLCs) be installed in an electrical enclosure. The application is six (6) fermenters that are controlled by the PLCs. The fermenters change raw materials into ethanol for gasoline. The PLCs will control temperature, time of cooking, and when chemicals are added to the four-story 25,000 gal. fermenters.

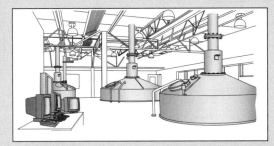
PLC-CONTROLLED PROCESS

Task:
Install four (4) PLCs in an enclosure that is located on the process floor. Install the PLCs in two columns, with each column having two rows. Follow all specifications to allow required airflow. Allow for raceway installation.

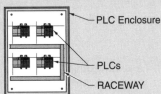

Reference Prints:

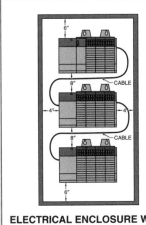
ELECTRICAL ENCLOSURE WITH VERTICALLY MOUNTED PLCs

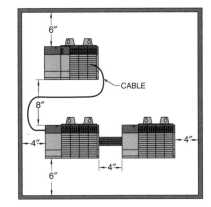
ELECTRICAL ENCLOSURE WITH HORIZONTALLY AND VERTICALLY MOUNTED PLCs

Refer to the CD-ROM "Prints" Chapter 1 Minimum PLC Airflow Clearances

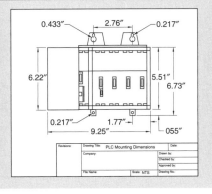

Refer to the CD-ROM "Prints" Chapter 1 PLC Mounting Dimensions

Example—Installing PLCs

Step 1: Type 1 large panel enclosure has been selected. The enclosure has the following standards ratings:
- UL 50 Listed, TYPE 1
- CSA C22.2 No. 40 CERTIFIED, TYPE 1
- NEMA STANDARD FOR TYPE 1
- IEC 60529, IP30

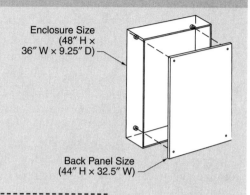

Step 2: Scribe a horizontal line and a vertical line to represent the top left corner of the PLC. Use the minimum distances required for proper airflow.

Panel clearance:
36.00″ − 32.50″ = 3.50″
3.50″ ÷ 2 = 1.75″ per side
Left side of PLC is 2.50″ in from left side of back panel.

Airflow clearance:
1.75″ + 2.50″ = 4.25″ from left wall

4.25″ – EXTRA DISTANCE AIDS IN CENTERING PLCs IN SPACE

Panel clearance:
48.00″ − 44.50″ = 4.00″
4.00″ ÷ 2 = 2.00″ per side
Top of PLC is 5.00″ down from top of back panel.

Airflow clearance:
2.00″ + 5.00″ = 7.00″ from top wall

7.00″ – EXTRA DISTANCE AIDS IN CENTERING PLCs IN SPACE

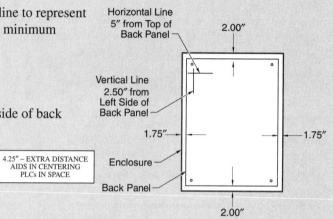

BACK PANEL ENCLOSURE CLEARANCES

Step 3: With the PLC lined up with the horizontal and vertical layout lines, center punch the four (4) mounting holes. The mounting holes are 2.75″ center-to-center horizontally.

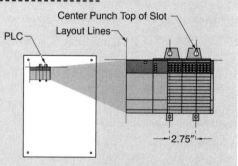

Step 4: Scribe the horizontal and vertical layout lines for PLC #2. Because a PLC is 9.25″ wide and the air space between PLCs is 4.00″, the left side of PLC #2 starts 13.25″ from the PLC #1 vertical layout line. (9.25″ + 4.00″ = 13.25″)

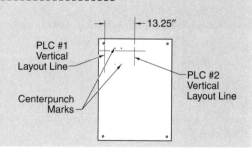

☑ Example—Installing PLCs

Step 5: With PLC #2 lined up with the horizontal and vertical layout lines, center punch the four (4) mounting holes. The mounting holes are 2.75″ center-to-center horizontally.

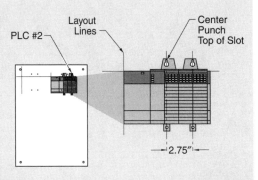

Step 6: Scribe the horizontal layout lines for the top PLC row raceway (2.00″ raceway). **Raceway distance (top):**
5.00″ + 6.25″ + 4.00″ = 15.25″ down from the top of the back panel.
Raceway distance (bottom):
5.00″ + 6.25″ + 4.00″ + 2.00″ = 17.25″ down from the top of the back panel.

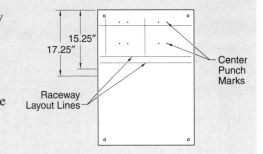

Step 7: Scribe a horizontal line and vertical line that represents the top left corner of PLC #3. Use the minimum distances required for proper airflow.
Airflow clearance:
1.75″ + 2.50″ = 4.25″ from left wall
Airflow clearance:
2.00″ from bottom of raceway

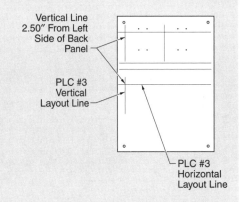

Step 8: With PLC #3 lined up with the horizontal and vertical layout lines, center punch the four (4) mounting holes. The mounting holes are 2.75″ center-to-center horizontally.

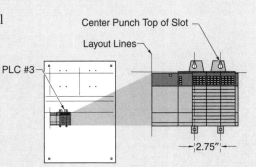

Example — Installing PLCs

Step 9: Scribe the vertical layout line for PLC #4. Because a PLC is 9.25″ wide and the air space between PLCs is 4.00″, the left side of PLC #4 starts 13.25″ from PLC #3 vertical layout line. (9.25″ + 4.00″ = 13.25″)

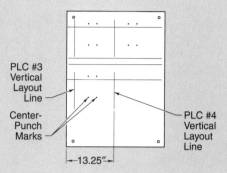

Step 10: With PLC #4 lined up with the horizontal and vertical layout lines, center punch the four (4) mounting holes. The mounting holes are 2.75″ center-to-center horizontally.

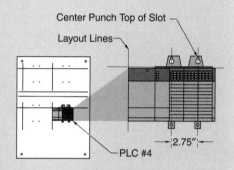

Step 11: Drill and tap all sixteen (16) center punch marks (four for each PLC). Drill with a #25 tap drill and tap with a #10-24 tap. Drill and tap the holes for the top and bottom raceways.

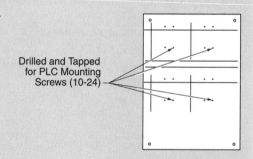

Step 12: Install back panel into enclosure. Mount raceways and PLCs to back panel.

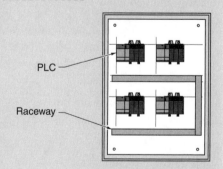

Example—Replacing Emergency Floodlight Fixtures

1-2

Scenario:
Example 1-2 is a troubleshooting activity requiring that four (4) emergency floodlight fixtures be replaced. The application is the kitchen and dining room of a Wendy's restaurant. There are two (2) emergency floodlight fixtures in the kitchen and two (2) in the dining room. The emergency floodlight fixtures presently in use in the Wendy's restaurant are no longer available.

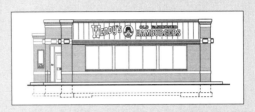

WENDY'S RESTAURANT - DINING ROOM AND KITCHEN EMERGENCY LIGHTING

Task:
Select four (4) emergency floodlight fixtures to replace the fixtures in the Wendy's restaurant.

Reference Print:

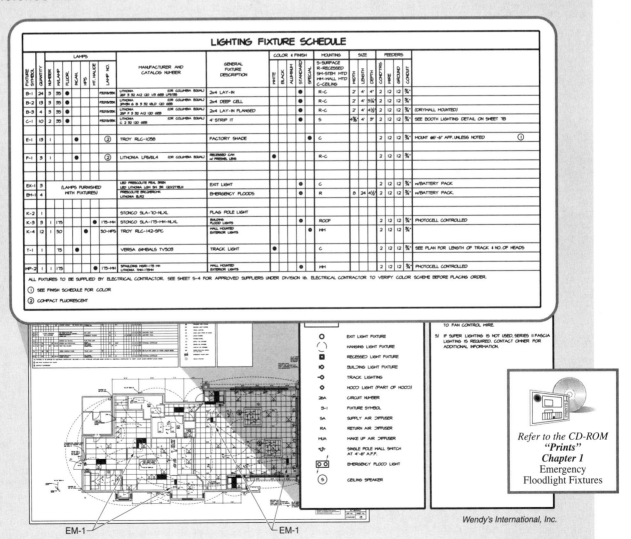

Wendy's International, Inc.

Refer to the CD-ROM "Prints" Chapter 1 Emergency Floodlight Fixtures

☑ Example—Replacing Emergency Floodlight Fixtures

Step 1: Identify the original emergency floodlight fixtures on the lighting fixture schedule (fixture EM-1), the legend, and the general notes.

Step 2: Examine the information on the print. Lamps are furnished with fixtures. The manufacturer and model is Prescolite ERC2. The description of the light fixtures is "Emergency Floods," and each fixture is 8″ W × 24″ L × 4.50″ D. The fixtures are standard in color and finish and have recessed battery packs.

Step 3: Evaluate possible replacement of emergency floodlight fixtures to determine which fixture fits the application requirements the best. Fixture A is a wall-mount fixture that is 5.50″ × 2.25″ in size. Fixture B is a ceiling-mount fixture with a recessed battery pack and is 24″ × 7.75″ × 4″ in size. Fixture C is a wall-mount fixture with exposed battery pack and is 11.25″ × 15.50″ × 4″ in size. Fixture D is a ceiling-mount fixture with a recessed battery pack and is 5″ × 7.75″ × 3.75″ in size.

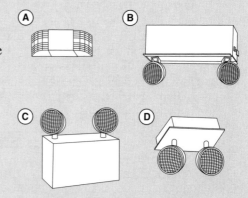

Step 4: Check Sheet 13 for light fixture mounting style. The emergency floodlight light fixtures are not shown on the walls, so the fixtures are ceiling mounted. The lighting fixture schedule states that the fixtures (battery packs) are recessed and provides the original dimensions. Fixture B is a recessed ceiling-mount fixture that best fits the application requirements.

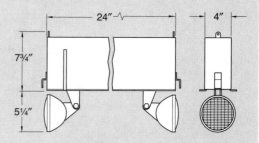

Review Questions and Activities
Printreading Fundamentals

Name _____   Date _____

True-False

T   F   1. Sets of prints can include site plans (plot plans), floor plans, elevations, sectional views, wiring diagrams, details, and schedules.

T   F   2. The basic types of print divisions are object, leader, hidden, cutting-plane, section, break, dimension, center, and extension.

T   F   3. Types of electrical schedules include fixtures, feeders, main switchboard, branch circuit panels, and transformers.

T   F   4. One of the most important pieces of information in the title block is the revision information.

T   F   5. The symbol for a detail drawing consists of a cutting-plane line drawn through the selected item and a circle divided in half by a horizontal line.

T   F   6. A schedule provides dimension information, identifies columns, and serves as a point of reference.

T   F   7. An architect's scale measures the size of a drawn object by using a specific scale.

T   F   8. A general note lists the codes, ordinances, and company policies that must be complied with during the project.

T   F   9. The CSI MasterFormat™ is a master list of numbers and titles used for organizing information about construction requirements, products, and activities in a standard sequence.

T   F   10. A cutting-plane line is a line that indicates the path through which an object will be cut so that its internal features can be seen.

Completion

_____   1. The ___ is the area of a print that contains important information about the contents of the print.

_____   2. A(n) ___ is a note that applies to a specific item in the drawing that the note appears with.

_____ 3. A(n) ___ is an agreed-upon method of displaying information on prints.

_____ 4. A(n) ___ is a letter or group of letters that represents a term or phrase.

_____ 5. Print numbering for a subdivision of electrical prints or a set of electrical prints begins with ___.

Multiple Choice

_____ 1. A ___ line is a line that connects a written description such as a dimension, note, or specification with a specific feature of a drawn object.
   A. dimension
   B. leader
   C. cutting-plane
   D. break

_____ 2. A ___ is a chart used to conserve space and display information in a concise and organized format.
   A. schedule
   B. flow chart
   C. pie graph
   D. title block

_____ 3. The ___ is an organization that develops standardized construction specifications.
   A. Construction Specifications Institute (CSI)
   B. American Institute of Architects (AIA)
   C. Associated General Contractors of America (AGC)
   D. Associated Specialty Contractors (ASC)

_____ 4. In a drawing on a print, a(n) ___ is identified by a symbol that has a number or letter inside a triangle, circle, or square.
   A. dimension
   B. elevation
   C. revision
   D. schedule

_____ 5. A ___ is a sentence or two that provides drawing information that does not fit within the space of the drawing.
   A. specification
   B. legend
   C. note
   D. schedule

Name _____   Date _____

Activity—Selecting Electrical Hardware 1-1

Scenario:
Activity 1-1 is an electrical hardware selection activity requiring that the devices, components, wiring, and location for an ice machine installation be identified. The application is a Wendy's restaurant kitchen that is under construction. The electrical subcontractor is in the building and is ready to wire the ice machine circuit.

Task:
Identify the load, voltage rating of required equipment, amount and size of wiring, and type of box and outlet to withdraw from stores. Also, identify the circuit number and location of the ice machine installation.

WENDY'S RESTAURANT — KITCHEN

Required Information for Wiring the Ice Machine:
1. Load in kilowatts: _____
2. Voltage: _____
3. Number of conductors including ground: ___
4. Wire size: _____
5. Ground size: _____
6. Conduit size: _____
7. Circuit number: _____
8. Specific connection type: _____
9. Plug/cord furnished by: _____
10. Circle the ice machine installation location on the enlarged view of the electrical plan.

Reference Prints:

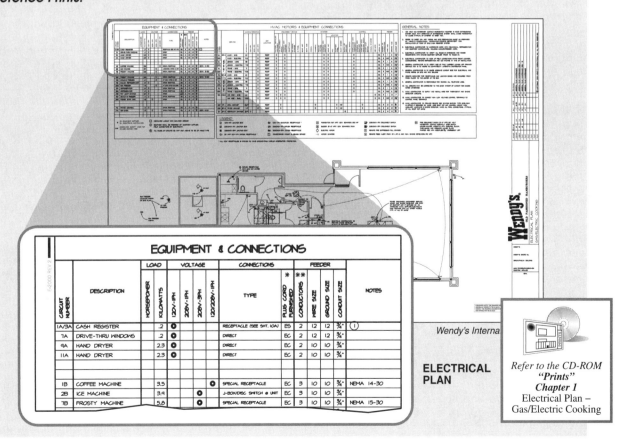

ELECTRICAL PLAN

Refer to the CD-ROM "Prints" Chapter 1 Electrical Plan – Gas/Electric Cooking

35

Activity—Selecting Electrical Hardware

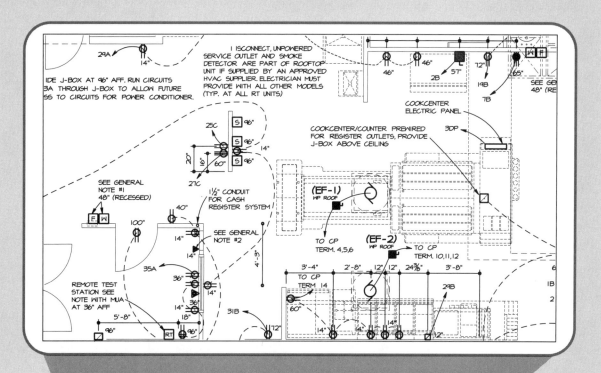

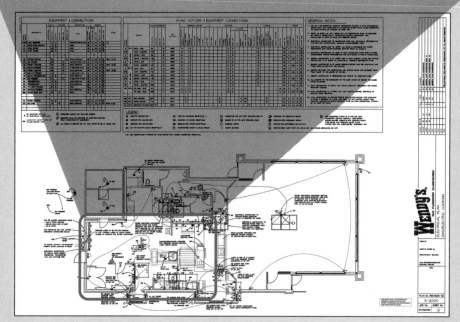

Wendy's International, Inc.

Name _____   Date _____

Activity—Installing an Automatic Lubricator 1-2

Scenario:
Activity 1-2 is an installation activity requiring that an automatic, low-pressure, centralized-system oil lubricator be installed. The application is an automatic rotating pallet wrapper that has a pallet with fifty (50) cases of toothpaste that must be cycled every one and one-quarter minutes (75 sec). The lubricator sends oil to the bearings and gears of the turntable and to each of the conveyor drive systems.

Task:
Install a lubricator to the center of a metal plate (stand) and connect conduit to the lubricator for connection to the pallet-wrapper control panel.

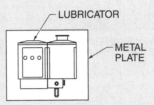

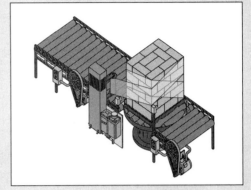

AUTOMATIC ROTATING PALLET WRAPPER

Reference Print

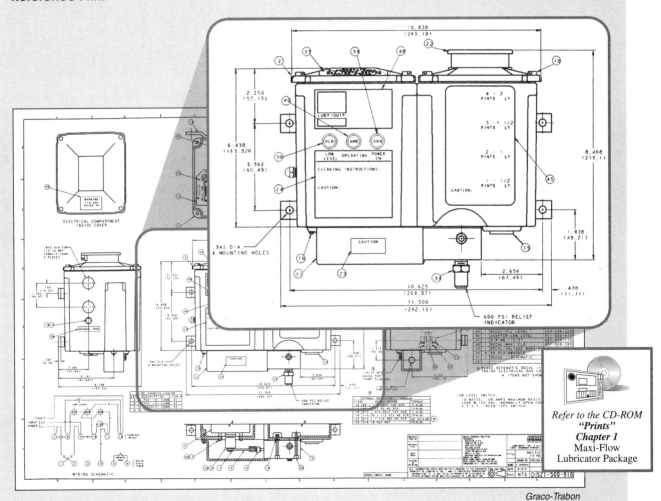

Refer to the CD-ROM "Prints" Chapter 1 Maxi-Flow Lubricator Package

Graco-Trabon

Activity—Installing an Automatic Lubricator

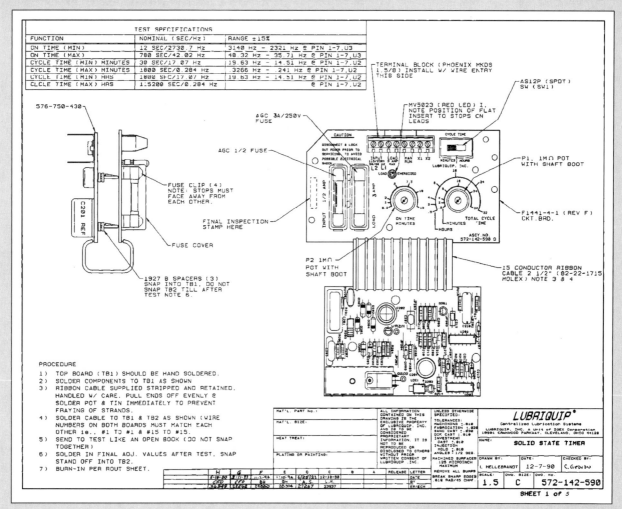

Graco-Trabon

Name _____ Date _____

REFERENCE PRINT #1 (Power & Systems Plan Basement)

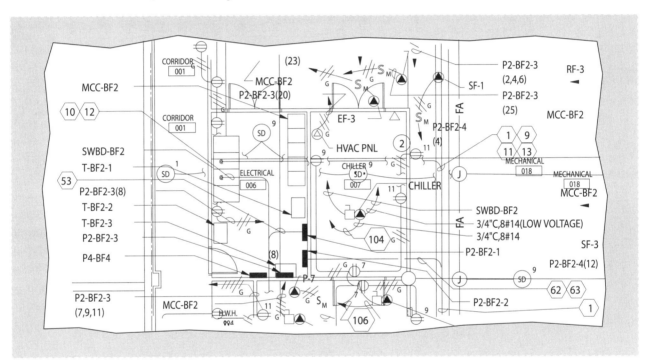

Questions – Reference Print #1

_____ 1. Room number 006 is the ___.

_____ 2. There are ___ entrances to Room 006.

T   F   3. There are no telephone outlets in the Chiller Room.

T   F   4. MCC-BF2 is located in the Electrical Room.

T   F   5. The exact location of the chiller is shown on the print.

_____ 6. There are ___ transformers in the Electrical Room.

_____ 7. MCC-BF2 has ___ sections.

_____ 8. The electrical feed for the Chiller comes from ___.

REFERENCE PRINT #2 (Concrete Water Tank – Electrical Plan)

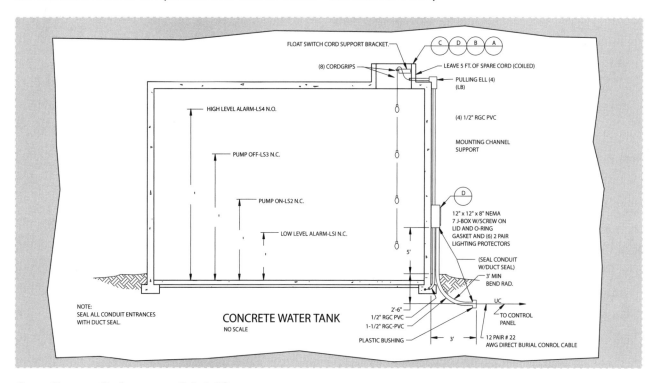

Questions – Reference Print #2

_____ 1. There are ___ float switches in the tank.

_____ 2. LS 1 is the low level alarm with ___ contacts.

_____ 3. The junction box mounted on the side of the tank is a NEMA ___ enclosure.

T   F   4. The drawing provides the exact measurements from the float switches to the bottom of the tank.

T   F   5. The distance between the bottom of the junction box and the ground is 5′.

_____ _____ 6. A(n) ___ pair ___ gauge direct burial control cable runs from the junction box to a control panel.

_____ _____ 7. A short length of ___ rigid ___ conduit protects the control cable where it exits the box.

ADVANCED CD-ROM PRINT QUESTIONS (Concrete Water Tank – Electrical Plan)

_____ 1. Detail ___ provides additional information about the junction box mounted on the side of the tank.

_____ _____ 2. The junction box mounted to the side of the tank is fastened with concrete anchors and ___″ ___ bolts.

_____ 3. Detail ___ provides additional information about grounding of the concrete foundation.

_____ 4. The drawing indicates that conduits are to be sealed with ___.

Residential and Commercial Electrical Symbols

Printreading for Installing and Troubleshooting Electrical Systems

Architectural electrical symbols are simplified drawings of electrical devices and components used on architectural floor plans, plot plans, and detail drawings to show the location of electrical equipment. Architectural symbols shown on plans and drawings are for location purposes only and are not intended to show how the device or component is to be wired or connected. The actual wiring and terminal connections of electrical devices and components are determined by the trade (electrical, HVAC, and/or plumbing) that will install the equipment in accordance with the National Electrical Code® (NEC®), state codes, local codes, and acceptable practices.

STANDARDS ORGANIZATIONS

Standards organizations develop symbols for consistent representation of specific devices and components used in electrical circuits. Symbols are also used to represent how devices and components are connected together in a circuit. Symbols are a universal language because a person does not have to speak a specific language to understand the symbol. Abbreviations, on the other hand, require an understanding of the language in which the abbreviation is written.

When symbols are connected together, diagrams and circuits are developed. Diagrams and circuits are used by architects, engineers, technicians, and maintenance personnel to create electrical and electronic prints of electrical systems.

Electrical symbols have been used on diagrams and prints since manufacturers began to create drawings of products. The symbols were used so that the products could be built the same way each time. The first electrical symbols bore a strong resemblance to the actual components represented. Symbols were typically developed by manufacturers and varied from manufacturer to manufacturer.

The electrical symbols used today have undergone considerable changes. **See Figure 2-1.** The development of new symbols is a continuing process as new devices and components are developed and existing ones are changed, consolidated, or expanded. Standards organizations develop new symbols for new technologies all the time and work to ensure that symbols and abbreviations are used consistently throughout industry.

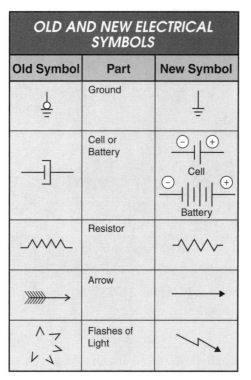

Figure 2-1. The electrical symbols used today have undergone considerable changes since they were first used in the late 1800s.

Standards organizations were developed to provide uniformity. For example, standards are written by committees of organizations that represent the manufacturers and consumers of materials such as electrical equipment. Standards organizations include trade associations, technical societies, government departments, and national, international, and private organizations. Standards organizations often work together in developing and promoting national and international standards.

Trade Associations

A *trade association* is an organization that represents the manufacturers of a specific type of product. For example, the Electronic Industries Alliance (EIA), formerly known as the Electronic Industries Association, is an organization whose members produce electrical and electronic devices and components. **See Figure 2-2.**

The National Electrical Manufacturers Association (NEMA) is a trade association for the electrical manufacturing industry. NEMA helps create electrical product standards that must be followed by all manufacturers. The Electrical Apparatus Service Association, Inc. (EASA) is an international trade organization for manufacturers of electromechanical products. EASA is a means for people working in the electrical field to stay up to date on state-of-the art technology.

Figure 2-2. Trade associations represent the manufacturers of electrical products.

Trade associations have been formed in most industries. For example, the American Petroleum Institute (API) is an organization of petroleum producers. API publishes many standards for the production and handling of petroleum products.

Technical Societies

A *technical society* is an organization that is composed of groups of engineers and technical personnel united by a professional interest, such as creating standards. The Institute of Electrical and Electronics Engineers (IEEE) is a technical society whose members are engaged in the electrical and electronics industry. Electrical and electronics industry standards are published by the IEEE and the American National Standards Institute (ANSI). For example, ANSI/IEEE C37.102-1988 (R1991) is a publication entitled, *Guide for AC Generator Protection*. **See Figure 2-3.**

TECHNICAL SOCIETIES

Institute of Electrical and Electronics Engineers (IEEE)
3 Park Avenue, 17th Floor
New York, NY 10016-5997
www.ieee.org

ASME International
Three Park Avenue
New York, NY 10016-5990
www.asme.org

Instrumentation, Systems, and Automation
67 Alexander Drive
Research Triangle Park, NC 27709
www.isa.org

Figure 2-3. Technical societies are organizations that develop standards and are composed of groups of engineers and technical personnel united by professional interest.

UNITED STATES GOVERNMENT DEPARTMENTS

U.S. Department of Defense Pentagon
Army Navy Drive & Fern Street
Arlington, VA 22202
www.defenselink.mil

U.S. Environmental Protection Agency (EPA)
Ariel Rios Building
1200 Pennsylvania Avenue, N.W.
Washington, DC 20460

Occupational Safety and Health Administration (OSHA)
200 Constitution Ave NW
Washington DC 20210
www.osha.gov

Figure 2-4. Besides the Department of Defense, other government departments that develop standards include the Departments of Commerce, Energy, Transportation, and Agriculture and the Environmental Protection Agency.

Technical societies in other areas include the ASME International, the Institute of Electrical and Electronic Engineers (IEEE), and the Instrumentation, Systems, and Automation Society. Each technical society promotes standardization in their respective industry.

> Members of the IEEE can join IEEE societies. There are 44 individual IEEE societies including societies for Aerospace and Electronic Systems, Education, Industry Applications, and Product Safety Engineering.

United States Government Departments

Among the departments of the United States government that develop materials related to standards, the work of the Department of Defense (DOD) is probably the best known. DOD standards are known as United States Military Standards. *United States Military Standards (Mil Standards)* are Department of Defense standards used by the armed forces, but are not restricted to the armed forces. **See Figure 2-4.**

Other government departments that develop standards include the Department of Commerce (DOC), the Department of Energy (DOE), the Department of Transportation (DOT), the Department of Agriculture (USDA) and the Environmental Protection Agency (EPA). The Occupational Safety and Health Administration (OSHA) is concerned with the development and enforcement of safety standards for construction and for industrial workers.

Personal protective equipment (PPE) required for electrical work is specified in NFPA 70E.

National and International Standards Organizations

Standards organizations are national and international organizations that work with governmental standards groups. National and international standards organizations develop standards or coordinate the development of standards among member groups. The most well-known national standards organization is ANSI. **See Figure 2-5.**

NATIONAL AND INTERNATIONAL STANDARDS ORGANIZATIONS

American National Standards Institute (ANSI)
11 West 42nd Street
New York, NY 10036
www.ansi.org

CSA International (CSA)
178 Rexdale Blvd.
Rexdale, ON M9W 1R3
www.csa.ca

International Organization for Standardization (ISO)
1, rue de Varambé, Case postale 56
CH-1211 Geneva 20, Switzerland

Figure 2-5. The most well-known national standards organization is the American National Standards Institute (ANSI), and the most well-known international organization is the International Organization for Standardization.

The *American National Standards Institute (ANSI)* is a standards-developing organization that adopts and copublishes standards that represent the needs of its members and other standards organizations from around the world. ANSI is always working to accredit programs that assess company or government conformance to national and/or international standards.

CSA International (CSA) is a product-testing organization (headquartered in Ontario, Canada) that certifies products meeting safety and performance levels set by standards. The CSA stamp is seen on many electrical devices, such as switches and receptacles, as well as electrical components, such as solenoids and heating elements.

The *International Organization for Standardization (ISO)* is an international standards-developing organization that develops standards for worldwide use. ANSI also manages United States participation with the ISO in international standards activities. The ISO is a nongovernmental international organization that is comprised of national standards institutions of over 90 countries.

Foreign national standards are developed by standards-setting organizations in many countries. The organizations often coordinate with their own government agencies. Standards written by foreign national standards organizations are identified by acronym. **See Figure 2-6.** For example, the *Standards Council of Canada (SCC)* is an organization that aids in the development and use of standards. European community standards are developed by the European Community for Standardization and are known as Euronorms (EN).

ORGANIZATION BY COUNTRY OF ORIGIN

Country	Acronym
Austria	ON
Belgium	NBN
Bulgaria	BDS
Canada	SCC
China	SAC
France	AFNOR
Germany	DIN
Hungary	MSZT
Italy	UNI
Japan	JISC
Poland	PKN
Romania	ASRO
Spain	AENOR
Sweden	SIS
United Kingdom	BSI
USA	ANSI

Figure 2-6. Foreign national standards organizations are identified by acronyms.

Private Organizations

Private organizations create standards that are an accumulation of an organization's knowledge and experience with materials, methods, and practices. Standards written by private organizations are often modified from other standards to meet the needs of the organization. Private organizations often impose stricter standards than government or international organizations. **See Figure 2-7.**

PRIVATE ORGANIZATIONS

National Fire Protection Association (NFPA)
1 Batterymarch Park
PO BOX 9101
Quincy, MA 02269-9191
www.nfpa.org

Underwriters Laboratories, Inc.® (UL®)
333 Pfingsten Road
Northbrook, IL 60062
www.ul.com

National Electrical Manufacturers Association (NEMA)
1300 N 17th St
Suite 1847
Rosslyn, VA 22209
www.nema.org

Figure 2-7. The NFPA, as a private organization, creates electrical standards that are assembled as the National Electrical Code® (NEC®) and safety standards that are known as 70E.

The *National Fire Protection Association (NFPA)* is a world leader in writing codes and standards for fire prevention and public safety. *Underwriters Laboratories, Inc.® (UL)* is a not-for-profit product-safety testing and certification organization. The *National Electrical Manufacturers Association (NEMA)* is an organization that develops technical standards and government regulations for electrical equipment.

The NFPA 70E, *Standard for Electrical Safety in the Workplace,* is a standard that covers the electrical safety requirements of workers when performing electrical work. The standard covers the safety requirements for the installation of all electrical equipment in residential, commercial, and industrial facilities.

RESIDENTIAL AND COMMERCIAL ELECTRICAL PRINTS

Architectural electrical symbols are used to show the location of lights, switches, receptacles, smoke detectors, power panels, and other electrical devices in residences, offices, or commercial facilities. Various prints are used to determine the location, type, and number of required devices to be installed in each room or commercial area. **See Figure 2-8.**

SYMBOLS AND COMPONENTS

A *symbol* is a graphic representation of a device, component, or object on a print. Electrical, electronic, mechanical, and fluid power symbols are used in ladder (line), wiring, and schematic diagrams. **See Figure 2-9.**

Misinterpretations and functional conflicts can result when symbols with multiple meanings are used. These conflicts typically occur when a drawing contains information for several trades. For example, the symbol for 3φ, 3-wire, delta circuit is a triangle. The triangle symbol is also used to represent a triangle in math and drafting, and direction of flow in fluid power. Notes, asterisks, or flags are used to clarify the intent of a symbol that might have multiple meanings. A tabular list of symbol meanings is generally provided with a drawing or set of prints.

A motor symbol can represent a variety of motors. A specification sheet is required to identify the specific motor being used.

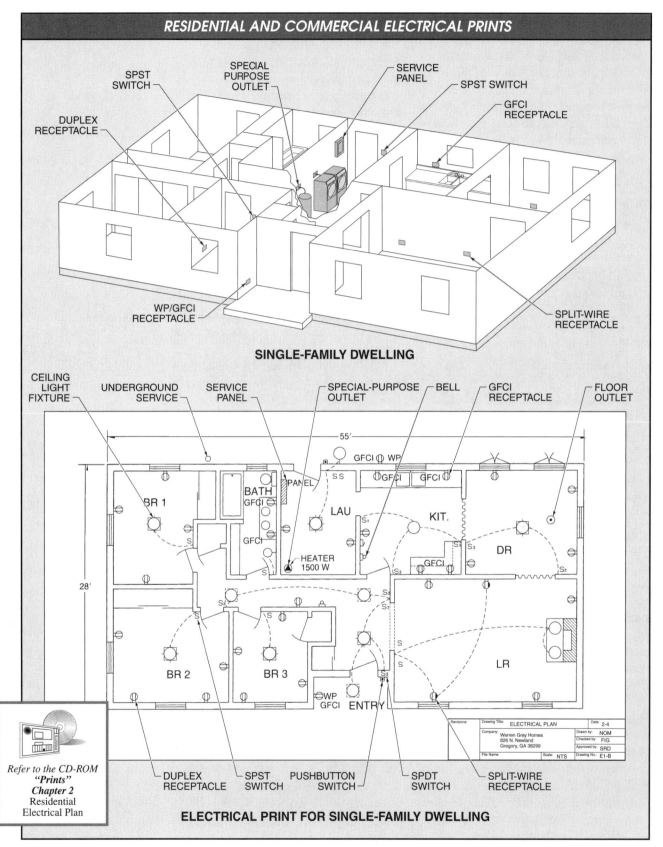

Figure 2-8. Electrical prints show the location of lights, switches, receptacles, smoke detectors, and power panels.

Chapter 2 — Residential and Commercial Electrical Symbols   47

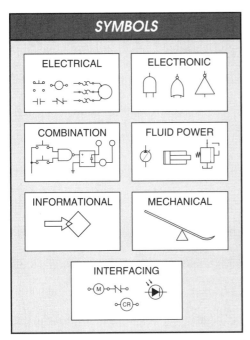

Figure 2-9. Symbols are graphic representations of devices, components, or objects on a print.

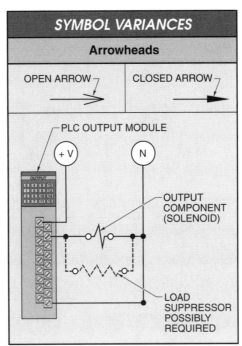

Figure 2-10. The use of a symbol for a specific application can be modified to account for non-standardized equipment or operations.

Lighting symbols indicate the type of lamp, shape of fixture, whether a fixture is wall-mounted or ceiling-mounted, and whether a fixture is flush or recessed.

Electrical symbols are drawn to a relative size and are designed so that their connection points coincide with the connection points of the next symbol. Terminal symbols (typically small circles) are added to some symbols to indicate proper placement of connections.

Arrowheads may be open or closed unless specifically indicated as open or closed by a standard. **See Figure 2-10.** Locations or paths for future or associated equipment are shown using short, dashed lines.

Lighting Symbols

On October 19, 1879, after many years of trying, Thomas Edison succeeded in developing the first lamp to produce a sustainable light. Because of the improvements constantly being made to light bulbs, lighting became the first major application of residential and commercial electricity distribution. Today, the basic incandescent light bulb designed by Edison is still used as a major source for producing light, although energy-efficient lamps, such as compact lamps, are becoming increasingly popular.

Lighting symbols are used on residential and commercial prints to show the location, type of lamp, and style of fixture (recessed or surface). **See Figure 2-11.** In addition to showing the location and type of lamp (incandescent, fluorescent, or HID), lighting symbols also show the type of required light-fixture mounting (ceiling or wall). Along with light fixtures, other items can also be mounted on ceilings or walls. Items such as fans and clocks must be connected at properly mounted outlets (boxes). **See Figure 2-12.**

LIGHTING

Device	Type	Symbol
LIGHTING	Wall Abbr = L	○ OR ⊸Ⓛ
	Ceiling Abbr = L	✦
	Recessed Ceiling Fixture Abbr = LAC	⟦✦⟧ Outline indicates shape of fixture
	Continuous Wireway for Fluorescent Lighting	▭○▭ Extend rectangle for length of installation
	Lighting Outlet with Lamp Holder Abbr = L	Ⓛ
	Lighting Outlet with Lamp Holder & Pull Switch Abbr = LPS	Ⓛ$_{PS}$
INCANDESCENT	Track Lighting Abbr = LTL	▭ ○ ○ ○ ▭
EXIT LIGHT	Abbr = LEX	⊗ Shaded areas denote faces (double-sided)
FLOOD LIGHT	Multiple Assembly Abbr = FL	⋎ ⋎ ⋎

Figure 2-11. Lighting symbols are used on residential and commercial prints to show the location, type of bulb, style of fixture, and type of fixture mounting.

MISCELLANEOUS OUTLETS

Device	Type	Symbol
FAN OUTLET	Abbr = F	Ⓕ Box listed as acceptable for fan support
JUNCTION BOX	Ceiling Abbr = J	Ⓙ
DROP CORD	Outlet Equipped with Adapter Abbr = D	Ⓓ
CLOCK OUTLET	Abbr = C	Ⓒ or 🕐 Ceiling Mount or ⊸Ⓒ⊸ 🕐 Wall Mount

Installing Ceiling Fan

Installing Clock

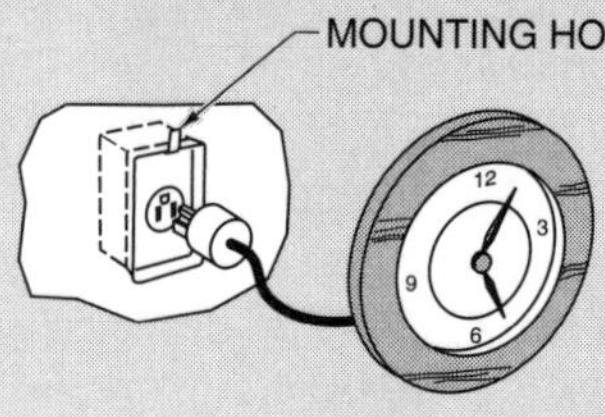

Figure 2-12. Various items can be mounted on ceilings or walls, such as fans and clocks, which must be connected to properly mounted outlets.

Incandescent Lamps. Incandescent lamps are the most widely used lamps in the world. An *incandescent lamp* is a gas-filled bulb that produces light using the flow of current through a tungsten filament. Incandescent lamps are available in a variety of shapes and sizes. **See Figure 2-13.** The outer glass of a bulb is available in various colors to allow a bulb to provide a color other than white light.

The shape of a bulb is designated by a letter or letters. The most common bulb shape is the "A" bulb used in most residential and commercial indoor lighting applications. It uses bulb wattages that range from 15 W to 200 W. Larger bulb sizes ranging from 150 W to 2000 W use a "PS" bulb, which has a longer neck than the A bulb. Higher-wattage PS bulbs used in commercial, school, industrial, and street lighting are being replaced, for the most part, with fluorescent and high-intensity discharge lighting.

Fluorescent Lamps. For large indoor-lighting requirements, fluorescent lamps are used. A *fluorescent lamp* is a low-pressure discharge lamp in which the ionization of mercury vapor transforms ultraviolet energy into light. Fluorescent lamps provide approximately double the light per watt of incandescent lamps and operate at much cooler temperatures. **See Figure 2-14.**

Standard fluorescent lamps vary from ⅝″ to 2⅛″ in diameter and from 6″ to 96″ in length. Fluorescent lamp bulbs are available that produce different shades of color. Standard bulbs have ratings of cool (pale blue-green whiteness), moderate (very pale yellow-whiteness), and warm (yellow-whiteness).

Compact Fluorescent Lamps. Compact fluorescent lamps (CFLs), are a new design of lamp that use less energy while providing the same amount of visible light. CFLs are also considered to have a longer life expectancy according to the ratings.

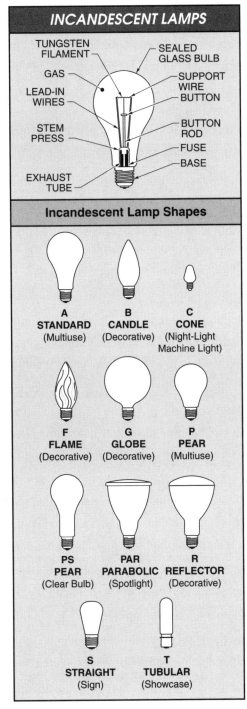

Figure 2-13. An incandescent lamp is a gas-filled bulb that produces light using the flow of current through a tungsten filament.

The shade of fluorescent lamp color chosen for an application (kitchen, office, or studio) depends on the type of color tinting desired.

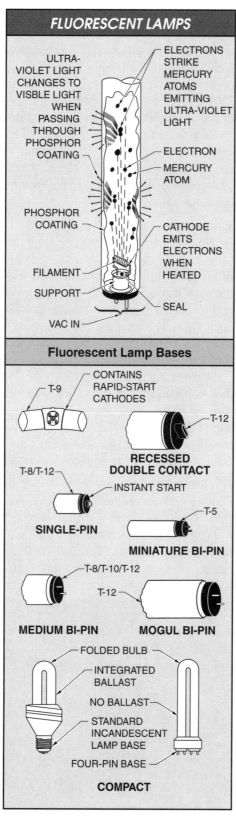

Figure 2-14. A fluorescent lamp is a low-pressure discharge lamp in which the ionization of mercury-vapor transforms ultraviolet energy into light.

High-Intensity Discharge Lamps. A *high-intensity discharge (HID) lamp* is a lamp that produces light from an arc tube similar to the tungsten filament of an incandescent bulb. HID lamps produce more light per watt then any other kind of bulb. **See Figure 2-15.**

The problem with most HID lamps is that they have poor color rendering. Color rendering is the appearance of a color when illuminated by a light source. Low-pressure sodium HID lamps produce a yellow to yellow-orange light, mercury-vapor and metal-halide HID lamps produce a light yellow light, and high-pressure sodium HID lamps produce a golden white light. However, HID lamps are still preferred for outdoor lighting and large-area lighting such as in sports stadiums and warehouses because HID lamps produce far more light per watt than incandescent or fluorescent bulbs.

Switch Symbols

Electrical circuits must be controlled at all times. A *switch* is a device used to control the flow of current in an electrical circuit. The most common types of switches used in lighting circuits are two-way, three-way, and four-way.

Two-Way Switches. A *two-way switch* is a single-pole, single-throw (SPST) switch that has an ON (closed) position and an OFF (open) position. Two-way switches are the typical switches used for light switches in residential and commercial buildings. **See Figure 2-16.**

Three-Way Switches. A *three-way switch* is single-pole, double-throw (SPDT) switch that is used to control lamps from two switch locations. A *four-way switch* is double-pole, double-throw (DPDT) switch that is used between two three-way switches when controlling a lamp from three or more switch locations. Other electrical switches, such as dimmers, are used for specialized functions. Dimmers control the brightness of a lamp. **See Figure 2-17.**

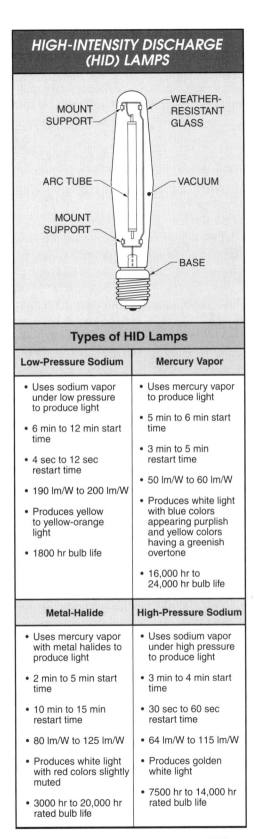

HIGH-INTENSITY DISCHARGE (HID) LAMPS

Types of HID Lamps

Low-Pressure Sodium	Mercury Vapor
• Uses sodium vapor under low pressure to produce light • 6 min to 12 min start time • 4 sec to 12 sec restart time • 190 lm/W to 200 lm/W • Produces yellow to yellow-orange light • 1800 hr bulb life	• Uses mercury vapor to produce light • 5 min to 6 min start time • 3 min to 5 min restart time • 50 lm/W to 60 lm/W • Produces white light with blue colors appearing purplish and yellow colors having a greenish overtone • 16,000 hr to 24,000 hr bulb life

Metal-Halide	High-Pressure Sodium
• Uses mercury vapor with metal halides to produce light • 2 min to 5 min start time • 10 min to 15 min restart time • 80 lm/W to 125 lm/W • Produces white light with red colors slightly muted • 3000 hr to 20,000 hr rated bulb life	• Uses sodium vapor under high pressure to produce light • 3 min to 4 min start time • 30 sec to 60 sec restart time • 64 lm/W to 115 lm/W • Produces golden white light • 7500 hr to 14,000 hr rated bulb life

Figure 2-15. A high-intensity discharge (HID) lamp is a lamp that produces light from an arc tube similar to the tungsten filament of an incandescent bulb.

TWO-WAY SWITCHES

Device	Type	Symbol
SINGLE-POLE	Abbr = SPST	S OR S_1
WITH PILOT LIGHT	Abbr = SPST	S_P
WEATHER-PROOF	Abbr = WP SW	S_{WP}
DOUBLE-POLE	Abbr = DPST	S_2

Two-Way Switch Circuit

Figure 2-16. A two-way switch is a single-pole, single-throw (SPST) switch that has an ON (closed) position and an OFF (open) position.

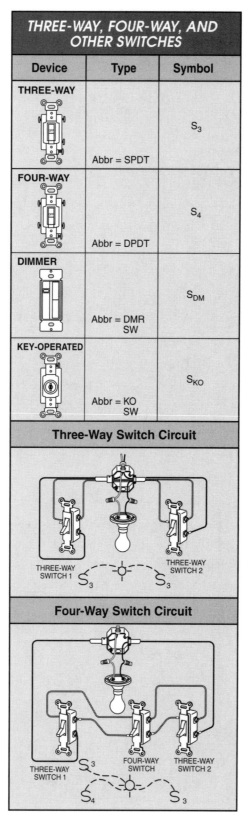

Figure 2-17. When three or more switch locations are required, one or more four-way switches are used between three-way switches.

Receptacle Symbols

Electrical energy is transformed into other forms of energy, such as heat, light, sound, and mechanical energy (linear or rotating) to produce work. When an electrical load is known, such as with an electric range, a dedicated (specific) symbol is used on a floor plan to show the location of the electric range outlet.

When electrical loads such as freestanding lamps, computers, and TVs are not known, general-purpose receptacles (duplex receptacles) are installed around a room to allow access to power as required. Duplex receptacles are the most common type of receptacles used for general power distribution throughout residential and commercial buildings. **See Figure 2-18.**

The three basic receptacle types are the standard, isolated, and ground-fault circuit interrupter (GFCI). The 15 A, 115 VAC standard receptacle is the most common type of receptacle. Some 15 A, 115 VAC convenience outlets also have a switch. **See Figure 2-19.** For higher current loads, a 20 A, 115 VAC receptacle is used. For current requirements over 20 A, special-purpose receptacles are typically specified on building prints.

Isolated Ground Receptacles. An *isolated ground receptacle* is a type of receptacle that provides a grounding path for each individual outlet that is not part of the normal grounding system used by standard receptacles. **See Figure 2-20.** A separate ground is used to minimize problems with sensitive electronic equipment, such as computers and medical equipment, that may be affected by unwanted electrical signals (electrical noise). Isolated ground receptacles are identified by an orange triangle on the face of the receptacle and/or an orange-colored face.

> Isolated ground receptacles create a noise-free grounding path to the service panel and should be used along with surge protectors to protect computers and other sensitive equipment.

RECEPTACLES

Device	Type	Symbol
SINGLE OUTLET	Wall Abbr = OUT.	—◯—
DUPLEX	Abbr = DX RCPT	═◯—
TRIPLEX	Abbr = TRI RCPT	═⊕═ OR ═◯═3
SPLIT-WIRED DUPLEX	Abbr = SPW RCPT	—●—
SINGLE SPECIAL-PURPOSE OUTLET	Abbr = SS OUT.	—◯△—
DUPLEX SPECIAL-PURPOSE	Abbr = DS RCPT	═◯△—
RANGE OUTLET	Abbr = R or RNG OUT.	═◯═R
SINGLE FLOOR OUTLET	Abbr = SF OUT.	▢ OR ⊡
DUPLEX FLOOR	Abbr = DF RCPT	▢

Figure 2-18. To power electrical loads, duplex receptacles are installed around a room to allow access to power as required.

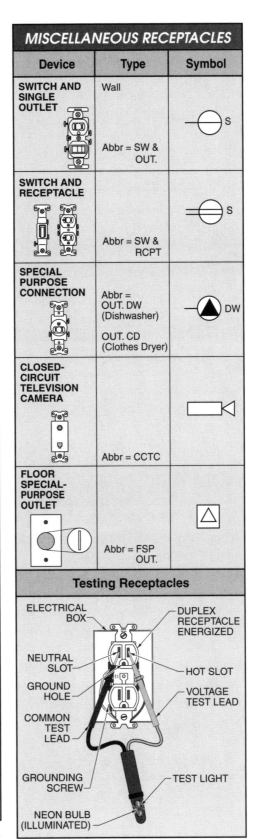

Figure 2-19. Various receptacles are used in residential and commercial buildings.

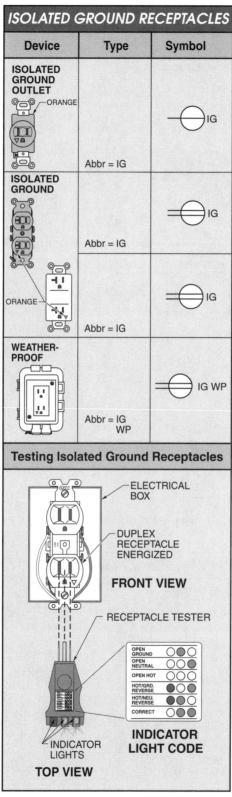

Figure 2-20. An isolated ground receptacle provides a grounding path for each individual outlet that is not part of the normal grounding system used by standard receptacles.

Ground-Fault Circuit Interrupter Receptacles. A *ground-fault circuit interrupter (GFCI)* is a type of receptacle designed to help protect people by detecting ground faults and quickly disconnecting the power from a circuit that has developed a ground fault. **See Figure 2-21.** A ground fault exists any time current takes a path it was not designed to take. A GFCI typically trips any time current is above the level that will cause a dangerous shock.

Power Symbols

In order to deliver power to lamps and receptacles, electrical power must be distributed throughout a building. Power is typically brought into a building from a utility company step-down transformer to a main power and lighting panel. Power and lighting panels are also called panelboards. **See Figure 2-22.**

The main panelboard, or service panel, in a residential or commercial building includes a main multiphase circuit breaker or fuses to protect the entire electrical system. Power from the service panel is broken up into branch circuits that make up the building's electrical system. Smaller circuit breakers or fuses protect each branch circuit.

Various lines are used on electrical prints to indicate whether the wiring for a system is hidden in the ceilings or walls or if the wiring is exposed. Many other electrical power system items are also found on electrical prints, such as emergency generators, transformers, and various pull boxes. **See Figure 2-23.**

> The three types of ground fault circuit interrupters are receptacle (used in place of standard receptacles), circuit breaker (GFCI circuit breaker installed in panel protects all circuits connected to GFCI), and portable (extension cord with built-in GFCI circuitry).

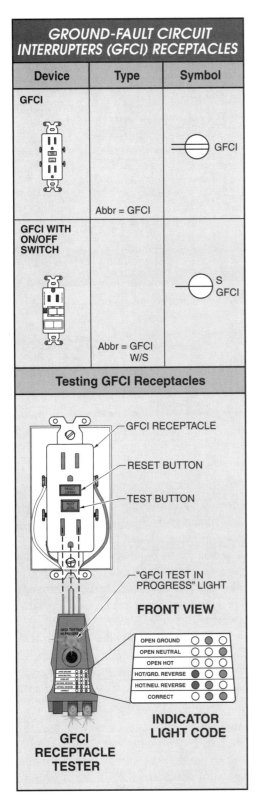

Figure 2-21. A ground-fault circuit interrupter (GFCI) is a type of receptacle designed to help protect occupants by detecting ground faults and quickly disconnecting the power from the circuit that has developed the ground fault.

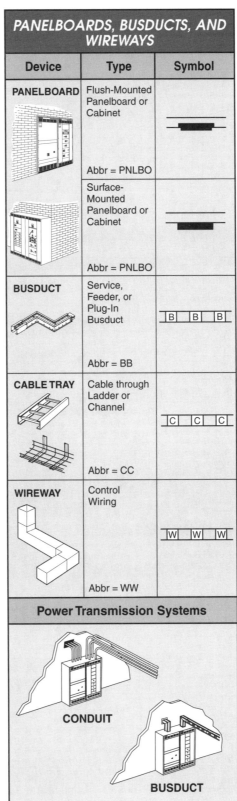

Figure 2-22. Electrical power from a utility company is distributed from panelboards throughout a building using busducts, wireways, and conduit.

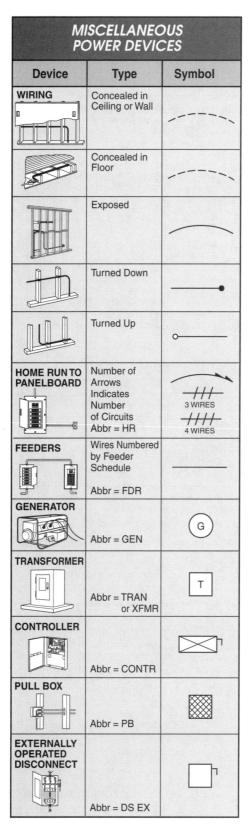

Figure 2-23. Many types of electrical power system items are found on electrical prints, such as emergency generators, transformers, and various pull boxes.

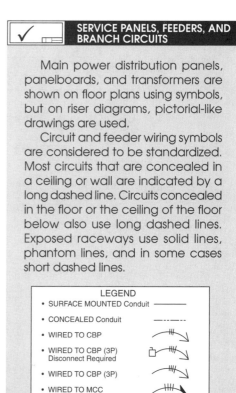

Sonny's Enterprises, Inc.

Electrical Panels. *A power panel* is a wall-mounted distribution cabinet used in large commercial and industrial buildings. A *service panel* is a power panel used for a residential structure. A *switchboard* is a power panel that is freestanding (not wall mounted). The individual circuit breakers or fuses in a power and lighting panel supply power to individual branch circuits. **See Figure 2-24.** A *branch circuit* is the portion of a distribution system between the circuit breaker or fuse and all receptacles, lamps, and loads connected to the circuit breaker or fuse.

Busducts. *A busduct* is the metal housing of a busbar distribution system and is available in prefabricated tees, elbows, crosses, and straight sections. Busducts also have specialty pieces for various types of functions. By bolting busduct sections together, electrical power is distributed throughout an area and easily available when needed. **See Figure 2-25.**

Chapter 2—Residential and Commercial Electrical Symbols

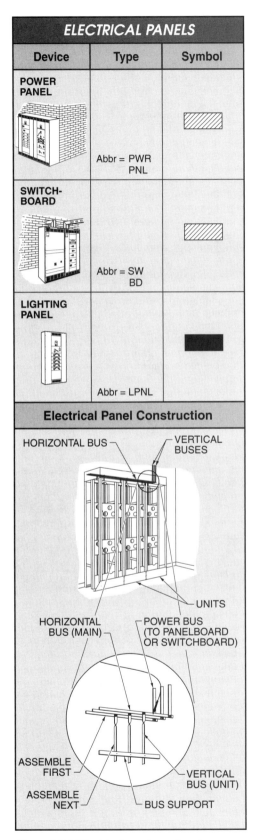

Figure 2-24. Electrical panels are found as power panels, service panels, and switchboards.

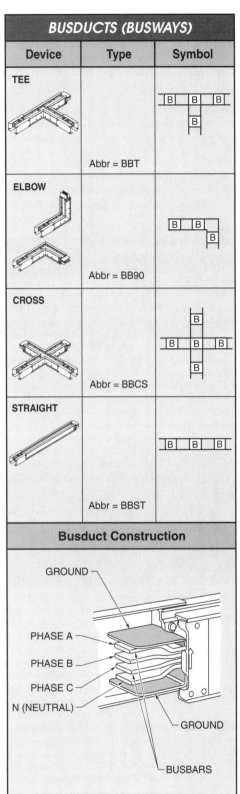

Figure 2-25. A busduct is a metal-enclosed section of a busbar distribution system and is available in prefabricated tees, elbows, crosses, and straight sections.

Signal Symbols

In addition to lighting and power circuits, electrical circuits also include specialized circuits for doorbells, sound systems, fire alarms, and other signaling devices. The main purpose of a signaling device is to provide a visual and/or audible warning signal. **See Figure 2-26.** Along with lamps and receptacles, signaling devices are also shown on electrical prints.

Alarms, chimes, horns, sirens, and bells are typically used to produce a loud warning signal designed to draw attention to an abnormal condition. Audible signals are used in applications that typically require a much faster response than a visual signal alone could provide. Other types of signaling systems, such as computer, telephone, and sound systems, transmit information for work and leisure. **See Figure 2-27.**

PLOT PLAN SYMBOLS

When a building is designed, the building floor plan is used to show the location of all electrical devices (lamps, outlets, power and lighting panels) inside the structure and on the outside walls. Buildings are located at fixed positions on a lot. A *plot plan* is an aerial view of one building lot and provides specific information about the lot. Plot plans identify the property lines, location of the structures on the lot, finished floor elevation, elevations and contours of the ground, utility locations, and easements. **See Figure 2-28.**

Private Property Symbols

All plot plans must have a North arrow for proper orientation and a point of beginning for surveying. The building property and adjacent properties include objects such as utilities, trees, hedges, and fences that must be considered when installing or servicing electrical systems. **See Figure 2-29.**

> Most cities require a plot plan to include the street address of the property.

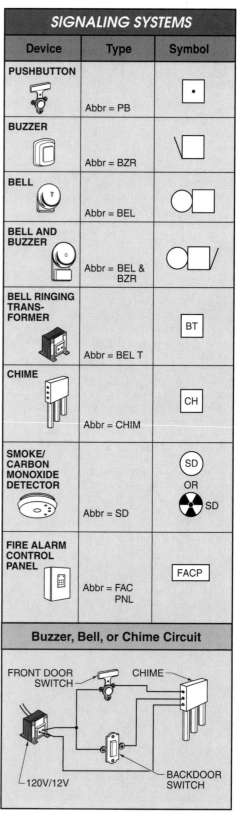

Figure 2-26. The main purpose of a signaling device is to provide a visible and/or audible warning signal.

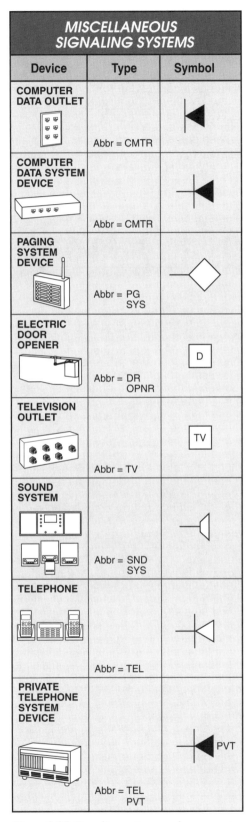

Figure 2-27. Signaling systems, such as computer systems, telephone systems, and sound systems, transmit information for work or leisure.

Figure 2-28. A plot plan is an aerial view of one building lot and provides specific information such as property line location, location of the structures on the lot, finished-floor elevations, elevations and contours of the ground, utility locations, and easements.

PRIVATE PROPERTY		
Device	**Type**	**Symbol**
NORTH	Abbr = N	N
POINT OF BEGINNING	Abbr = POB	
UTILITY METER OR VALVE	Abbr = UMR	
MAILBOX	Abbr = MLBX	
TREE	Abbr = TRE	
BUSH	Abbr = BSH	
HEDGE ROW	Abbr = HDGRW	
FENCE	Abbr = FNC	—•—•—

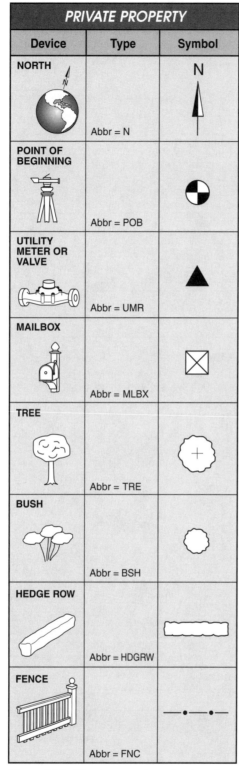

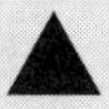

Figure 2-29. *All private property plot plans have a North arrow as part of the plan for proper orientation and include objects such as placement of utilities, trees, hedges, and fences that must be considered when installing or servicing electrical systems.*

Public Property Symbols

Symbols are used on plot plans to show the location, type of objects, and electrical devices located around a building. Many times the objects are part of public property but are indicated on a private property plot plan. Public property objects typically found on private property plot plans are light standards, street sign location, fire hydrants, and utility holes. **See Figure 2-30.**

PUBLIC PROPERTY		
Device	**Type**	**Symbol**
LIGHT STANDARD	Abbr = LT STD	
TRAFFIC SIGNAL	Abbr = TFC SIG	
STREET SIGN	Abbr = ST SGN	
FIRE HYDRANT	Abbr = FHY	
MANHOLE	Abbr = MH	M

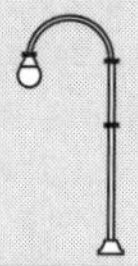

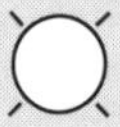

Figure 2-30. *Symbols are used on plot plans to show the location, type of objects, and electrical devices located around a building that are part of public property, but are indicated on the private property plot plan.*

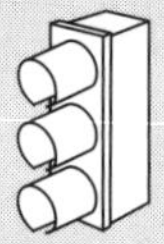

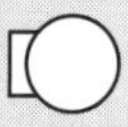

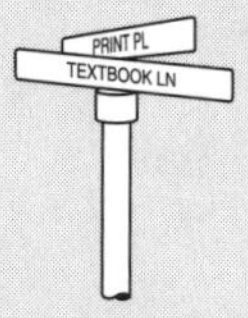

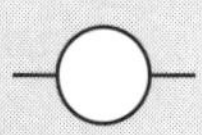

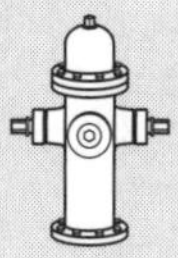

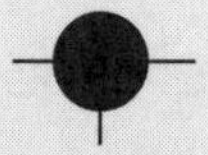

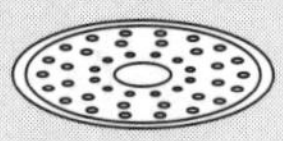

ABOVEGROUND AND UNDERGROUND DISTRIBUTION AND LIGHTING SYMBOLS

Electrical power is distributed aboveground, underground, or a combination of both. **See Figure 2-31.** A wide variety of electrical equipment is required to distribute power to a building from utility power lines. Electrical power distribution systems contain single- and three-phase transformers (utility), single- and three-phase transformers (property owner), meter sockets and meters, and wiring with or without conduit.

In the past, electrical distribution and lighting systems distributed electricity aboveground to save costs and provide easier access for repairs and expansion. **See Figure 2-32.** Today, underground electrical distribution and power for lighting is preferred to help prevent power outages caused by storms.

Underground power-distribution systems provide additional safety and protection when the location of underground lines are clearly marked and known. Underground distribution systems can be damaged by improper digging. Whether aboveground or underground, all electrical devices and equipment are shown on plot plans using symbols. **See Figure 2-33.**

Standard notes found on underground distribution and lighting prints include the following:

1. To provide space for fire protection, minimum clearances are used for transformer foundations:
   a. 10′ from windows (horizontal and vertically)
   b. 5′ from buildings or structures
   c. 10′ from door (entrance), fire escape, and ventilation ducts
2. Concrete pad can be poured or precast.
3. All conduits shall be positioned before positioning pad.
4. Conduit shall be rigid nonmetallic per the NEC®.
5. Where vehicles are present, transformer shall have appropriate barrier.

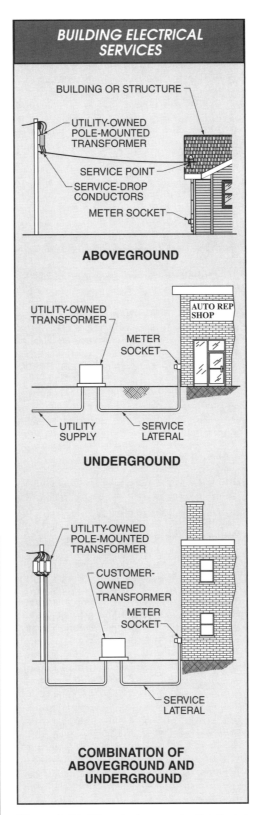

Figure 2-31. Electrical power is distributed to residential and commercial buildings above ground, below ground, or a combination of above and below ground.

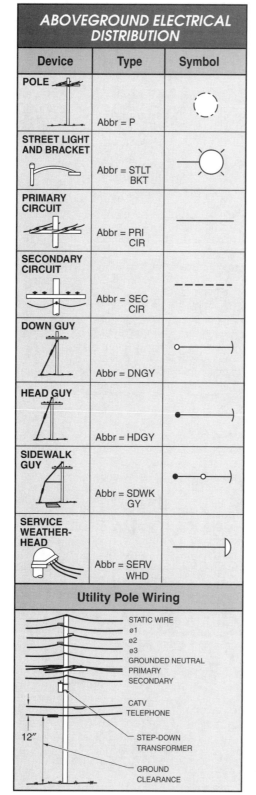

Figure 2-32. Electrical distribution and lighting systems are installed aboveground because aboveground installation saves money and provides easier access for repairs and expansion.

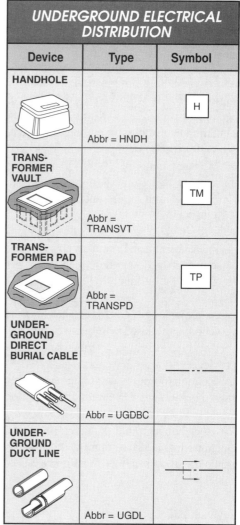

Figure 2-33. Underground electrical distribution and power for lighting is preferred because it prevents power outages caused by storms.

According to the NFPA, electrical distribution and lighting equipment is involved in an estimated 32,000 home fires annually. These 32,000 fires cause annual averages of
- 220 deaths
- 950 injuries
- $674 million in property damage

Overall, electrical distribution and lighting equipment is annually involved in
- 9% of all home fires
- 8% of all home fire deaths
- 6% of all home fire injuries

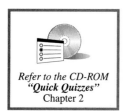

Refer to the CD-ROM "Quick Quizzes" Chapter 2

☑ Example—Filling Out a Commercial Bill of Materials

Step 5: Count the number of standard 3-way switches.
Bedroom #1 = 0
Bedroom #2 = 0
Bedroom #3 = 0
Hallway = 2
Total = 2

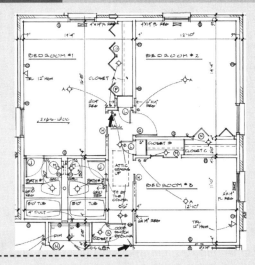

Step 6: Count the number of standard 4-way switches.
Bedroom #1 = 0
Bedroom #2 = 0
Bedroom #3 = 0
Hallway = 0
Total = 0

Step 7: Count the number of dimmer switches.
Bedroom #1 = 0
Bedroom #2 = 0
Bedroom #3 = 0
Hallway = 0
Total = 0

Step 8: Count the number of standard 2-way switches with receptacles.
Bedroom #1 = 0
Bedroom #2 = 0
Bedroom #3 = 0
Closets = 0
Hallway = 0
Total = 0

Step 9: Count the number of standard porcelain lamp fixtures with pull chains.
Bedroom #1 = 0
Bedroom #2 = 0
Bedroom #3 = 0
Closets = 0
Hallway = 0
Total = 0

Step 10: Count the number of standard porcelain lamp fixtures without pull chains.
Bedroom #1 and Closet = 2
Bedroom #2 Closet = 2
Bedroom #3 Closet = 2
Hallway = 2
Hallway Closet = 1
Total = 9

Example—Filling Out a Commercial Bill of Materials

Step 11: Count the number of special electrical devices (door switches).
  Bedroom #1 Closet = 0
  Bedroom #2 Closet = 0
  Bedroom #3 Closet = 0
  Hallway Closet = 1
    Total = 1

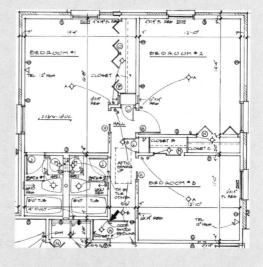

Step 12: Count the number of special electrical devices (telephone jacks).
  Bedroom #1 = 1
  Bedroom #2 = 0
  Bedroom #3 = 1
  Hallway = 0
    Total = 2

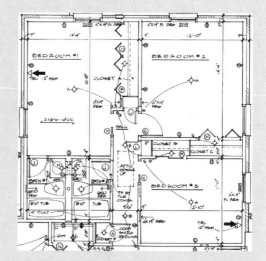

Chapter 2 — Residential and Commercial Electrical Symbols 67

☑ Example—Troubleshooting Game Room Electrical Circuits

2-2

Scenario:
Example 2-2 is a troubleshooting activity requiring that an electrical problem in a game room be diagnosed. The circuit breaker for branch circuit B-9 keeps tripping. Game room personnel swapped the games on circuit B-9 with the games on circuit B-10, but B-9 still has the tripping problem.

Task:
Troubleshoot circuit B-9's electrical problem.

Reference Print:

GAME ROOM — GAMES

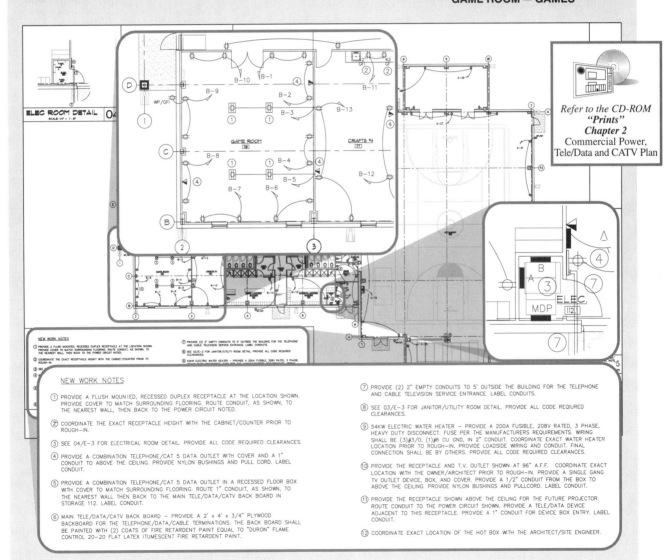

Refer to the CD-ROM "Prints" Chapter 2 Commercial Power, Tele/Data and CATV Plan

NEW WORK NOTES

① PROVIDE A FLUSH MOUNTED, RECESSED DUPLEX RECEPTACLE AT THE LOCATION SHOWN. PROVIDE COVER TO MATCH SURROUNDING FLOORING. ROUTE CONDUIT, AS SHOWN, TO THE NEAREST WALL, THEN BACK TO THE POWER CIRCUIT NOTED.

② COORDINATE THE EXACT RECEPTACLE HEIGHT WITH THE CABINET/COUNTER PRIOR TO ROUGH-IN.

③ SEE 04/E-3 FOR ELECTRICAL ROOM DETAIL. PROVIDE ALL CODE REQUIRED CLEARANCES.

④ PROVIDE A COMBINATION TELEPHONE/CAT 5 DATA OUTLET WITH COVER AND A 1" CONDUIT TO ABOVE THE CEILING. PROVIDE NYLON BUSHINGS AND PULL CORD. LABEL CONDUIT.

⑤ PROVIDE A COMBINATION TELEPHONE/CAT 5 DATA OUTLET IN A RECESSED FLOOR BOX WITH COVER TO MATCH SURROUNDING FLOORING. ROUTE 1" CONDUIT, AS SHOWN, TO THE NEAREST WALL THEN BACK TO THE MAIN TELE/DATA/CATV BACK BOARD IN STORAGE 112. LABEL CONDUIT.

⑥ MAIN TELE/DATA/CATV BACK BOARD — PROVIDE A 2' x 4' x 3/4" PLYWOOD BACKBOARD FOR THE TELEPHONE/DATA/CABLE TERMINATIONS. THE BACK BOARD SHALL BE PAINTED WITH (2) COATS OF FIRE RETARDENT PAINT EQUAL TO "DURON" FLAME CONTROL 20-20 FLAT LATEX ITUMESCENT FIRE RETARDENT PAINT.

⑦ PROVIDE (2) 2" EMPTY CONDUITS TO 5' OUTSIDE THE BUILDING FOR THE TELEPHONE AND CABLE TELEVISION SERVICE ENTRANCE. LABEL CONDUITS.

⑧ SEE 03/E-3 FOR JANITOR/UTILITY ROOM DETAIL. PROVIDE ALL CODE REQUIRED CLEARANCES.

⑨ 54KW ELECTRIC WATER HEATER — PROVIDE A 200A FUSIBLE, 208V RATED, 3 PHASE, HEAVY DUTY DISCONNECT. FUSE PER THE MANUFACTURERS REQUIREMENTS. WIRING SHALL BE (3)#3/0, (1)#6 CU GND, IN 2" CONDUIT. COORDINATE EXACT WATER HEATER LOCATION PRIOR TO ROUGH-IN. PROVIDE LOADSIDE WIRING AND CONDUIT. FINAL CONNECTION SHALL BE BY OTHERS. PROVIDE ALL CODE REQUIRED CLEARANCES.

⑩ PROVIDE THE RECEPTACLE AND T.V. OUTLET SHOWN AT 96" A.F.F. COORDINATE EXACT LOCATION WITH THE OWNER/ARCHITECT PRIOR TO ROUGH-IN. PROVIDE A SINGLE GANG TV OUTLET DEVICE, BOX, AND COVER. PROVIDE A 1/2" CONDUIT FROM THE BOX TO ABOVE THE CEILING. PROVIDE NYLON BUSHINGS AND PULLCORD. LABEL CONDUIT.

⑪ PROVIDE THE RECEPTACLE SHOWN ABOVE THE CEILING FOR THE FUTURE PROJECTOR. ROUTE CONDUIT TO THE POWER CIRCUIT SHOWN. PROVIDE A TELE/DATA DEVICE ADJACENT TO THIS RECEPTACLE. PROVIDE A 1" CONDUIT FOR DEVICE BOX ENTRY. LABEL CONDUIT.

⑫ COORDINATE EXACT LOCATION OF THE HOT BOX WITH THE ARCHITECT/SITE ENGINEER.

Example—Troubleshooting Game Room Electrical Circuits

Step 1: Examine receptacles on Circuit B-9 for obvious problems such as missing or loose screws, loose wires, and indications of shorting, such as burn marks.
* Look for any visible circuit overloading.

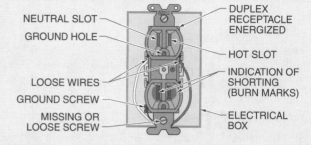

Step 2: Using the electrical print for the game room, identify the receptacle branch circuits in the game room.

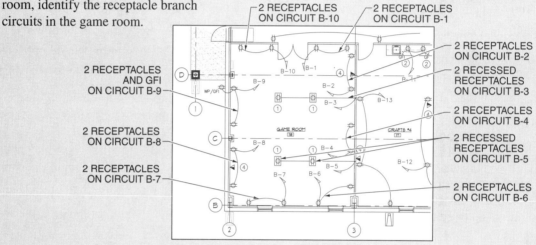

Step 3: Of all the branch circuits in the game room, identify the branch circuit with the additional device.

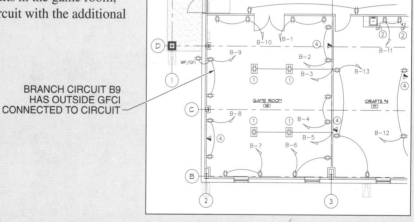

Step 4: Determine the probable cause of branch circuit B-9's tripping problem.

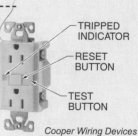

Cooper Wiring Devices

Industrial Electrical and Electronic Symbols
Printreading for Installing and Troubleshooting Electrical Systems

Industrial electrical prints include information about machine connections to power supplies, type and size of power supply conduits and wiring, balancing machine loads on building electrical systems, shielding control circuits, and grounding various electrical systems for proper machine and system operation.

INDUSTRIAL EQUIPMENT

Industrial electrical prints indicate the general location of transformers, panelboards, junction boxes, circuit breakers, busways, and switches. Schematic drawings indicate the various electrical loads, circuits, and demands on the electrical system of a building. Schematic drawings and specifications for machines indicate the various power quality issues a specific machines electrical system may have. **See Figure 3-1.** Besides typical communication and alarm system information, industrial electrical prints will have prints indicating the machine communication (logic) system for automated processes.

Power Sources

All electrical circuits must have a power source. The power source may be chemical (battery), magnetic (generator), heat (thermocouple), or solar (photovoltaic cell). The source of power used depends on the application and the amount of power required.

Battery. A *battery* is a chemical power source that produces electricity through a chemical reaction between battery plates and battery acid. A battery may consist of a single cell or a combination of cells. When individual cells are connected in series, the voltage output of the battery is increased. When individual cells are connected in parallel, the current output of the battery is increased. Typical battery voltage ratings are 1.5 V, 6 V, 9 V, 12 V, 24 V, and 36 V. Batteries are a common source of electricity, especially in portable equipment and hand tools. **See Figure 3-2.**

Batteries are rated in ampere-hours (AH). Ampere-hours represent the amount of amps a battery can deliver in 1 hr before being discharged.
Size D = 20,500 mAH
Size C = 8350 mAH
Size AA = 2850 mAH
Size 9 V = 625 mAH

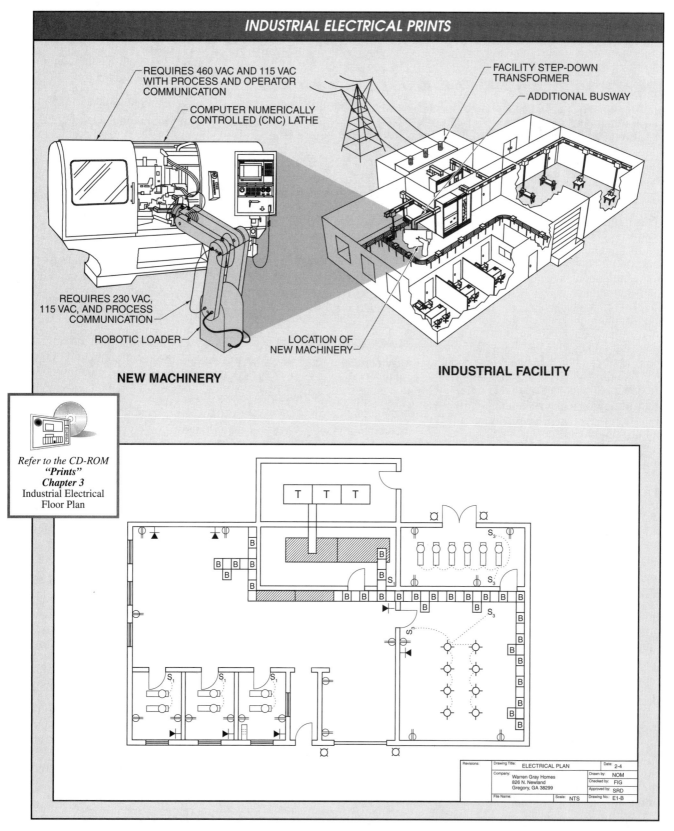

Figure 3-1. All industrial equipment comes with electrical drawings that show how the equipment connects to building systems and process controls.

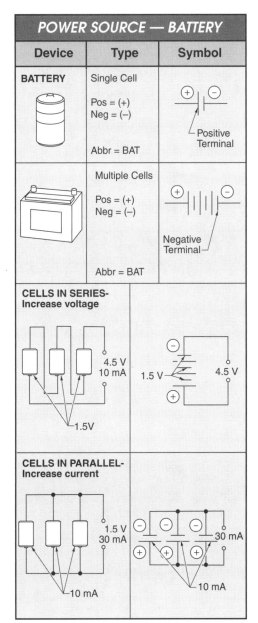

Figure 3-2. Cells or batteries convert chemical energy into electrical energy. Cells in series increase voltage and cells in parallel increase current.

Generator. A *generator* is a magnetic power source that produces electricity when a rotating conductor cuts magnetic lines of force. Magnetic lines of force are produced by the magnetic field that is present between the north and south poles of a magnet. A magnetic field can be produced by permanent magnets or by electromagnets. **See Figure 3-3.**

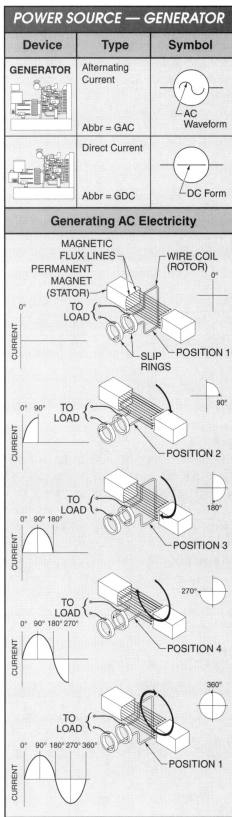

Figure 3-3. Generators convert mechanical energy into electrical energy.

A magnetic field is produced when electric current passes through a wire coil. The voltage produced by a generator depends on the strength of the magnetic field and the speed of rotation. The stronger the magnetic field and the faster the rotation, the higher the produced voltage.

The output of a generator ranges from a few volts to several thousand volts. A force such as a steam, water, or combustion engine rotates the generator. Generators are the most common way to produce electricity.

Thermocouple. A *thermocouple* is a heat power source that produces electricity when two different metals that are joined together are heated. The amount of voltage produced depends on the temperature. The higher the temperature, the higher the voltage produced. However, the amount of electricity produced is very small. Thermocouples produce an analog voltage that is typically used to provide a measurement of temperature. **See Figure 3-4.**

Photovoltaic Cell. A *photovoltaic cell* is a solar power source that produces electricity when light strikes the surface of the cell. The light energy penetrates the cell and forces the electrons to move between the cell terminals. Photovoltaic cells typically produce only about 0.5 V at 10 mA but can be interconnected like battery cells to deliver high voltage and current outputs. **See Figure 3-5.**

Photovoltaic cells are typically used for solar power systems and as power sources in calculators, photographic equipment, and satellite communication systems. The applications for photovoltaic cells grow every year.

POWER SOURCE — THERMOCOUPLE

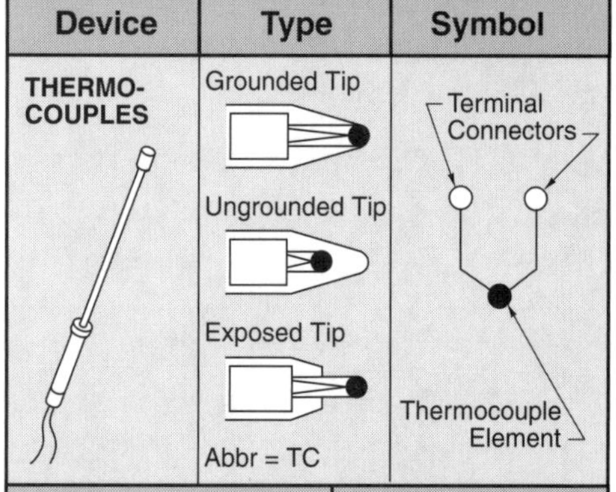

Device	Type	Symbol
THERMO-COUPLES	Grounded Tip / Ungrounded Tip / Exposed Tip / Abbr = TC	Terminal Connectors / Thermocouple Element

Thermocouple Type and Material	Maximum Temp. Range
B – Platinum & Rhodium	32°F to 3092°F
C – Tungsten & Rhenium	–32°F to 4208°F
E – Nickel-Chromium & Copper-Nickel	–328°F to 1652°F
J – Iron & Copper-Nickel	32°F to 1382°F
K – Nickel-Chromium & Nickel-Aluminum	–328°F to 2282°F
N – Nickel-Chromium-Silicon & Nickel-Silicon-Magnesium	–450°F to 2372°F
R – Platinum-Rhodium & Platinum	32°F to 2642°F
S – Platinum-Rhodium & Platinum	32°F to 2642°F
T – Copper & Copper-Nickel	–328°F to 662°F

Thermocouple and Thermowell Assembly

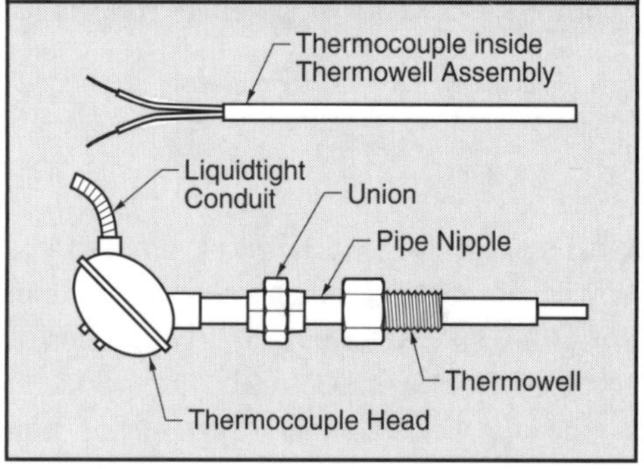

Figure 3-4. Thermocouples convert heat energy into electrical energy by using two dissimilar metals joined at one end that, when heated, allow electrons to flow from one metal to the other.

SOLAR POWER PRINTS

Photovoltaic cells are connected together to form modules, and modules are placed together to create installation panels. When a group of panels is connected together, an array is created.

Drawings must include all component locations and descriptions, whether they are device or architectural prints.

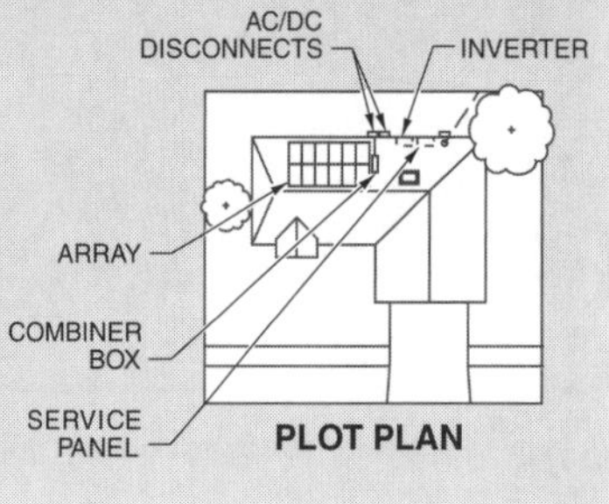

PLOT PLAN: AC/DC DISCONNECTS, INVERTER, ARRAY, COMBINER BOX, SERVICE PANEL

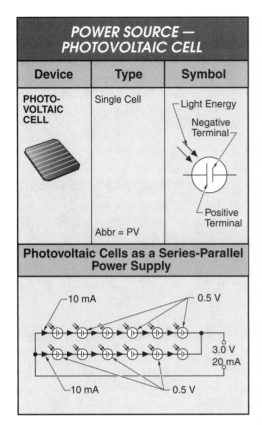

Figure 3-5. Photovoltaic cells convert light energy into electrical energy. Photovoltaic cells produce DC voltage in direct proportion to the amount of light energy striking the cell surface.

DIRECT CURRENT AND ALTERNATING CURRENT

Electricity may be either direct current (DC) or alternating current (AC). DC is produced by batteries, photovoltaic cells, thermocouples, and DC generators. All DC power sources have a definite positive and negative side. DC is used primarily in applications that require a portable power supply, such as flashlights, power tools, and automobiles.

DC is also used for applications that require electronic circuits. In most electronic circuit applications, DC is supplied by a rectified AC power supply. *Rectification* is the process of changing AC electricity into DC electricity. Appliances with high-current electronic circuits, such as computers, VCRs, and TVs, operate on DC that has been rectified from AC. **See Figure 3-6.**

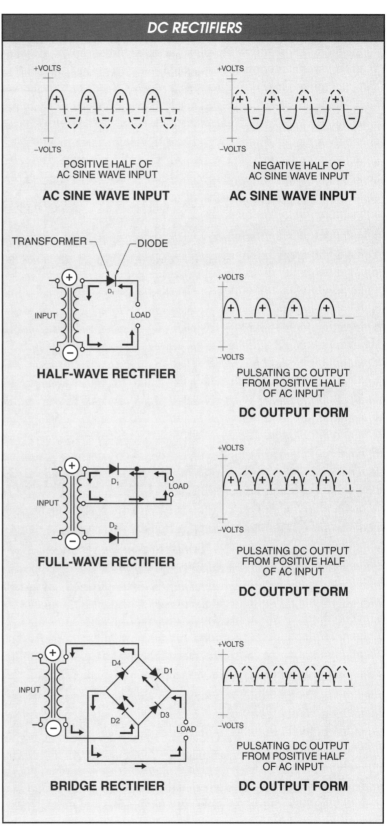

Figure 3-6. Rectifiers change AC electricity into various forms of DC electricity depending on the type of rectifier used.

AC electricity is produced by an AC generator. Small AC generators are typically known as alternators. AC electricity does not have definitive positive and negative poles; the current reverses in direction of flow once per cycle (every $\frac{1}{120}$ of a second). AC electricity is the most common form of electricity. Any application that does not require DC uses AC. Power distribution systems use AC because transformers operate on AC.

Generated voltage is stepped up to allow large amounts of power to be transmitted over great distances. Voltage at the point of use is stepped down to provide low and safe voltage levels for consumers. Typical consumer usages include lighting circuits, heating circuits, and many types of motor applications.

A generator that is connected to an electrical system must be installed with a special device called a transfer switch. A transfer switch is a heavy-duty double-pole switch that disconnects utility power from a system when the generator is in use.

Disconnects and Overcurrent Protection Devices

A *disconnect switch (disconnect)* is a switch that removes electrical power from motors and machines. **See Figure 3-7.** Disconnects are used to manually remove power from or apply power to a circuit. With few exceptions, disconnects are required in all circuits that control large loads over 10 A. Overcurrent protection is added to protect the circuit conductors, control apparatus, and loads from short circuits, grounds, and excessive current levels. A disconnect of some type is used in all motor, high-intensity discharge (HID) lighting, and heating applications.

One disconnect is used for DC, 1ϕ, low-voltage, and 115 VAC circuits. Two disconnects in the same unit are used for 1ϕ, high-voltage, and 230 VAC circuits. Three disconnects are used in the same unit for all 3ϕ circuits. Caution must be used when a disconnect is opened because the input (line) side of the disconnect still has power, even when the switching element is opened and the fuses are removed.

	DISCONNECTS	
Device	**Type**	**Symbol**
DISCONNECT SWITCH 	Single-Pole Single-Throw Abbr = SPST	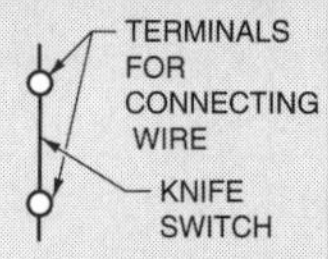 TERMINALS FOR CONNECTING WIRE / KNIFE SWITCH
	Double-Pole Single-Throw Abbr = DPST	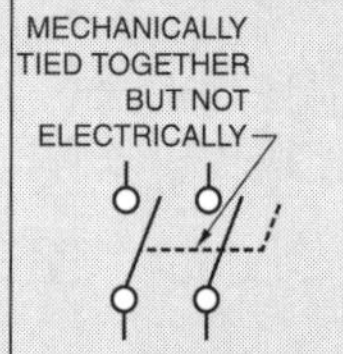 MECHANICALLY TIED TOGETHER BUT NOT ELECTRICALLY
	Three-Pole Single-Throw Abbr = 3PST	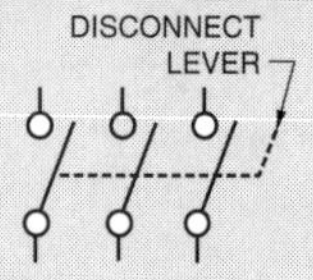 DISCONNECT LEVER
DISCONNECT WITH BREAKERS 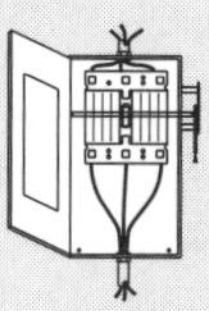	Three-Phase Abbr = DISC SW/CB	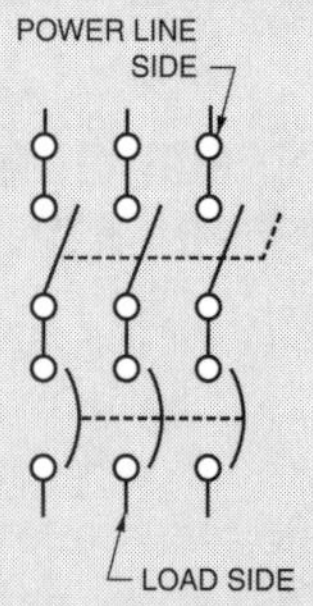 POWER LINE SIDE / LOAD SIDE
DISCONNECT WITH FUSES 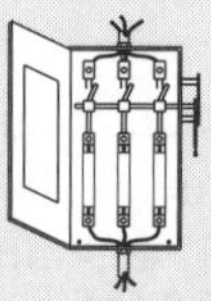	Three-Phase Abbr = DISC SW/FU	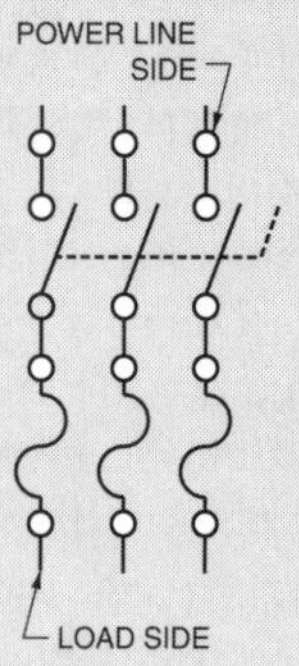 POWER LINE SIDE / LOAD SIDE

Figure 3-7. To apply or remove power from a circuit, the National Electrical Code® (NEC®) requires a means of disconnecting a motor and/or controller from a circuit using disconnects.

Chapter 3—Industrial Electrical and Electronic Symbols   83

An *overcurrent protection device (OCPD)* is a circuit breaker or fuse that provides overcurrent protection to a circuit. **See Figure 3-8.** OCPDs remove power and provide protection for all components connected to the output (load) side of a disconnect switch.

A *thermal overload* is a device that detects the amount of current flowing in a motor circuit by sensing the heat generated by the current flow. The greater the current flow, the higher the temperature. Thermal overloads are set at a preset temperature to operate thermal overload contacts that open the motor starter control circuit. Thermal overloads are also known as heaters.

Disconnect switches include a point at which power can be locked out. Per OSHA standards, equipment must be locked out and tagged out before any maintenance or service is performed. **See Figure 3-9.** *Lockout* is the process of removing the source of power and installing a lock that prevents the power from being turned ON until the lock is removed. *Tagout* is the process of placing a danger tag on the source of electrical power, which indicates that the equipment cannot be operated until the person who placed the tag removes it.

OVERCURRENT PROTECTION DEVICES

Device	Type	Symbol
CIRCUIT BREAKER	Single-Pole Circuit Breaker Abbr = SPCB	CIRCUIT BREAKER ELEMENT
	Double-Pole Circuit Breaker Abbr = DPCB	
	Three-Pole Circuit Breaker Abbr = 3PCB	
FUSE	Single Abbr = FU	FUSE ELEMENT OR
THERMAL OVERLOAD	Three-Phase Abbr = OL	THERMAL ELEMENT (HEATER)

Figure 3-8. Overcurrent protection devices (OCPDs) interrupt power and provide protection for all components connected to the output (load) side of an OCPD when a predetermined value of current has been exceeded.

One OCPD is used for DC, 1ϕ, low-voltage, and 115 VAC circuits. Two OCPDs are used in the same unit for 1ϕ, high-voltage, and 230 VAC circuits. Three OCPDs are used in the same unit for all 3ϕ circuits.

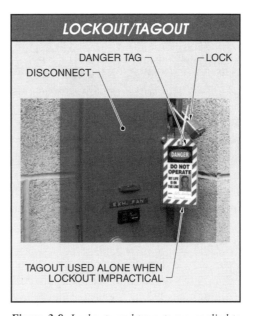

Figure 3-9. Lockouts and tagouts are applied to equipment to prevent injury from energized circuits and to prohibit equipment operation during installation and maintenance.

OSHA Standards:
1910.147—*Lockout and Tagout*
1910.300 to 1910.399—*Electrical*

Contacts

All switches, including relays, contactors, and motor starters, use contacts to start or stop the flow of current in a circuit. A *contact* is the conducting part of a switch or relay that operates with another conducting part to make or break a circuit. The contact position, number of poles, number of throws, and type of break are used to describe the type of contact. **See Figure 3-10.**

Contact life is the number of times the contacts of a switch or relay can be actuated before malfunctioning. Contacts are designed for either low-power switching, such as with relays, or for high-power switching, such as with contactors and motor starters. Contact life is reduced when contacts are used with the incorrect type of load or with switch loads that develop destructive arcs.

Contact position is the position of the contacts before the contacts are activated. Normally open (NO) contacts are open and close when they are activated. Normally closed (NC) contacts are closed and open when they are activated.

The activation of contacts can be accomplished manually (by a person), mechanically (by a machine), or automatically (by a PLC or controller). Typically, contacts are held in the normal position by a spring. Any switch with a spring has a normal position. For example, most doorbell (pushbutton) switches contain a spring, thus the switch contacts have a normal position (NO). Switches used to turn the lights ON in a room (toggle switches) do not contain a spring; the switch contacts do not have a normal position.

Poles. A *pole* is the number of isolated circuit contacts that are used to activate individual circuits. A single pole contact carries current through only one circuit at any one time. A double pole contact carries current through two circuits simultaneously. The two circuits are mechanically connected so that both circuits open or close at the same time while remaining electrically insulated from each other. The symbol for an internal mechanical connection in a switch is a dashed line connecting the poles together. **See Figure 3-11.**

CONTACTS

Device	Type	Symbol
CONTACTS SMALL	Normally Open Abbr = NO	MECHANICAL CONTACT (OPEN) SOLID-STATE CONTACT (OPEN)
LARGE	Normally Closed Abbr = NC	CONTACT CLOSED
SWITCH	Single-Pole Single-Throw Single-Break Abbr = SPST	POLE TERMINALS
	Single-Pole Single-Throw Double-Break Abbr = SPST	BREAK BOTH SIDES
	Single-Pole Double-Throw Single-Break Abbr = SPDT	
	Single-Pole Double-Throw Double-Break Abbr = SPDT	CONTACT CLOSED
	Double-Pole Single-Throw Single-Break Abbr = DPST	MECHANICALLY CONNECTED
	Double-Pole Double-Throw Single-Break Abbr = DPDT	COMMON TERMINALS
	Double-Pole Double-Throw Double-Break Abbr = DPDT	
	Multiple Contact Rotary	COMMON TERMINAL

Figure 3-10. A contact is a conducting part of a switch or relay that operates with the pole to make or break a circuit.

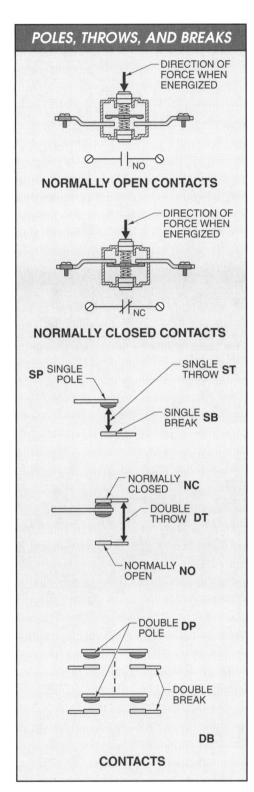

Figure 3-11. The pole is the number of isolated circuit contacts that are used to activate individual circuits, with the throw being the number of closed contact positions per pole and the break being the number of separate places on the contacts where the circuit opens and closes.

Throws. A *throw* is the number of closed contact positions per pole. Throws denote the number of different circuits that each individual pole is capable of controlling. When a pole can control only one circuit, the pole is a single-throw. When the pole can control two circuits, the pole is a double-throw.

Breaks. All contacts are either single-break or double-break. A *break* is the number of separate places on the contacts where the circuit opens and closes. A *single-break contact* is a contact that breaks the electrical circuit in one place. A *double-break contact* is a contact that breaks the electrical circuit in two places.

Control Switches

A *control switch* is a switch that controls the flow of current in a circuit. Switches can be activated manually, mechanically, or automatically. Once activated, the switch operates the contacts. The contacts are used to start or stop the flow of current.

Manually operated control switches include pushbuttons, selector switches, and foot switches. A *manually operated switch* is a switch that is activated by a person and is the most common type of switch used in control circuits. Even in automatically operated circuits, manual switches are typically included to override the automatic switches. The ability to override automatic switches is necessary when troubleshooting and in the event of an emergency. Typical uses of manually operated switches include turning lights and equipment off and on. **See Figure 3-12.**

Manually operated switches are represented differently than mechanically operated and automatically operated switches. In all manually operated switches, the operator is drawn above the terminals for NO switches and below the terminals for NC switches.

Capacitive touch sensors are developed for any product that uses mechanical switches and are manufactured using traces on standard two-layer PCBs.

Limit switches are the only type of mechanically operated switches. A *limit switch* is a switch that detects the physical presence of an object and is typically used as a safety device. Typical uses include detection to determine when a door is closed (automobiles, microwaves, and ovens), when safety guards are in place, and when machine parts are in the proper position or products are properly placed. **See Figure 3-12.** In all mechanically operated switches, the operator is drawn below the terminals for NO switches and above the terminals for NC switches.

MANUALLY OPERATED CONTROL SWITCHES

Device	Type	Symbol
PUSHBUTTON	Single-Circuit Normally Open Abbr = PB or PB-NO	MANUAL OPERATION TERMINALS
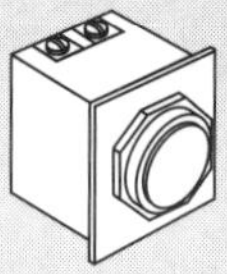	Single-Circuit Normally Closed Abbr = PB or PB-NC	
OR	Double-Circuit Normally Open and Closed Abbr = PB or PB-NO/NC	
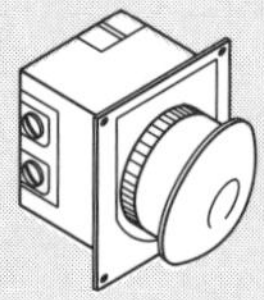	Mushroom Head Abbr = PB or PB-NO/NC	MUSHROOM OPERATOR
	Maintained Abbr = PB	MECHANICAL LINK
	Illuminated R = Red G = Green A = Amber B = Blue Abbr = PB/LT	R = RED R LAMP INSIDE OPERATOR
SELECTOR SWITCH 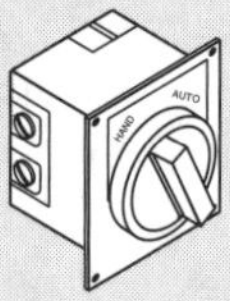	Two-Position Abbr = SS or SELT SW	JOG RUN A1 A2 (J, R; A1: X, blank; A2: blank, X)
	Three-Position Abbr = SS or SELT SW	STOP JOG RUN A1 A2 (J, S, R; A1: X, blank, blank; A2: blank, blank, X) J = Jog R = Run X = Contacts closed Blank = Contacts open
FOOT SWITCH 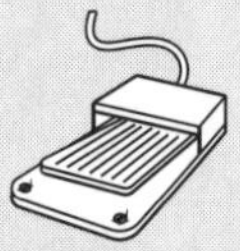	Normally Open Abbr = FTS	FOOT OPERATOR
	Normally Closed Abbr = FTS	

Figure 3-12. Manually operated switches require a person to activate them.

MECHANICALLY OPERATED CONTROL SWITCHES

Device	Type	Symbol
LIMIT SWITCH (MECHANICAL)	Normally Open Abbr = LS	MECHANICAL OPERATOR
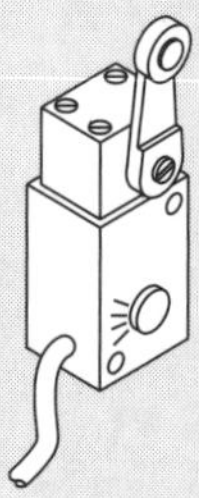	Normally Open, Held Closed Abbr = LS	OPERATOR IN CLOSED POSITION
	Normally Closed Abbr = LS	
	Normally Closed, Held Open Abbr = LS	OPERATOR IN OPEN POSITION
LIMIT SWITCH (SOLID-STATE) 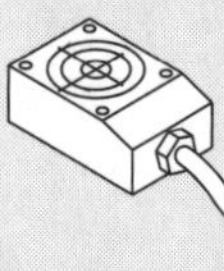	Normally Open Abbr = PROX	SOLID-STATE OR
	Normally Closed Abbr = PROX	OR

Figure 3-13. Mechanically operated switches (limit switches) detect the physical presence of an object by converting mechanical motion into an electrical signal or by using magnetism to detect the presence or absence of an object using a proximity switch (solid-state limit switch).

An *automatically operated switch* is a switch that maintains the operation of electrical circuits with little or no manual input. Automatically operated control switches include photo electric, temperature, pressure, flow, and level switches. Typical uses include keeping ovens and rooms at a set temperature, basements dry, tanks full, pressure in a system, and products moving. **See Figure 3-14.** In all automatically operated switches, the operator is drawn below the terminals for NO switches and above the terminals for NC switches.

Relays and Timers

A *relay* is an interface device that controls one electrical circuit by opening and closing contacts with another low-voltage circuit. **See Figure 3-15.** Relays multiply the number of available output contacts and/or permit a low current to switch a high current. The two basic types of relays are electromechanical and solid-state.

An *electromechanical relay* is a relay with multiple contacts that are typically used in a control circuit to switch low currents. An electromechanical relay uses a movable armature to open or close its relay contacts. The armature is moved by the electromagnetic field that is created when power is applied to the coil of the relay. The contacts of electromechanical relays can be NO, NC, or any combination of both.

A *solid-state relay* is a relay that uses electronic switching devices, such as SCRs and triacs, in place of mechanical contacts to switch current flow. The electronic switching device is controlled by power being applied to or removed from the input terminal of the relay.

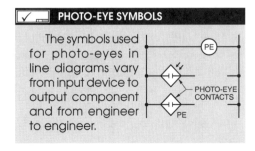

PHOTO-EYE SYMBOLS

The symbols used for photo-eyes in line diagrams vary from input device to output component and from engineer to engineer.

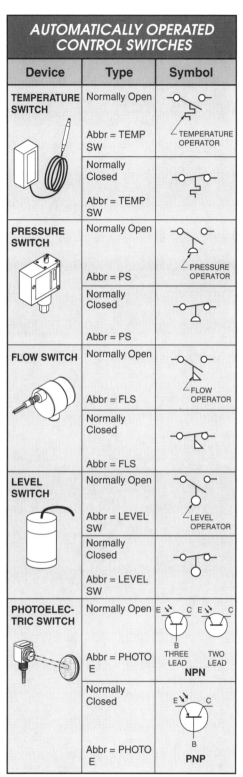

Figure 3-14. Automatically operated switches maintain the operation of electrical circuits with little or no manual input, such as opening or closing contacts by responding to temperature changes, pressure changes, movement of a fluid, or the level of a substance.

RELAYS AND TIMERS		
Device	Type	Symbol
CONTROL RELAY (MECHANICAL)	Coil Abbr = CR	CR
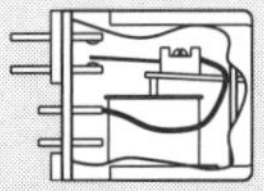	Normally Open Contact Abbr = NO	─┤ ├─
	Normally Closed Contacts Abbr = NC	─┤/├─
CONTROL RELAY (SOLID-STATE)	Input (Coil) Abbr = SSR	◇SSR◇
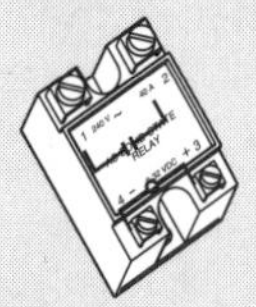	Normally Open Output Abbr = NO	◇─┤ ├─◇ SSR
	Normally Closed Output Abbr = NC	◇─┤/├─◇ SSR
TIMER, ON DELAY	Coil Abbr = TR	TR
	Normally Open, Timed Closed Abbr = NOTC	TC ─┤ ├─ OR ─o/o─ TIMED CLOSED
	Normally Closed, Timed Open Abbr = NCTO	TIMED OPEN TO ─┤/├─ OR ─o/o─ POINTS IN THE DIRECTION OF TIME RELAY
TIMER, OFF DELAY	Coil Abbr = TR	TR
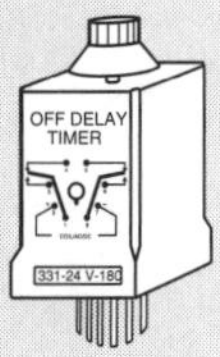	Normally Open, Timed Open Abbr = NOTO	TO ─┤ ├─ OR ─o/o─
	Normally Closed, Timed Closed Abbr = NCTC	TC ─┤/├─ OR ─o/o─ POINTS IN THE DIRECTION OF TIME DELAY

Figure 3-15. Relays are interface devices that open and close contacts using a low-voltage control circuit. Timers use a preset time period to delay starting and stopping.

Relays control loads that are typically rated at less than 10 A. Relays are used to directly switch lamps, solenoids, small heating elements, and motors of less than 1 HP. When controlling high-current loads, relays are used to control the contactors of lights and heating elements or the motor starters of motors. High-current loads are always connected to contactors or motor starters, which are designed to handle the large currents required by the load.

A *timer* is a control device that uses a preset time period as part of the control function. Timers have mechanical contacts or solid-state switching devices. Depending on the type of timer used, the time period can be applied at different intervals of circuit operation. Time-delay relays are available in timing ranges of a few milliseconds to hundreds of hours.

Like relays, timers can be used to directly switch low-current loads. However, timers are typically used to indirectly control high-current loads by switching relays, contactors, or motor starters ON.

> **TIMER MODES OF OPERATION**
>
> The output of an on-delay timer is de-energized until the end of the time delay, then the output is energized. The output of an interval timer is energized until the end of the time delay, then the output is deenergized. The output of an off-delay timer is energized when power is removed and the output stays energized until the end of the timer delay. The output of a single shot timer is energized (1 time) for the duration of the time delay.

Contactors and Motor Starters

A *contactor* is an electrically operated switch that uses a low control-circuit current to energize or de-energize a high-current-load circuit. **See Figure 3-16.** Contactors are devices that turn high-current loads ON and OFF and that do not require overload protection. Contactors are used to switch lights, heating elements, and solenoids. These loads require overcurrent protection, but not overload protection.

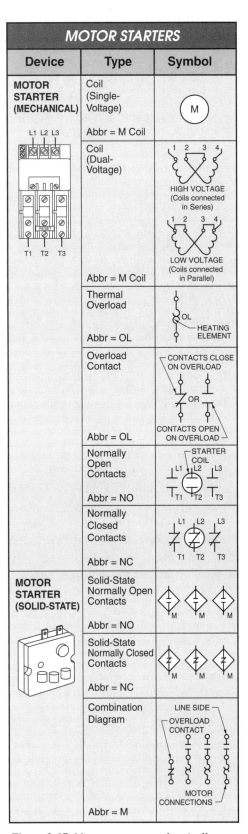

Figure 3-16. Contactors are devices that turn high-current loads such as lights, heating elements, and solenoids ON and OFF.

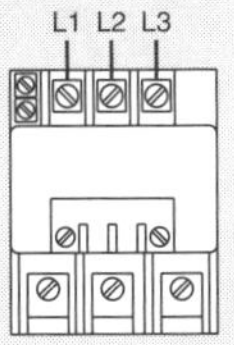

A *motor starter* is an electrically operated switch (contactor) for use with motors that includes 3φ motor overload protection. **See Figure 3-17.** Overload protection, in addition to overcurrent protection, is required with all motors. Overload protection is provided by heaters or electronic circuits, which are included as part of the motor starter. Overload protection protects the motor from overheating when the motor is running. Overcurrent protection is provided by fuses and circuit breakers that protect the motor during start-up or when a short occurs.

Both overload and overcurrent protection are required with motor circuits because motors draw a different amount of current when starting than when running. Motors draw approximately two to five times more current when starting than when running. Overload protection is selected based on the running current rating of the motor. Overcurrent protection is selected based on the starting current rating of the motor and is designed to react in a few milliseconds. Overload protection devices (heaters) are designed to react after 5 seconds or more.

Figure 3-17. Motor starters are electrically operated motor START/STOP switches (contactors) that include 3φ motor overload protection.

Resistors

Electrical resistance is the opposition to electron flow of any material. A *resistor* is a component with a specific amount of electrical resistance. **See Figure 3-18.** Resistors are classified by resistance value (in ohms) and power dissipation (in watts). Resistors are used for dividing voltage, dropping voltage, limiting current, and developing heat.

The two basic types of resistors are fixed and variable. A *fixed resistor* is a resistor with a set value, such as 250 Ω. A *variable resistor* is a resistor with a set range of values, such as 0 Ω to 1000 Ω.

Variable resistors are either linear or nonlinear. In a linear variable resistor, 1° of shaft rotation results in the same change of resistance regardless of the position of the shaft. In a nonlinear variable resistor, 1° of shaft rotation results in a different change of resistance for each position of the shaft. Fixed and variable resistors are available in a wide range of values ranging from 1 Ω to more than 10,000 Ω.

Variable resistors are used in applications where it is necessary to make an electrical change in a circuit. By varying the resistance of a circuit, the power output of the circuit is changed. Small variable resistors (potentiometers) are used to change the volume output of speakers, brightness of a TV screen, or the amount of screen contrast. Large variable resistors (rheostats) are used to change the heat output of heating elements (such as in an electric space heater or range) or to control the brightness of a lamp.

Resistance wire is wire with a fixed resistance that is designed not to melt while providing high heat. Resistance wire has a high enough resistance so that the wire turns bright red and gives off heat when energized. Resistance wire is used to heat rooms, dry clothes, and make toast.

The resistance of a variable resistor can be changed manually or automatically. A thermistor, which is heat sensitive, automatically changes in resistance with a change in temperature. **See Figure 3-19.**

RESISTORS

Device	Type	Symbol
RESISTOR	General	—/\/\/—
	Abbr = RES	R₁ = RESISTOR 1
	Variable	DIAGONAL LINE INDICATES VARIABLE
	Abbr = RES or POT	
	Adjustable	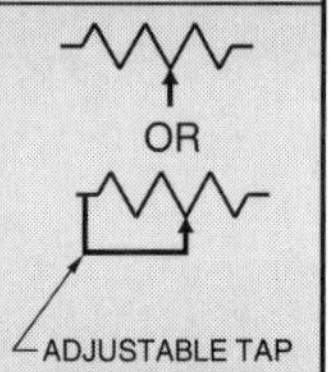 OR — ADJUSTABLE TAP CONNECTED TO ONE OF THE INPUTS
	Abbr = RES or RH	
	Tapped	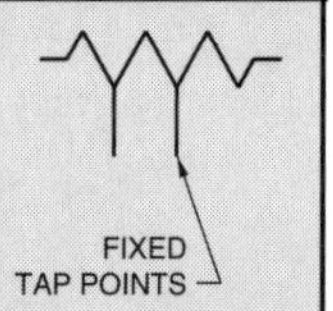 FIXED TAP POINTS
	Abbr = RES	

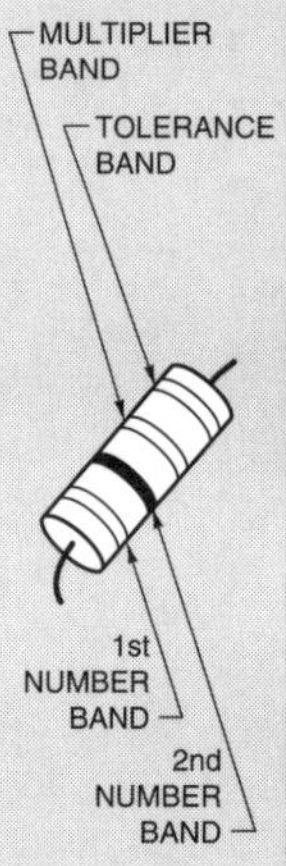

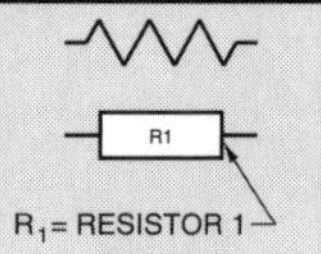

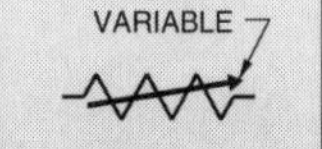

Resistor Color Codes

Color	Digit 1st	Digit 2nd	Multiplier	Tolerance*
Black (BK)	0	0	1	0
Brown (BR)	1	1	10	—
Red (R)	2	2	100	—
Orange (O)	3	3	1000	—
Yellow (Y)	4	4	10,000	—
Green (G)	5	5	100,000	—
Blue (BL)	6	6	1,000,000	—
Violet (V)	7	7	10,000,000	—
Gray (GY)	8	8	100,000,000	—
White (W)	9	9	1,000,000,00	—
Gold (Au)	—	—	0.1	5
Silver (Ag)	—	—	0.01	10
None	—	—	0	20

* in %

Figure 3-18. Resistors are electrical devices that have a specific amount of electrical resistance.

The higher a resistor's wattage rating (heat dissipation), the larger the resistor's physical size. Resistor wattage ratings are 1/4 W, 1/2 W, 1 W, 2 W, 5 W and up.

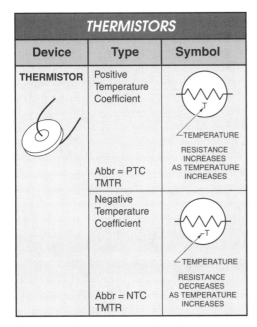

Figure 3-19. A thermistor is a heat-sensitive resistor that, with a change in temperature, either increases or decreases in resistance depending on whether the thermistor has a positive or negative coefficient.

The two types of thermistors are the positive temperature coefficient thermistor and the negative temperature coefficient thermistor. A *positive temperature coefficient (PTC) thermistor* is a thermistor whose resistance value increases with an increase in temperature. A *negative temperature coefficient (NTC) thermistor* is a thermistor whose resistance value decreases with an increase in temperature.

Capacitors

Capacitance is the ability of an object to store energy in the form of an electrical charge. Capacitance is the property of an electrical device that permits the storage of electrically separated charges when potential differences exist between the conductors.

A *capacitor* is an electronic device used to store an electrical charge. **See Figure 3-20.** Capacitors can take on many different forms, but all capacitors have two electrodes, often in the form of plates that are separated by a dielectric material. A *dielectric material* is a medium in which an electrical field is maintained with little or no outside energy supply.

Capacitors are used as filters in AC circuits, to block DC voltages in electronic circuits, and to improve torque in motors. Capacitors can be fixed or variable. A *fixed capacitor* is a capacitor that has one value of capacitance. A *variable capacitor* is a capacitor that varies in capacitance value.

Fixed Capacitors. Fixed capacitors are classified according to the dielectric used. Fixed capacitors include ceramic, electrolytic, mica, and paper. Ceramic capacitors use ceramic as the dielectric. Ceramic capacitors typically range in size from 500 pF to 0.01 µF.

Electrolytic capacitors use a paste as the dielectric. Electrolytic capacitors provide more capacitance for a specific size than any other type of capacitor. Because of this, electrolytic capacitors are used when a very high capacitance is required. Electrolytic capacitors can reach 100,000 µF of capacitance or more.

Mica capacitors use mica as the dielectric. Mica capacitors are small in size and have a very low capacitance rating. However, they have extremely high voltage ratings. Because of this, mica capacitors are often used in high-voltage circuits. They typically range from 1 pF to about 0.1 µF. Paper capacitors use paper or a plastic film as the dielectric. Paper capacitors have a low capacitance rating, typically 0.001 µF to 1 µF.

VISUAL INSPECTION OF CAPACITORS

Electrolyte failure in a capacitor causes the production of hydrogen gas, which creates pressure in the capacitor housing. The most common method of identifying electrolyte failure in a capacitor is a visual inspection. When hydrogen gas is building up in a capacitor, the vent (the small impression on the top) will begin to bulge.

CAPACITORS		
Device	Type	Symbol
CAPACITORS  Abbr = CAP	Fixed	 —│├—
	Variable	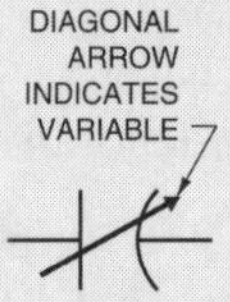 DIAGONAL ARROW INDICATES VARIABLE —│╱├—
	Electrolytic Polarized Abbr = CAP	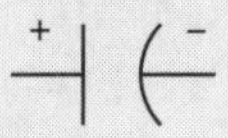 + —│├— -

Ceramic Capacitor Color Codes

Color	Number	Multiplier	Tolerance*	
			Over 10 pF*	10 pF or less†
Black	0	1	20	2.0
Brown	1	10	1	—
Red	2	100	2	—
Orange	3	1000	—	—
Yellow	4	—	—	—
Green	5	—	—	—
Blue	6	—	—	—
Violet	7	—	5	0.5
Gray	8	0.01	—	0.25
White	9	0.1	10	1.0

* in %
† in pf

MICA Capacitor Color Codes

Color	Number	Multiplier	Tolerance*	Voltage†
None	—	—	20	500
Black	0	1	—	—
Brown	1	10	1	100
Red	2	100	2	200
Orange	3	1000	3	300
Yellow	4	10,000	4	400
Green	5	100,000	5 (EIA)	500
Blue	6	1,000,000	6	600
Violet	7	10,000,000	7	700
Gray	8	100,000,000	8	800
White	9	1,000,000,000	9	900
Gold	—	0.1	5 (JAN)	1000
Silver	—	0.01	10	2000

* in %
† in volts

Figure 3-20. A capacitor is an electronic device used to store an electrical charge.

Variable Capacitors. Variable capacitors typically use air as the dielectric and include both movable and stationary metal plates. When the movable plates are fully meshed together with the stationary plates, the capacitance is at maximum. Moving the plates apart lowers the capacitance. Variable capacitors are used as tuning capacitors in radio receivers. The capacitance is varied to tune in different stations.

Diodes

A *diode* is a semiconductor device that allows current to flow in only one direction by offering very high opposition to current in one direction and very low opposition to current in the other. **See Figure 3-21.** Diodes are also known as rectifier devices because diodes are used to change AC electricity into pulsating DC electricity. Diodes are rated according to type, voltage level, and current capacity.

The most common types of diodes include rectifier, zener, tunnel, photoconductive, and light-emitting. Rectifier diodes are used in very low-current electronic circuits and very high-power electrical circuits. Rectifier diodes are available in current ranges of a few milliamps to over 1000 A.

Zener diodes are designed to operate in a reverse-biased mode as voltage regulators without being damaged. When reverse-biased, a zener diode allows varying amounts of reverse current to flow through it but still maintains a relatively constant voltage drop across itself.

Tunnel diodes are designed so that the current flowing through the diode decreases with an increase in applied voltage for a specified range of forward voltage. Tunnel diodes operate as amplifiers or oscillators in electronic circuits and are commonly used in logic circuits.

Photoconductive diodes conduct current when energized by light. Current increases with the intensity of the light. Light-emitting diodes (LEDs) emit light when forward-biased. LEDs are typically used as visual indicators because of their durability and long life.

DIODES

Device	Type	Symbol
DIODE	A = Anode C(K) = Cathode Abbr = D or DIO	A ▶︎— C PATH OF CURRENT FLOW
	Bridge Rectifier Abbr = D or DIO	AC INPUT CONNECTIONS AC +DC −DC AC DC OUTPUT CONNECTIONS
DIODE (ZENER)	Abbr = DZ or DIO	A ▶︎— C
DIODE (TUNNEL)	Abbr = DT or DIO	A ▶︎— C
DIODE (PHOTOCON-DUCTIVE)	Abbr = Photo D or Photo DIO	LINES INDICATE LIGHT INPUT A ▶︎— C
LIGHT EMITTING DIODE (LED)	Abbr = LED	LINES INDICATE LIGHT OUTPUT A ▶︎— C

Figure 3-21. A diode is a semiconductor device that allows current to flow in one direction only and is typically used as a rectifier to change AC electricity into pulsating DC electricity.

Thyristors

A *thyristor* is a solid-state switching device that turns current ON when it receives a quick pulse of control current at its gate. **See Figure 3-22.** Thyristors are designed to be either ON or OFF. In the ON position, thyristors allow current to flow. In the OFF position, thyristors block the flow of current in a circuit.

THYRISTORS

Device	Type	Symbol
SILICON-CONTROLLED RECTIFIER	SCR Abbr = SCR	DIRECTION OF CURRENT FLOW A — G K(C) A = Anode G = Gate K (C) = Cathode
	Anti-Parallel SCR Abbr = SCR	USED IN AC CIRCUITS
DIAC	Abbr = DIAC	T — T
TRIAC	Abbr = TRIAC	MT2 — G MT1 MT1 = Main Terminal 1 MT2 = Main Terminal 2 G = Gate

Figure 3-22. Thyristors are solid-state switching devices such as silicon-controlled rectifiers (SCRs), diacs, and triacs that turn current ON by using a quick pulse of control current to a gate.

Once a thyristor is switched ON, no further control current is required to keep it in that state. A thyristor is switched OFF only when the current flowing through the thyristor reaches zero. Thyristors can switch currents from a few milliamperes to over 1000 A. The three basic types of thyristors are the silicon-controlled rectifier (SCR), diac, and triac.

A *silicon-controlled rectifier (SCR)* is a three-terminal semiconductor thyristor that is normally open until a signal applied to the gate terminal switches the SCR into the conducting state for one direction. SCRs are used to control the flow of current in DC circuits. When used in AC circuits, SCRs rectify the AC current in addition to controlling current flow. Two SCRs can be used to switch alternating current.

A *diac* is a thyristor that triggers in either direction when the breakover voltage of the diac is exceeded. Diacs are typically used in trigger circuits to maintain nearly equal firing delay angles for each half-cycle of an AC supply voltage. A delay angle is produced because the diac must first reach its breakover voltage level before allowing current to flow.

Diacs function like an open switch (in both directions) until a set voltage level is reached. Once the voltage level has been reached, diacs conduct current. Diacs that are conducting current continue to do so until the current is reduced to a minimum level.

A *triac* is a three-terminal semiconductor thyristor that is triggered into conduction in either direction through a small amount of current to its gate terminal. Triacs are used to control the flow of current in AC circuits. Triacs vary the amount of power applied to a load. For example, a solid-state lamp dimmer uses a triac to vary the brightness of the lamp. Triacs are typically used in low-current (less than 50 A) AC circuits. Antiparallel SCRs are used for high-current AC circuits.

ISOLATION DIODES FOR MULTIPLE POWER SUPPLIES

In a circuit with a back-up power supply, the voltage of the primary power supply is set approximately 0.2 V higher than the back-up supply. This causes the diode of the back-up supply to be reverse biased, which allows the primary power supply to deliver current to the load. When the output voltage of the primary supply decreases by more than 0.2 V, the condition is reversed and only the back-up supply delivers load current.

Transistors

A *transistor* is a solid-state device that is used as a switch or as a signal amplifier. **See Figure 3-23.** Transistors are three-terminal devices made of semiconductor material and are part of almost every electronic circuit. Transistors are used to switch or to increase current flow in DC circuits.

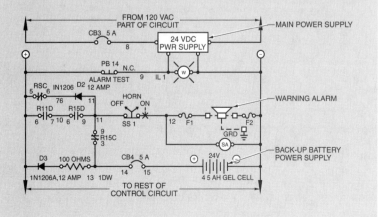

Figure 3-23. Transistors are solid-state devices that are used as switches or as signal amplifiers.

Power must be applied to all three terminals in order for a transistor to operate. Two of the terminals are used as the switching element. The other terminal is used as the control terminal. A very small current change at the control terminal produces a very large current change in the switching element.

Power transistors have a change ratio that is typically about 1 to 20. For example, a 2 mA input can control a 40 mA output. Signal transistors have a change that is typically about 1 to 400. For example, a 250 µA input can control a 100 mA output. In a transistor, the emitter and collector are used as the switching terminals, and the base is used as the control terminal. A small change in the base circuit results in a much higher change in the emitter to collector circuit.

A *phototransistor* is a transistor that controls the amount of current flowing through the emitter to the base junction based on the amount of light encountered. Phototransistor current flow increases with an increase in light. A phototransistor is a combination of a photodiode and a transistor.

Digital Logic Gates

A *digital logic gate* is a circuit that performs a special logic operation such as AND, OR, NOT, NAND, or NOR. Digital logic gates are used in most electronic circuits. Digital circuits operate on binary signals (zeros and ones). **See Figure 3-24.**

International Rectifier

Integrated circuit (IC) packages of digital logic gates are manufactured in various sizes.

DIGITAL LOGIC GATES

Device	Boolean Expression	Symbol
AND GATES	$Y = A \cdot B$ ($\cdot$ = AND)	
OR GATES	$Y = A + B$ (+ = OR)	
NOT (INVERTER) GATES	$Y = \overline{A}$ (− = NOT)	
NOR GATES	$Y = \overline{A + B}$	
NAND GATES	$Y = \overline{A \cdot B}$	
EXCLUSIVE (XOR) GATES	$Y = A \oplus B$ ($\oplus$ = EXCLUSIVE)	

Truth Tables

AND Inputs		Output	OR Inputs		Output
A	B	Y	A	B	Y
0	0	0	0	0	0
0	1	0	0	1	1
1	0	0	1	0	1
1	1	1	1	1	1

NOT Inputs	Output	NOR Inputs		Output
A	Y	A	B	Y
0	1	0	0	1
1	0	0	1	0
		1	0	0
		1	1	0

NAND Inputs		Output	XOR Inputs		Output
A	B	Y	A	B	Y
0	0	1	0	0	0
0	1	1	0	1	1
1	0	1	1	0	1
1	1	0	1	1	0

Figure 3-24. Digital logic gates use binary signaling (zeros and ones) to perform AND, OR, NOT, NOR, NAND, and XOR operations.

Binary signals have only two states. A binary signal is either high (1) or low (0). A high signal is typically 5 V, but can range from 2.4 V to 5 V. A low signal is typically 0 V, but can range from 0 V to 0.8 V.

The five basic logic gates are AND, OR, NOT (inverter), NOR, and NAND. An AND gate has two or more inputs and one output. When both or all inputs are ON, the output is ON. An OR gate has two or more inputs and one output. When either or all inputs are ON, the output is ON. A NOT gate has one input and one output. The output of a NOT gate is always opposite the input. A NOR gate has two or more inputs and one output. When any input is ON, the output is OFF. A NAND gate has two or more inputs and one output. When both or all inputs are ON, the output is OFF.

In addition to the five basic logic gates, additional logic gates are also available. For example, an exclusive OR gate is another type of logic gate. An exclusive OR gate has two inputs and one output. When one (and only one) input is ON, the output is ON. Exclusive OR circuits are used when only one condition can be true, such as with forward and reversing circuits.

Coils

A *coil* is a winding of insulated conductors arranged to produce a magnetic field. **See Figure 3-25.** Coils are also known as inductors or chokes. Coils are made by winding a length of wire conducting material around a core of air or iron. The conductor is typically a copper wire coated with a thin layer of enamel insulation. The iron core concentrates the magnetic field. When current flows through the coil, the coil develops a magnetic field and the magnetic field opposes a change of current in the circuit in which the coil is used.

Inductance is the opposition to a change of current in an AC circuit. The amount of inductance in a coil depends on the current in the wire, the number of turns in the coil, and the applied frequency. As frequency increases, the inductance of the coil increases. Increasing the number of turns also increases the coil's inductance. Inductance is present in any AC circuit that includes a coil, such as a solenoid, motor, transformer, or lighting ballast. Without the effect of inductance created by a coil of wire, the coil of wire would act like a short circuit when connected to a power supply.

COILS

Device	Type	Symbol
COIL 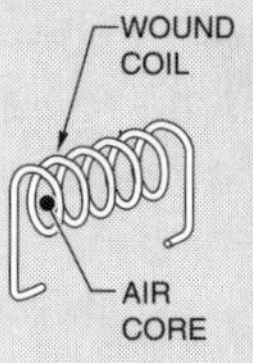	Air Core Abbr = L or INDT	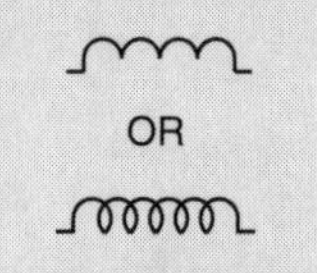
	Iron Core Abbr = L or INDT	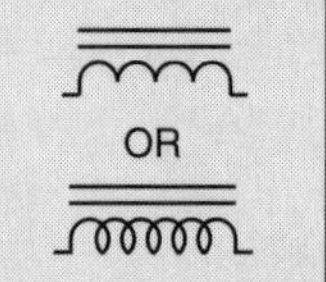
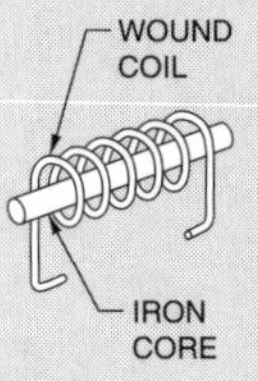	Tapped Core Abbr = L or INDT	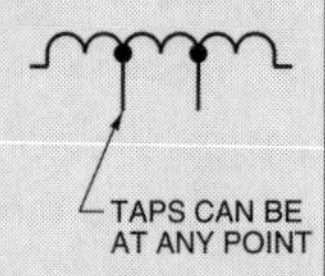
	Adjustable Core Abbr = L or INDT	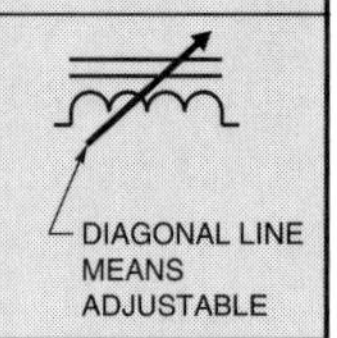

Figure 3-25. Coils are made by winding a length of insulated conducting material (wire) around a core of air or iron.

Solenoids

A *solenoid* is an electrical output component that converts electrical energy into linear mechanical force. A solenoid has an electromagnet that moves an iron plunger when energized. The plunger transmits the force created by the solenoid into useful mechanical work. The moving plunger can be used to control electrical contacts, as in contactors and motor starters, or can be used directly to move nonelectric devices, such as the spool of a valve. **See Figure 3-26.**

Chapter 3—Industrial Electrical and Electronic Symbols 97

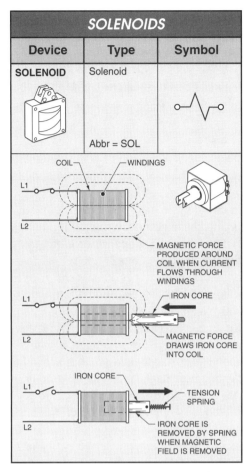

Figure 3-26. Solenoids are electrical output components (loads) that convert electrical energy into linear mechanical force by using an electromagnet to move an iron plunger.

> The K rating (K-1, K-4, K-9, K-13, K-20, K-30, or K-40) of a power transformer represents the extra ability of the transformer to power nonlinear loads, such as computers, which produce harmonics that result in heat buildup.

Transformers

A *transformer* is an electrical interface device that has no moving parts and is designed to change AC electricity from one voltage level to another voltage level. **See Figure 3-27.** Voltage is changed through the interaction of the two magnetic fields in the transformer. Supply voltage is applied to the primary-side coil. Loads are connected to the secondary-side coil.

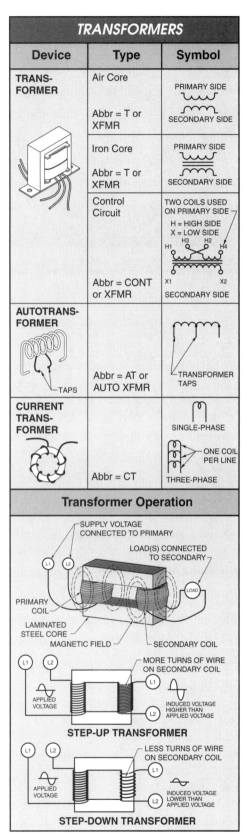

Figure 3-27. A transformer changes AC electricity from one voltage level to another voltage level.

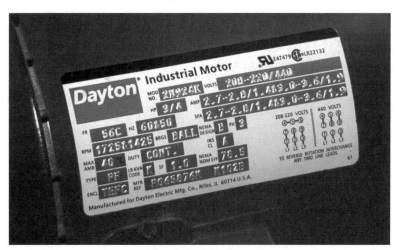

A motor nameplate consists of abbreviated and coded information that lists the electrical, operating, environmental, and mechanical ratings of the motor. The nameplate typically also includes the motor's wiring diagram.

A *step-up transformer* is a transformer in which the secondary-coil output voltage is greater than the primary-coil input voltage. A *step-down transformer* is a transformer in which the secondary-coil output voltage is less than the primary-coil input voltage. Transformers can step up or step down voltage but cannot change the amount of power. The power is the same (minus about 2% for heat loss) on both sides of a transformer. Power is maintained because when voltage is stepped up, the current is proportionally stepped down, and when voltage is stepped down, the current is proportionally stepped up.

An *autotransformer* is a transformer that changes voltage level using the same common coil for both the primary and secondary sides. Taps on the secondary side of an autotransformer allow for various voltages to be accessed.

A *current transformer* is a transformer in which a coil is placed around a wire that carries current. The current-carrying wire is the primary side of the transformer and the coil is the secondary side. A current transformer measures the current flowing through the wire by measuring the strength of the magnetic field around the wire. Current transformers are used to measure the current flowing through AC power lines and are typically connected to current monitors that are set to activate contacts when a predetermined amount of current is reached. Contacts are used in control circuits of current transformers to remove power when an overload occurs.

Motors

A *motor* is a component that develops a rotating mechanical force (torque) on a shaft, which is used to produce work. Electric motors convert electrical energy into rotating mechanical energy. The rotary motion is developed by the opposition of magnetic fields within the motor.

There are two types of magnetic fields in a motor that produce rotation. The two magnetic fields are the stationary field and the rotating field. The stationary field is known as the field in DC motors and the stator in AC motors. The rotating field is known as the armature in DC motors and the rotor in AC motors. Armatures and rotors are connected to the shafts of the motors.

The four types of DC motors are the series, shunt, compound, and permanent-magnet motors. **See Figure 3-28.** A *series motor* is a DC motor that has the stationary field connected in series with the armature and produces the highest torque of all DC motors. A *shunt motor* is a DC motor that has the stationary field connected in parallel with the armature and is used where constant or adjustable speed is required. A *compound motor* is a DC motor that has one stationary field connected in series with the armature and the other stationary field connected in parallel with the armature. Compound motors are used when high starting torque and constant speed are required. A *permanent-magnet motor* is a DC motor that has armature connections but no field connections. Permanent-magnet motors are used where a motor runs only for short periods of time.

AC motors are the most common type of motor. **See Figure 3-29.** Three-phase AC motors are available with ratings from fractional horsepower to several hundred horsepower.

Chapter 3—Industrial Electrical and Electronic Symbols **99**

DC MOTORS

Device	Type	Symbol
DIRECT CURRENT MOTORS	Armature Abbr = ARM	A1 —(ARM)— A2 OR (A) BRUSHES
	Series Field Abbr = FLD	S1 ⌒⌒⌒ S2 OR S1 ⌒⌒⌒⌒ S2 S1 = Series Connection 1 S2 = Series Connection 2
	Shunt Field Abbr = FLD	F1 ⌒⌒⌒ F2 OR F1 ⌒⌒⌒⌒⌒ F2 F1 = Field Connection 1 F2 = Field Connection 2
	Interpoles or Compensating Field Abbr = INTER	F1 ⌒⌒ F2 OR F1 ⌒⌒⌒⌒ F2
DC SERIES MOTOR	Field and Armature Abbr = SERIES	+ S1 ⌒⌒ S2 — A1 (⊙) A2 S1 & S2 = Series Field A1 & A2 = Armature
DC SHUNT MOTOR	Field and Armature Abbr = SHUNT	+ F1 — A1 ⌒⌒ F2 — A2 F1 & F2 = Shunt Field A1 & A2 = Armature
DC COMPOUND MOTOR	Field(s) and Armature Abbr = COMP	+ F1 S1 ⌒⌒ S2 — A1 ⌒⌒ F2 A2 S1 & S2 = Series Field F1 & F2 = Shunt Field A1 & A2 = Armature
DC PERMANENT MAGNET MOTOR	Armature Abbr = P.M.	+ A1 (N S) A2 N = North Field Pole S = South Field Pole

Figure 3-28. DC motors have a stationary field (field) and a rotating field (armature).

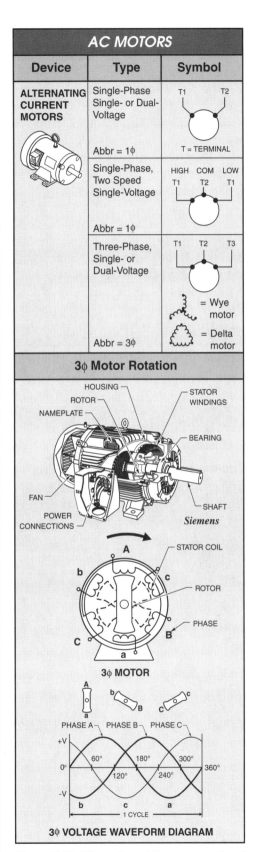

Figure 3-29. Three-phase AC motors are found from ¼ HP to hundreds of horsepower in size.

Lights, Alarms, and Meters

In some electrical circuits, a visual or audible indication of circuit conditions must be provided. A light is used to indicate the presence (but not the amount) or absence of voltage in part of a circuit. Alarms are used when the application calls for an audible indication. A meter is used when an indication of the amount of an electrical property is required.

A *pilot light* is an electrical component that provides a visual indication of the presence or absence of power in a circuit. **See Figure 3-30.** For example, lights are added to indicate power is ON in computers, VCRs, monitors, ovens, and other home appliances. In industrial facilities, lights are added to many applications and machines.

An *alarm* is an electrical component that provides an audible signal that is used as a safety feature, typically to indicate a problem or a potential problem. **See Figure 3-31.** Where there can be a definite safety problem, such as a fire, alarms should always be added. When there is a condition that can be a potential safety problem, such as a truck backing up, alarms should be considered.

Permanently connected meters must be added to electrical circuits when electrical properties need to be measured on a regular basis. **See Figure 3-32.** Permanently connected meters must be considered when monitoring or troubleshooting a circuit on an ongoing basis is important.

LIGHTS

Device	Type	Symbol
LIGHT	Pilot Light Abbr = L	LETTER INDICATES COLOR OR A = Amber G = Green R = Red B = Blue
	Indicator Light Abbr = I	
	Neon, AC Abbr = L	SOLID DOT INDICATES NEON GAS
	Neon, DC Abbr = L	INDICATES POSITIVE TERMINAL INDICATES NEGATIVE TERMINAL

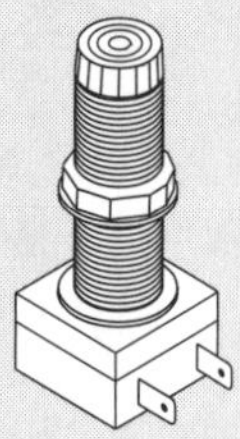

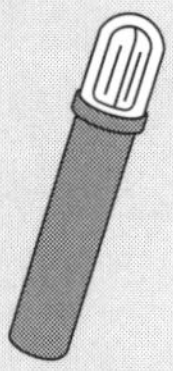

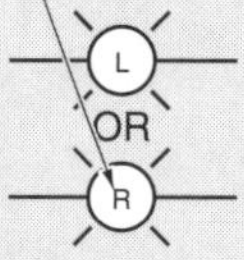

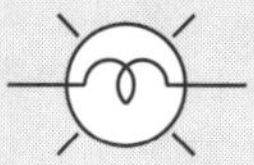

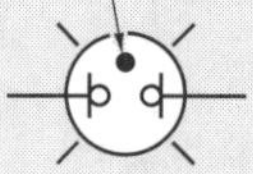

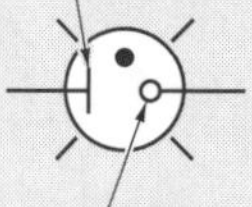

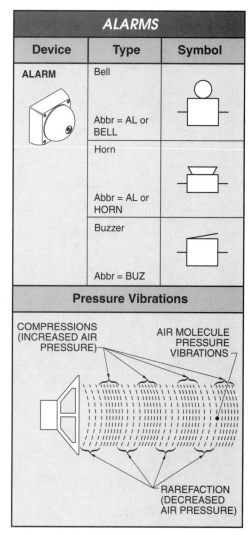

Figure 3-30. A pilot light is an electrical component that typically provides a visual indication of the presence or absence of power in a circuit.

ALARMS

Device	Type	Symbol
ALARM	Bell Abbr = AL or BELL	
	Horn Abbr = AL or HORN	
	Buzzer Abbr = BUZ	

Pressure Vibrations

Figure 3-31. Sound is energy that consists of pressure vibrations in the air that are produced by or originate from vibrating objects such as bells, horns, and buzzers.

General Wiring

General wiring is the wiring used to connect electrical components in a circuit. **See Figure 3-33.** Electrical circuits are made of interconnected components, with wire typically being used to make the connections. Using and tracing the wiring circuit is required when installing new equipment, troubleshooting, and when trying to understand the circuit.

Wire size determines the amount of current that a wire can safely carry. The larger the diameter of the wire, the greater the current-carrying capacity. Copper wire can carry a higher current than aluminum wire for a given wire size.

Wire is typically coated with an insulating material that can come in many different colors. The different colors are used to identify wires in wire harnesses. For example, black is used as a hot wire, white is a neutral wire, red is a switched wire, and green is a ground wire.

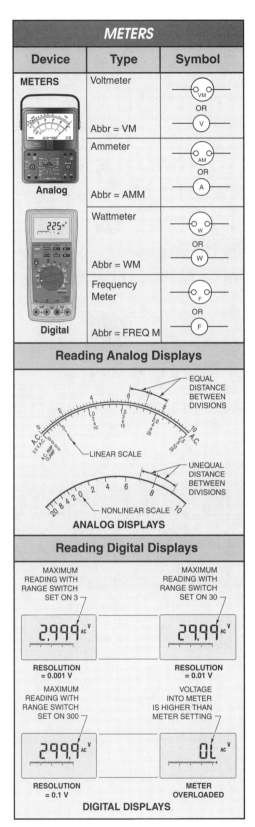

Figure 3-32. Permanently connected meters are used when the electrical properties of a circuit are required on a regular basis.

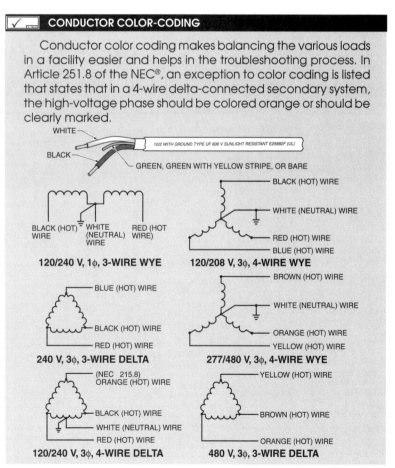

CONDUCTOR COLOR-CODING

Conductor color coding makes balancing the various loads in a facility easier and helps in the troubleshooting process. In Article 251.8 of the NEC®, an exception to color coding is listed that states that in a 4-wire delta-connected secondary system, the high-voltage phase should be colored orange or should be clearly marked.

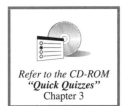

Refer to the CD-ROM
"Quick Quizzes"
Chapter 3

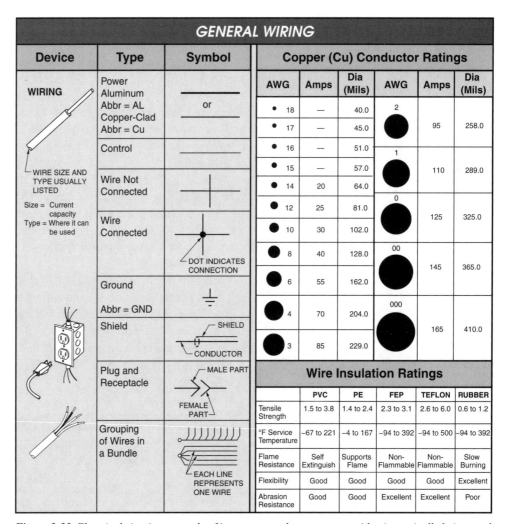

Figure 3-33. *Electrical circuits are made of interconnected components, with wire typically being used to make the connections.*

☑ Example—Ordering Parts for a Compactor Motor Control Enclosure

3-1

Scenario:
Example Activity 3-1 is an installation activity requiring a customer of a machine manufacturer to supply some parts for a purchased waste compactor.

Task:
Order the parts still needed by the customer for the installation of the waste compactor.

* Typically, the machine manufacturer supplies the control panel. The customer typically supplies all other electrical equipment necessary to install the machine.

DISCONNECT AND POWER ENCLOSURE

Reference Prints:

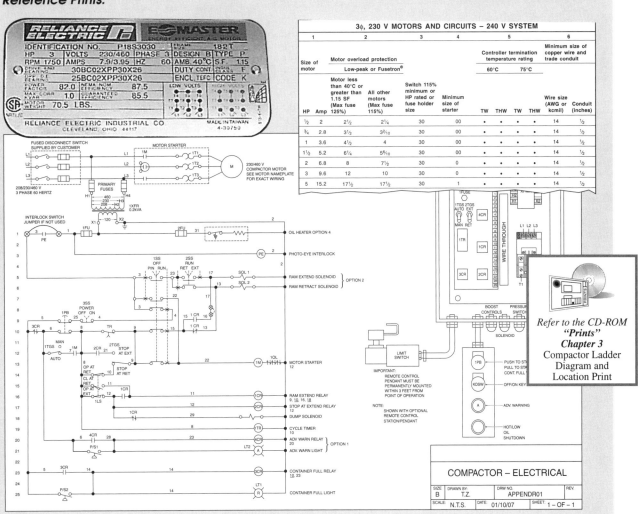

Refer to the CD-ROM "Prints" Chapter 3 Compactor Ladder Diagram and Location Print

☑ Example—Ordering Parts for a Compactor Motor Control Enclosure

Step 1:
Find the prints for the waste compactor. Determine which items are still needed for installation.

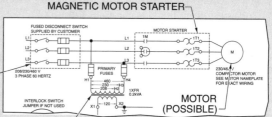

Step 2: Determine the type of disconnect switch needed.

DISCONNECT MUST BE SUPPLIED BY THE CUSTOMER BECAUSE ONLY THE CUSTOMER KNOWS THE VOLTAGE TO BE USED (208 V, 230 V, OR 460 V)

AVAILABLE POWER IN FACILITY FOR THE WASTE COMPACTOR IS 230 V, 3φ, 4-WIRE (L1, L2, L3, AND GRD)

FUSED DISCONNECT IS INDICATED

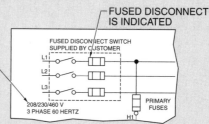

Step 3: Determine the motor needed.

WHEN DETERMINING MOTOR SIZE, ALWAYS GO WITH THE LARGER HP FOR POWER SELECTION REQUIREMENTS

MOTOR IS 3 HP

MOTOR WILL HAVE 182T FRAME

MOTORS CAN BE SUPPLIED BY THE MACHINE MANUFACTURER (2.5 HP TO 3 HP), BUT THE CUSTOMER TYPICALLY USES A MOTOR MANUFACTURER; USING THE SAME MOTOR MANUFACTURER KEEPS THINGS CONSISTENT IN A FACILITY, MAKING THE MAINTENANCE DEPARTMENT'S JOB EASIER

MOTOR CAN BE WIRED FOR 240 V

Step 4: Determine power requirements for the power circuit (fused disconnect) and the control circuit (transformer).

TRANSFORMER LOAD
200 VA ÷ 230 V = 0.87 A

```
230 V MOTOR          7.9 A
TRANSFORMER LOAD   + 0.87 A
                     8.77 A
```

A 30 AMP DISCONNECT WITH 10 AMP FUSES WILL WORK

30 A DISCONNECT SWITCH REQUIRED

10 A FUSES (L1, L2, AND L3)

14 AWG WIRE CAN BE USED
3-14 AWG WIRE CAN BE USED IN ½″ CONDUIT

		3φ, 230 V MOTORS AND CIRCUITS – 240 V SYSTEM			
1	2	3	4	5	6
Size of motor	Motor overload protection			Controller termination temperature rating	Minimum size of copper wire and trade conduit
	Low-peak or Fusetron®			60°C / 75°C	
	Motor less than 40°C or greater than 1.15 SF (Max fuse 125%)	All other motors (Max fuse 115%)	Switch 115% minimum or HP rated or fuse holder size	Minimum size of starter	Wire size (AWG or kcmil) / Conduit (inches)
HP	Amp				TW / THW / TW / THW
½	2	2½	2¼	30	00	• • • • 14 ½
¾	2.8	3½	3²⁄₁₀	30	00	• • • • 14 ½
1	3.6	4½	4	30	00	• • • • 14 ½
1½	5.2	6¼	5⁶⁄₁₀	30	00	• • • • 14 ½
2	6.8	8	7½	30	0	• • • • 14 ½
3	9.6	12	10	30	0	• • • • 14 ½
5	15.2	17½	17½	30	1	• • • • 14 ½

Step 5: Determine the type and size of motor starter to be used.

MAGNETIC MOTOR STARTER

NEMA SIZE 0

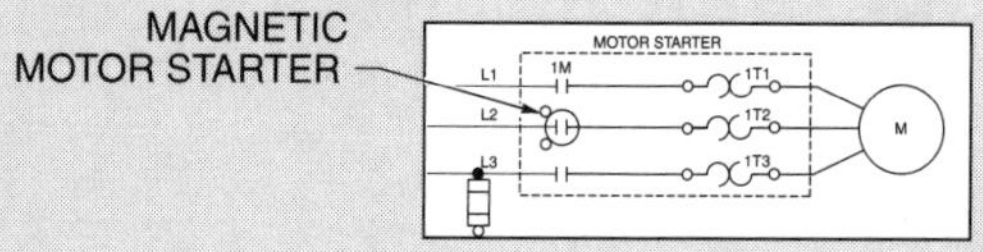

MOTOR STARTER IS TYPICALLY CUSTOMER-SUPPLIED BECAUSE THE CUSTOMER KNOWS EXACTLY WHICH MOTOR AND ENCLOSURE ARE BEING USED

1½	5.2	6¼	5⁶⁄₁₀	30	00	• • • • 14 ½
2	6.8	8	7½	30	0	• • • • 14 ½
3	9.6	12	10	30	0	• • • • 14 ½
5	15.2	17½	17½	30	1	• • • • 14 ½

Name _____ Date _____

Activity—Ordering Parts for a Pump Motor Control Enclosure

3-1

Scenario:
Activity 3-1 is an installation activity requiring a customer of a pump manufacturer to supply some parts for a bank of pumps that was purchased. Your company is using 480 V power supplies for new machinery and pump installations.

Task:
Order the parts still needed by the customer for the installation of the pumps.

* Typically, the machine manufacturer supplies the control panel. The customer typically supplies all other electrical equipment required to install the machinery.

Saftronics, Inc.
PUMP #1

MOTOR STARTER IS TYPICALLY CUSTOMER-SUPPLIED BECAUSE THE CUSTOMER KNOWS EXACTLY WHICH MOTOR AND ENCLOSURE ARE BEING USED

WHEN DETERMINING MOTOR SIZE, ALWAYS GO WITH THE LARGER HP FOR POWER SELECTION REQUIREMENTS

MOTORS CAN BE SUPPLIED BY THE MACHINE MANUFACTURER, BUT THE CUSTOMER TYPICALLY USES A MOTOR MANUFACTURER; USING THE SAME MOTOR MANUFACTURER KEEPS THINGS CONSISTENT IN A FACILITY, MAKING THE MAINTENANCE DEPARTMENT'S JOB EASIER

Reference Prints:

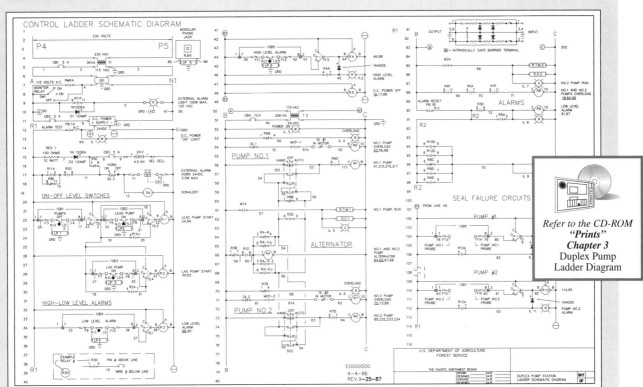

Refer to the CD-ROM "Prints" Chapter 3 Duplex Pump Ladder Diagram

109

Activity—Ordering Parts for a Pump Motor Control Enclosure

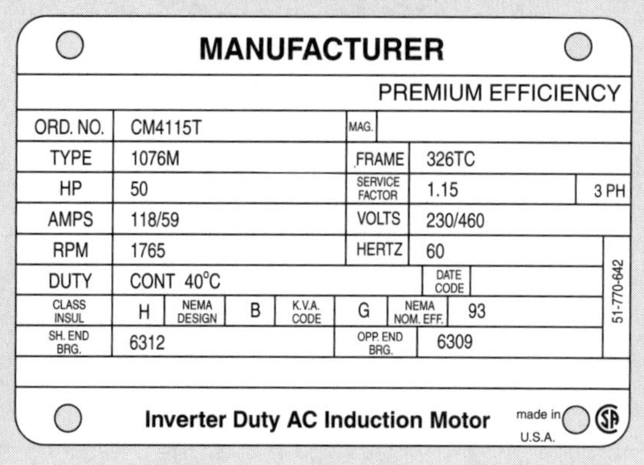

MANUFACTURER
PREMIUM EFFICIENCY

ORD. NO.	CM4115T	MAG.	
TYPE	1076M	FRAME	326TC
HP	50	SERVICE FACTOR	1.15 — 3 PH
AMPS	118/59	VOLTS	230/460
RPM	1765	HERTZ	60
DUTY	CONT 40°C	DATE CODE	
CLASS INSUL	H	NEMA DESIGN B	K.V.A. CODE G	NEMA NOM. EFF. 93
SH. END BRG.	6312	OPP. END BRG.	6309

51-770-642

Inverter Duty AC Induction Motor — made in U.S.A.

3φ, 460 V MOTORS AND CIRCUITS – 480 V SYSTEM

1		2		3	4	5				6	
Size of motor		Motor overload protection Low-peak or Fusetron®				Controller termination temperature rating				Minimum size of copper wire and trade conduit	
						60°C		75°C			
HP	Amp	Motor less than 40°C or greater than 1.15 SF (Max fuse 125%)	All other motors (Max fuse 115%)	Switch 115% minimum or HP rated or fuse holder size	Minimum size of starter	TW	THW	TW	THW	Wire size (AWG or kcmil)	Conduit (inches)
½	1	1¼	1⅛	30	00	•	•	•	•	14	½
¾	1.4	1⁶⁄₁₀	1⁶⁄₁₀	30	00	•	•	•	•	14	½
1	1.8	2¼	2	30	00	•	•	•	•	14	½
1½	2.6	3²⁄₁₀	2⁸⁄₁₀	30	00	•	•	•	•	14	½
2	3.4	4	3½	30	00	•	•	•	•	14	½
3	4.8	5⁶⁄₁₀	5	30	0	•	•	•	•	14	½
5	7.6	9	8	30	0	•	•	•	•	14	½
7½	11	12	12	30	1	•	•	•	•	14	½
10	14	17½	15	30	1	•	•	•	•	14	½
15	21	25	20	30	2	•	•	•	•	10	½
20	27	30	30	60	2	•	•	•		8	¾
									•	10	½
25	34	40	35	60	2	•	•	•		6	1
									•	8	¾
30	40	50	45	60	3	•	•	•		6	1
									•	8	¾
40	52	60	60	100	3	•	•	•		4	1
									•	6	1
50	65	80	70	100	3	•	•	•		3	1¼
									•	4	1
60	77	90	80	100	4	•	•	•		1	1¼
									•	3	1¼
75	96	110	110	200	4	•	•	•		1/0	1½
									•	1	1¼
100	124	150	125	200	4	•	•	•		3/0	2
									•	2/0	1½
125	156	175	175	200	5	•	•	•		4/0	2
									•	3/0	2
150	180	225	200	400	5	•	•	•		300	2½
									•	4/0	2
200	240	300	250	400	5	•	•	•		500	3
									•	350	2½
250	302	350	325	400	6	•	•	•		4/0-2/φ*	2-2*
									•	3/0-2/φ*	2-2*
300	361	450	400	600	6	•	•	•		300-2/φ*	2-1½*
									•	4/0-2/φ*	2-2*

* two sets of multiple conductors and two runs of conduit required

Activity—Troubleshooting a Duplex Pump Electrical System   3-2

Scenario:
Activity 3-2 is a troubleshooting activity requiring that an electrician understand the operation of a multiple-pump control system with a 480 V power supply.

Task:
Determine the operation of the multiple-pump control system to be able to troubleshoot the power and control circuits.

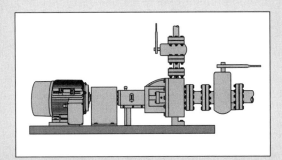

PUMP #1

Required Troubleshooting Information:
1. Control relay R5 has how many normally open (NO) contacts? __2__
2. Control relay R5 has how many normally closed (NC) contacts? __1__
3. The control systems (line 52 to line 76) for Pump #1 and Pump #2 are at what voltage? __24V__
4. What device is used as the alarm for Pump #1 seal failure? __RED LIGHT__
   a. What is the voltage rating of Pump #1 seal failure alarm? __24V__
5. What type of switch is SS1? __SELECTOR__
   a. Does SS1 have SB or DB contacts? _____
   b. Is SS1 an SP, DP, 3P, or 4P type of switch? __4__
   c. How many normally open (NO) contacts does SS1 have? _____
   d. How many normally closed (NC) contacts does SS1 have? _____

Reference Print:

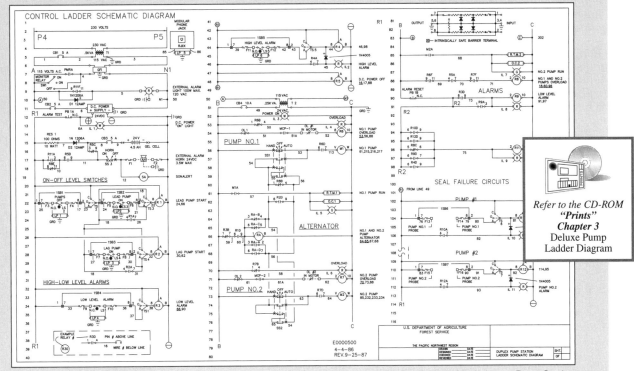

Refer to the CD-ROM "Prints" Chapter 3 Deluxe Pump Ladder Diagram

U. S. Department of Agriculture — Forest Service

Activity—Troubleshooting a Duplex Pump Electrical System

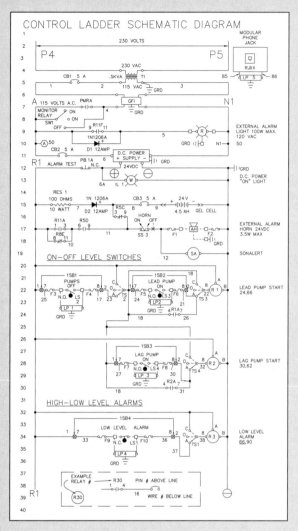
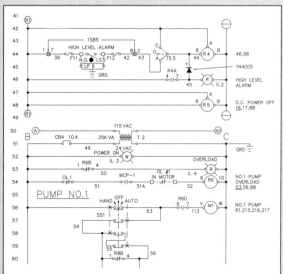
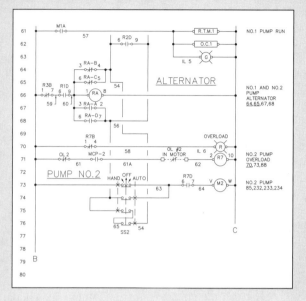
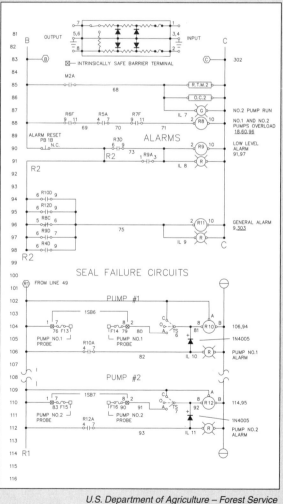

U.S. Department of Agriculture – Forest Service

Electrical Drawings and Plans

Printreading for Installing and Troubleshooting Electrical Systems

Drawings, plans, and diagrams are used to convey facts, ideas, directions, measurements, and information on the location and operation of devices, components, and electrical circuits. Drawings, plans, and diagrams use standard symbols to add meaning and simplify the large amount of information provided on prints. Because there is a large amount of technical information that must be conveyed in different ways at different stages, there are numerous types of drawings, plans, and diagrams used.

The major types of drawings are pictorial drawings, application drawings, location drawings, detail drawings, assembly drawings, instructional drawings, elevation drawings, and sectional drawings. The major types of plans are plot plans, floor plans, foundation plans, structural plans, and utility plans.

PRINTS

Drawings, plans, and diagrams are often generically referred to as "prints." Typically, a drawing, plan, or diagram must be on a border-ruled background that includes a title block to be classified as a print. **See Figure 4-1.** The difference between drawings, plans, and diagrams is the application and amount of information being presented.

Drawings show large amounts of detail of an individual device or small object through a picture type of drawing and often include dimensions, specifications, location, and/or part numbers. Plans show large objects or areas, such as a floor plan, and directly relate to other plans, as electrical floor plans relate to a given set of house prints. Diagrams typically show how electrical components are interconnected and used.

Drawings and plans are used with structures, while diagrams are typically used with electrical equipment. Drawings, plans, and diagrams all relate to each other to provide information that, when used together, shows what is required to select, install, and maintain electrical equipment.

Learning the various types of drawings, plans, and diagrams along with all the associated symbols is important when trying to understand printreading in the electrical field. Mastery in understanding the individual types of drawings, plans, and diagrams associated with a particular trade or field is required to perform work in that trade or field.

116   Printreading for Installing and Troubleshooting Electrical Systems

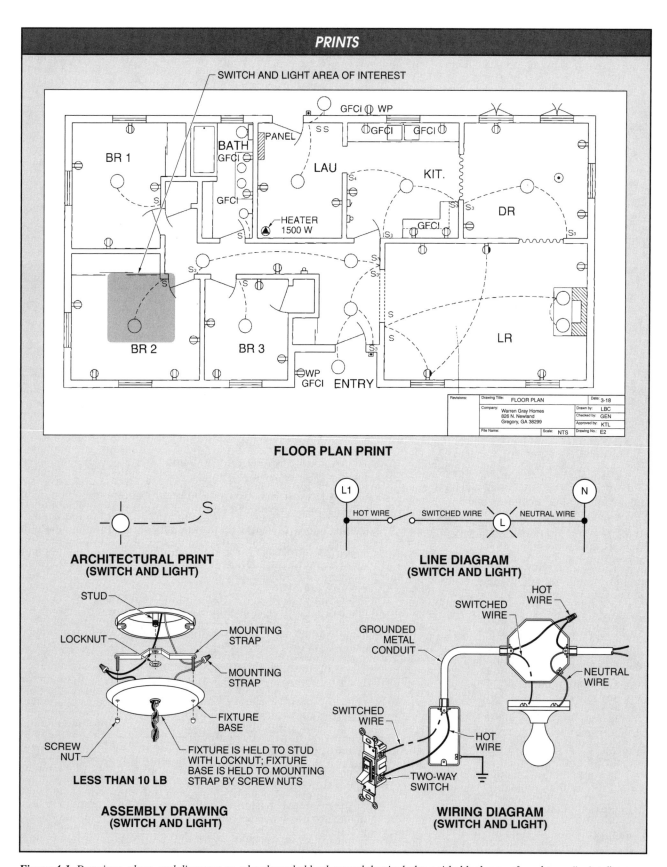

Figure 4-1. Drawings, plans, and diagrams on a border-ruled background that includes a title block are referred to as "prints."

For example, a residential electrician must master reading floor plans, site plans, and detail drawings that are associated with individual component location and mounting. An industrial electrician must master reading ladder (line) diagrams and wiring diagrams. A service technician must master reading schematic diagrams, block diagrams, and interconnecting diagrams. A maintenance person must master all the different types of drawings, prints, and diagrams because the problem may be located any place at any level (equipment, component, or connecting wires). Most individuals working in the electrical field master the types of prints that they often use for a specific job and develop a general understanding of the other print types that they do not typically use.

DRAWINGS

A *drawing* is an assembly of lines, dimensions, and notes used to convey general or specific information as required by the application and use. Drawings such as pictorial drawings, application drawings, location drawings, and assembly drawings display enough information to produce a visual picture of what a device or component looks like, how it can be used or how it fits into a system, and where its major parts are located. Pictorial, application, and location drawings typically do not include dimensions but may include part numbers and some installation and/or mounting information.

Drawings such as detail drawings, elevation drawings, and sectional drawings are used to provide as much detail as required for a clear understanding of the type, size, and dimensions of the device, component, or object. Detail, elevation, and sectional drawings are often used to show what surrounds the object being shown as well as to show the relationship between the object and the system.

Pictorial, application, location, detail, assembly, elevation, and sectional drawings are drawn as viewed from the front. The *front view* is the view when looking directly at an object from the same height as the object. **See Figure 4-2.** Plans, such as floor plans, plot plans, site plans, and utility plans, are drawn to be viewed from directly above (plan view).

Any drawing, such as a sectional drawing, can be customized to show as much detail as required. Sectional drawings are drawn to be viewed head-on and as if the object were sliced open so the actual internal composition and construction of the object can be seen. Any drawing can include dimensions, part numbers, special notes, and/or any additional information as required to convey the information needed to do the specified work.

> Traditionally, draftsmen worked at large drawing boards and used pencils, pens, compasses, protractors, triangles, T-squares, and other drafting tools to prepare a drawing. Today, pencils have been replaced by the computer mouse and a drafting board with a monitor, and drafting tools are found in the menu box.

Pictorial Drawings

The old saying "a picture is worth a thousand words" has merit when trying to show technical information and details. A *pictorial drawing* is any three-dimensional drawing that resembles a picture. The picture can be drawn to highlight information and details required to locate connecting points, install or order a part, show the location of switches, fuses, or other parts, or provide a clear picture of how the device looks or fits into a larger system.

A pictorial drawing is used to show the actual layout and position of all devices and components used in an application. In a pictorial drawing, devices and components are placed as near to their actual positions as possible. The objects in a pictorial drawing are typically drawn with great detail but can also be drawn in general outline form. **See Figure 4-3.**

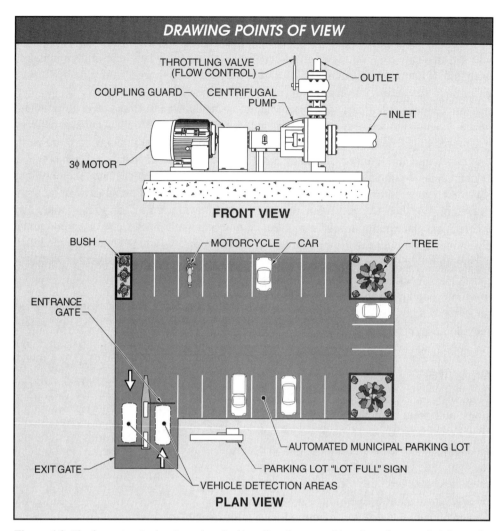

Figure 4-2. The front view is the view when looking directly at an object from the same height as the object, and the plan view is the view from directly above.

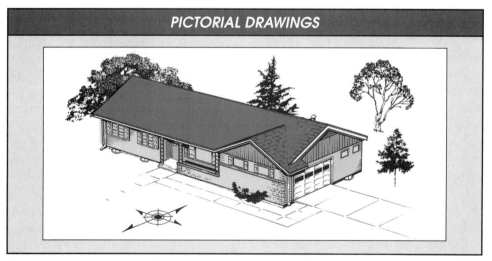

Figure 4-3. A pictorial drawing is any drawing that resembles a picture, with information and details to locate object connecting points, install a part, show location of switches, and/or provide a clear picture of how a device looks or fits into a larger system.

Orthographic Drawings

When an object is drawn as a two-dimensional drawing the object is drawn using orthographic projection. An *orthographic projection* is a type of drawing where all faces (front, top, bottom, and sides) of an object are projected onto flat planes that generally are at 90° to one another. Building drawings typically include orthographic views (elevations), which are front views of the sides of a building. **See Figure 4-4.**

Orthographic projection drawings include the dimensions and details required to convey the technical information of the object drawn. Typically, large orthographic drawings of buildings or large machines do not include many dimensions. To avoid cluttering the drawing, only basic dimensions such as total length, width, and depth are included. Plans for buildings and large objects typically include multiple drawings that show all the dimensions.

Smaller objects drawn with orthographic views (typically front, right side, and top views) include all dimensions needed to convey the required technical data and provide a better picture of the object. **See Figure 4-5.**

For example, industrial pushbuttons typically have a lockout means to prevent the pushbutton (start or run) from being pressed during downtime or repairs. A stop pushbutton will have a locking attachment that holds it in the open condition (pushed in). Pictorial drawings provide a clear image of what the lockout device looks like and how the device is applied to a pushbutton. However, pictorial drawings do not provide information on the physical size of the device or component.

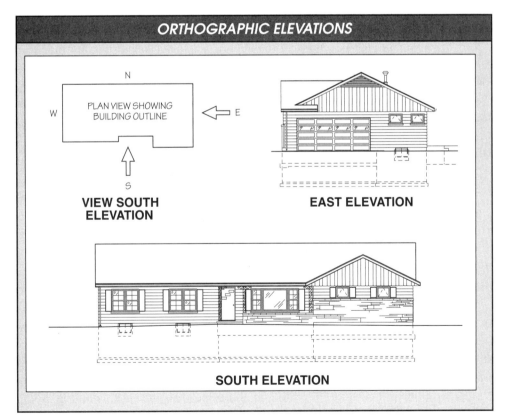

Figure 4-4. Orthographic projections of a building are known as elevations. Building prints typically include orthographic views, which are views of the sides of the building.

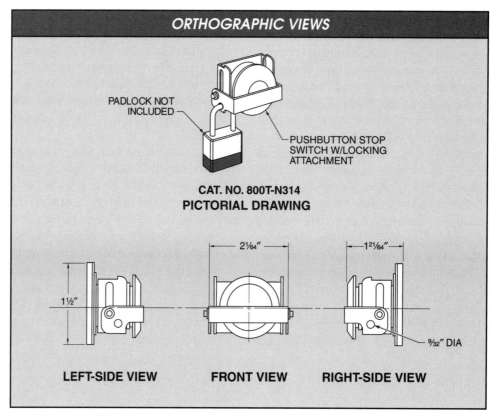

Figure 4-5. Objects drawn using orthographic views include all dimensions needed to convey the required technical data and provide a complete picture of the object.

Application Drawings

An *application drawing* is a type of drawing that shows the use of a specific piece of equipment or product in an application. Application drawings show product use and are not intended to indicate component connections, wiring, dimensions, or actual size or shape. Application drawings present ideas on how to use a product in problem solving and are used by manufacturers to promote products. Application drawings are also used during troubleshooting to present ideas for the use of new or different components.

For example, a *Hall effect sensor* is a type of sensor that detects the proximity of a magnetic field. Hall effect sensors are used in many applications requiring a small magnetically operated sensor that can be used as a switch. Hall effect sensors contain no moving parts, which allows them to be used in applications that are harsh to mechanical switches, such as in a beverage dispensing gun. **See Figure 4-6.** A dispensing gun can be completely submersed in water for easy cleaning and requires almost no maintenance because of the solid-state switching.

An application drawing with a Hall effect sensor (switch) can be used in a machine to indicate if the machine is level or has a tilt (in degrees). **See Figure 4-7.** Typically a magnet is installed with the Hall effect sensor and the switch is only activated when the machine is level. To fully activate the sensor, the magnet must be directly over the Hall effect sensor.

> Application drawings are used to present the concept of equipment usage. Architects and engineers must determine if the equipment will work in a system and must guarantee that the intended installation meets all code requirements.

Chapter 4—Electrical Drawings and Plans 121

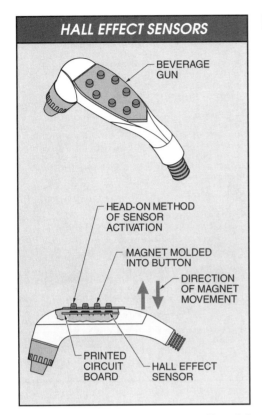

Figure 4-6. Hall effect sensors are small, sealed devices that contain no moving parts and are used in applications that are harsh to mechanical switches.

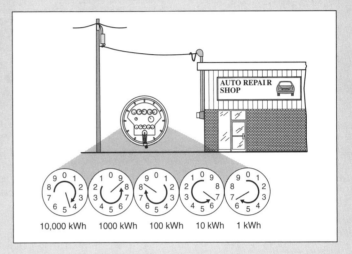

Note: Hands of adjacent dials rotate in opposite directions.

Service entrance meters record the kilowatt hours of electricity used. A small motor inside the meter operates at speeds proportional to the rate of power usage. One kilowatt hour (1 kWh) is 1000 W of energy used in one hour.

From right to left the dials represent units, tens, hundreds, thousands, and ten thousands in kWh. To read a service meter, begin with the units dial and record the highest number the hand has passed. Continue the process until readings have been taken from all five dials.

Location Drawings

A *location drawing* is a type of drawing used to show the position of switches, buttons, terminal connections, and other features found on a device or component. For example, location drawings are often used in installation and operational manuals to show where to connect external wires and position indicating lamps, switches, and displays. **See Figure 4-8.**

Typically, location drawings do not include dimensions. They include arrows and callouts showing the location of an object and a brief overview of what the object is used for. Location drawings can include simple callouts that state only what a switch (device) or component is, such as a "Run Lamp", or the callouts can provide as much information as space allows, such as "Caution: Run Lamp is on when electric motor drive and motor are running."

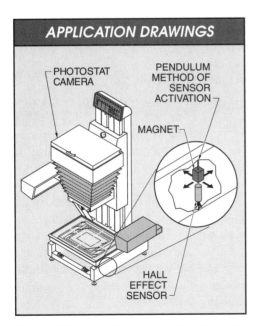

Figure 4-7. An application drawing shows the use of a specific piece of equipment or product in an application, but does not indicate component connections, wiring, dimensions, or actual size or shape.

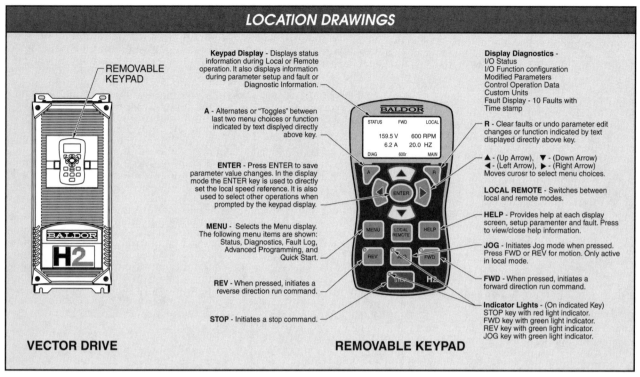

Figure 4-8. A location drawing is used to position switches, buttons, terminal connections, and other features found on a device or component.

Detail Drawings

A *detail drawing* is a type of drawing that provides all information needed to produce a part. Detail drawings include primary, auxiliary, section, or other views of an object. Detail drawings are typically used during the construction and/or assembly of devices and components and are typically shown on service bulletins. **See Figure 4-9.**

Dimensions and other information can be found on detail drawings. Detail drawings are drawn as orthographic projections to show one or more side views of an object.

Assembly Drawings

An *assembly drawing* is a type of drawing that shows as closely as possible the way individual parts or components are placed together to produce a finished piece of equipment or result. Assembly drawings must show as much information as possible to allow for proper assembly of an object but must not be overwhelming in their detail. When used, dimensions must be kept to a minimum. Dimensions are typically shown on separate detail drawings that easily relate back to the assembly drawing.

Assembly drawings may be one individual drawing or a series of drawings showing different steps or stages of an assembly process. The more complex an assembly, the greater the number of

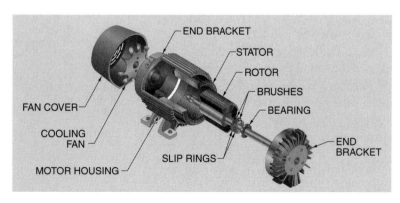

Motor assembly drawings are used to determine bolt or screw placement and the direction in which the rotor is pressed onto the shaft.

individual drawings required to convey the information. **See Figure 4-10.** For example, a *motor control center (MCC)* is a sheet metal enclosure that houses and protects fuses or circuit breakers, motor starters, overloads, and wiring.

Assembly drawings need to convey all information required to assemble the product without being too complicated or too simple. Assembly drawings that are too complicated or too simple end up costing time, material, and customers.

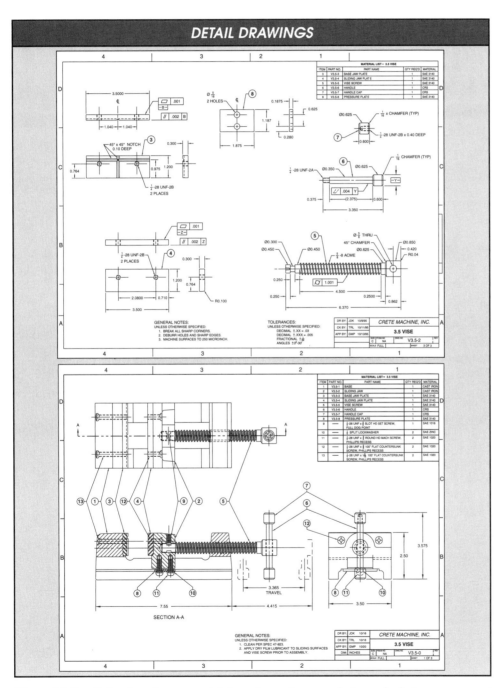

Figure 4-9. Detail drawings show as much information and as many features of a device or component as possible.

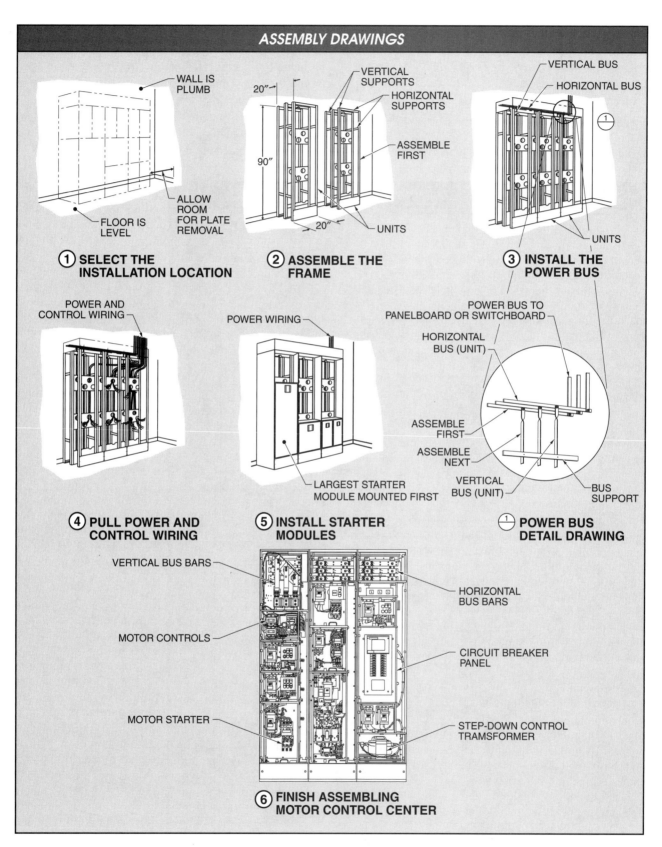

Figure 4-10. An assembly drawing shows as closely as possible the way individual parts or components are placed together to produce a finished piece of equipment.

Instructional Drawings

An *instructional drawing* is a type of drawing that is intended to indicate how to do work using the simplest and/or safest method. Instructional drawings are typically drawn as pictorial drawings, with arrows and callouts indicating the required work.

For example, there are two drive methods used in facilities to move material along a conveyor. **See Figure 4-11.** The two methods of conveyor operation are direct drive and roller drive. With the direct drive method, the material rides directly on a driven belt. The direct drive method is used for light loads. With the roller drive method, a belt underneath drives the rollers, and the material rides on top of them. The roller drive method is used for heavy loads.

> Instructional drawings are typically used in classroom presentations, manufacturer product manuals, assembly booklets, product specification sheets, and "how to" books. Instructional drawings go into great detail about how to assemble, install, and adjust electrical equipment.

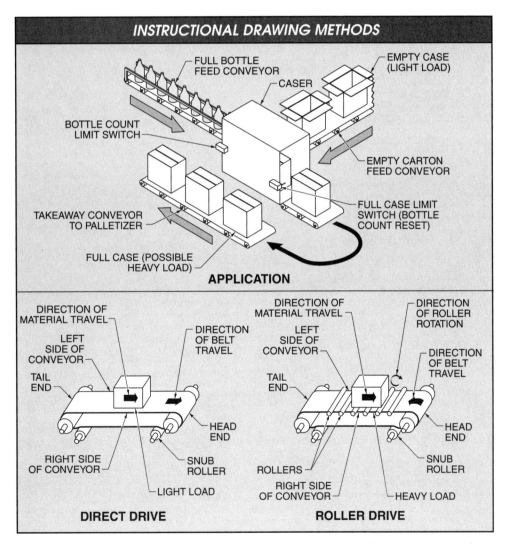

Figure 4-11. The two drive methods used in facilities to move material along a conveyor are direct drive and roller drive.

When a conveyor belt is not tracking properly, it will drift to one side and become damaged by rubbing against the stationary parts of the conveyor. **See Figure 4-12.** To realign a conveyor belt, one side of the belt snub roller is adjusted forward or to the rear. A good instructional drawing will show, using as few words as possible, how the conveyor belt is tracked.

Elevation Drawings

An *elevation drawing* is an orthographic projection of a structure's vertical surfaces. Elevation drawings are used to define the architectural style, structural materials, and features of a building or structure. An elevation drawing indicates the location and style of windows, shutters, doors, chimneys and other visible features, such as ramps and cables. **See Figure 4-13.**

Although dimensions are typically kept to a minimum on elevation drawings, vertical and overall dimensions are used to indicate building or structure height. When features are underground, such as foundations, these features are drawn using dashed lines.

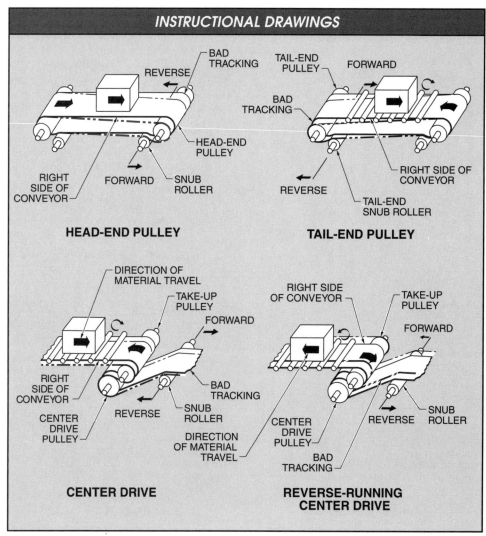

Figure 4-12. Instructional drawings are typically drawn as pictorial drawings with arrows and callouts and are intended to indicate how to do work using the simplest and/or safest method.

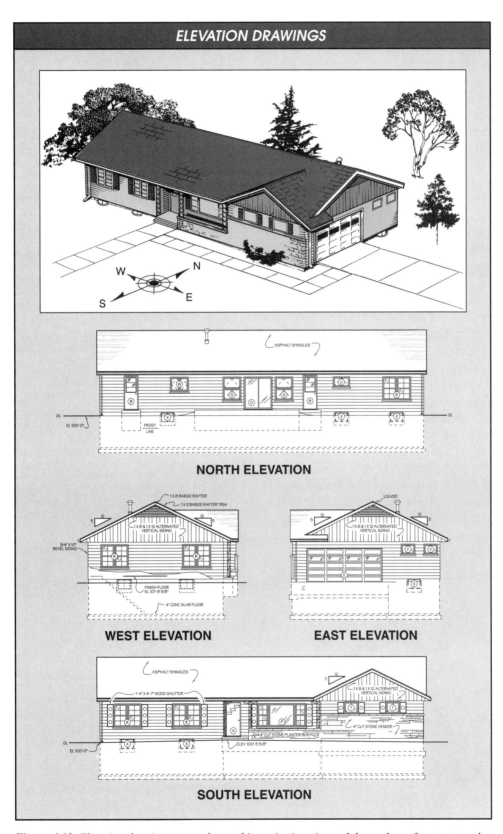

Figure 4-13. Elevation drawings are orthographic projection views of the surface of a structure that indicate the location and style of building windows, shutters, doors, chimneys and other visible features of a structure.

Sectional Drawings

A *sectional drawing* is a type of drawing that indicates the internal features of an object. An imaginary cutting plane is passed through the object perpendicular to the line of sight. The portion of the object between the cutting plane and the observer is removed, revealing the internal features of the object. **See Figure 4-14.**

In the orthographic section view, a right-side view (Section A-A) and a front view of a motor end bell are shown. The cutting plane line in the front view indicates the location and path where the end bell is cut through and the direction of sight for viewing the object after it is cut.

The views of sectional drawings may be pictorial or orthographic. Section lines are drawn on all surfaces cut by the cutting plane. General-purpose section lines are typically used and are drawn $\frac{1}{10}''$ (2.5 mm) apart at an incline of 45°. Section lines are inclined in either direction unless the features of the object dictate that an angle other than 45° be used. The angle of the section lines must not match the angle of any lines defining the shape of the object. Specific types of section lines are used to identify specific types of material, such as steel, aluminum, wood, or insulation.

In a typical pictorial sectional drawing, general-purpose section lines are used to indicate the type of material that the cutting plane cuts. **See Figure 4-15.** As the cutting plane passes through various parts, section lines are drawn at different angles to indicate the various parts. Shading is also used in place or as part of section lines to identify separate pieces on sectional drawings.

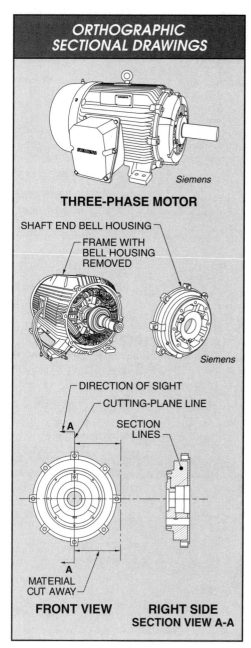

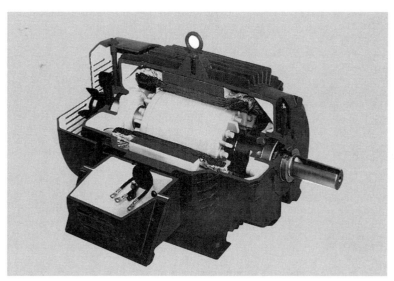

Rockwell Automation/Reliance Electric
When drawing pictorial sectional drawings, it can help to look at the real item with the relevant area cut away to see which areas of the drawing will receive section lines.

Figure 4-14. An orthographic sectional drawing uses the orthographic views of an object to indicate the cutting-plane line and resulting section view.

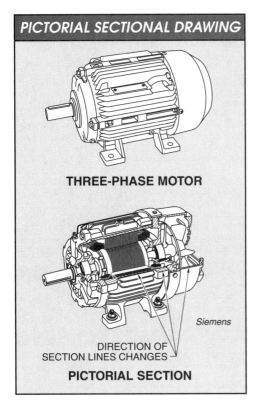

Figure 4-15. A pictorial section-view drawing typically uses general-purpose section lines at various angles to indicate the different parts and materials used.

PLANS

A *plan* is a drawing of an object as it is viewed from above. Plans are two-dimensional drawings designed to indicate the location of objects. **See Figure 4-16.** The type of plan being used depends on the type of object to be drawn.

For example, plot plans indicate the location of buildings, trees and shrubs, driveways, sidewalks, and other important physical features located on or adjacent to a property. Floor plans indicate the internal rooms of a structure and the location of doors, windows, electrical outlets, switches, light fixtures, cabinets, and other features. Plans include dimensions and callouts to indicate the intent and size of each room in a building or structure on a piece of property.

Plot (Site) Plans

A plot (site) plan is a type of drawing that indicates an entire property, with buildings

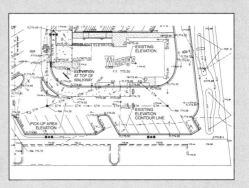

GRADING PLANS

Grading plans provide building site elevations plus utility and paving information required in the construction of buildings. Grading plans are used by excavating and paving contractors to determine cut and fill requirements for parking areas and to determine pavement slope for proper drainage. Drawing scales for grading plans are 1″ = 20′ or larger.

and other structures drawn in their proper locations on the property. In addition to indicating the location and size of buildings, sidewalks, driveways, and other structures, plot plans also show the locations of utilities, such as sewer pipes, water pipes, electrical runs, gas lines, communication cables, and other hidden or important items. **See Figure 4-17.** When utilities are not indicated in detail on a plot plan or more information about the utilities is required, a separate utility plan is used.

Plot plans are used by surveyors to locate the exact position of a building on a piece of property and to ensure that all easements and boundaries are met. To ensure proper building orientation, all plot plans must include a direction compass pointing north. A plot plan is typically the first plan used by workers on a construction site because a plot plan is used to locate the exact place to begin excavation.

In addition to plot plans, other closely related plans are used to provide data about a property. A *survey plan* is a type of drawing that is prepared by a licensed surveyor or civil engineer that accurately provides land contour information, dimensions, and other important feature information about a piece of property and adjacent properties. A plot plan and survey plan can be the same plan, with the difference being that a survey plan ultimately becomes a legal document.

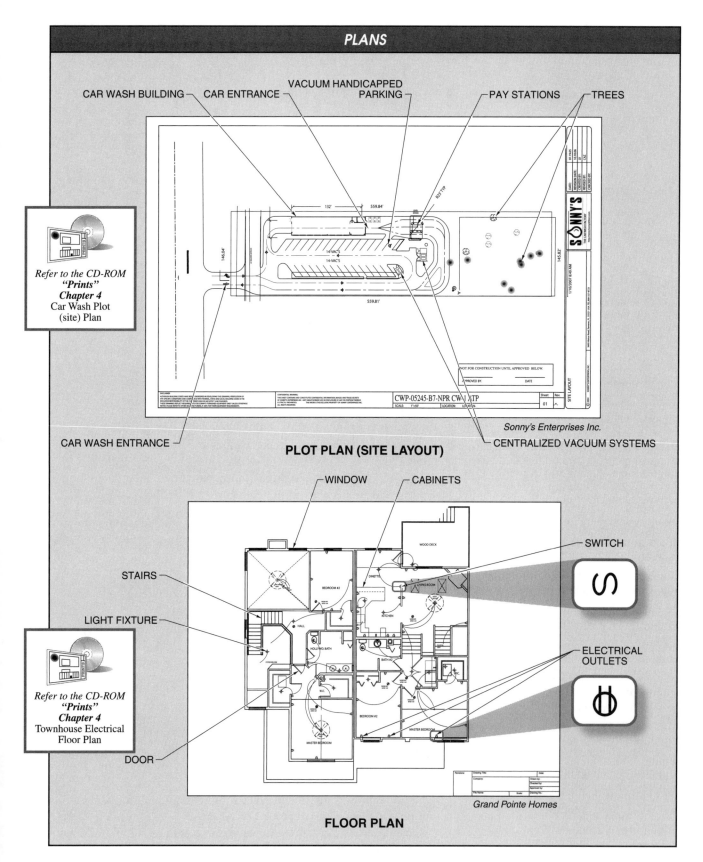

Figure 4-16. A plan is a two-dimensional drawing that indicates the location of objects as seen from a bird's eye view.

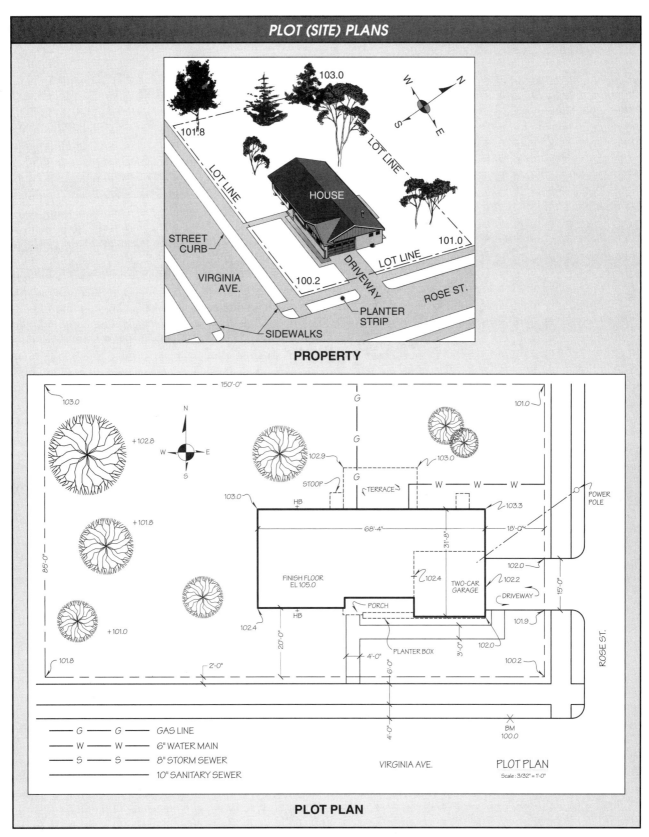

Figure 4-17. A plot (site) plan indicates the entire property, with buildings, other structures, and utilities drawn in their proper locations.

132 Printreading for Installing and Troubleshooting Electrical Systems

Because survey plans become legal documents, survey plans must be prepared by a licensed surveyor or civil engineer approved by the city, county, or state. A *landscape plan* is a type of drawing that indicates land contours along with buildings and other structural information so that landscaping designs can be created.

Floor Plans

A *floor plan* is a type of drawing that shows exterior walls, all room partitions, doors, windows, fireplaces, stairs, bathrooms, cabinetry, and any fixtures or appliances. Each floor (level) of a building or structure has its own floor plan. **See Figure 4-18.**

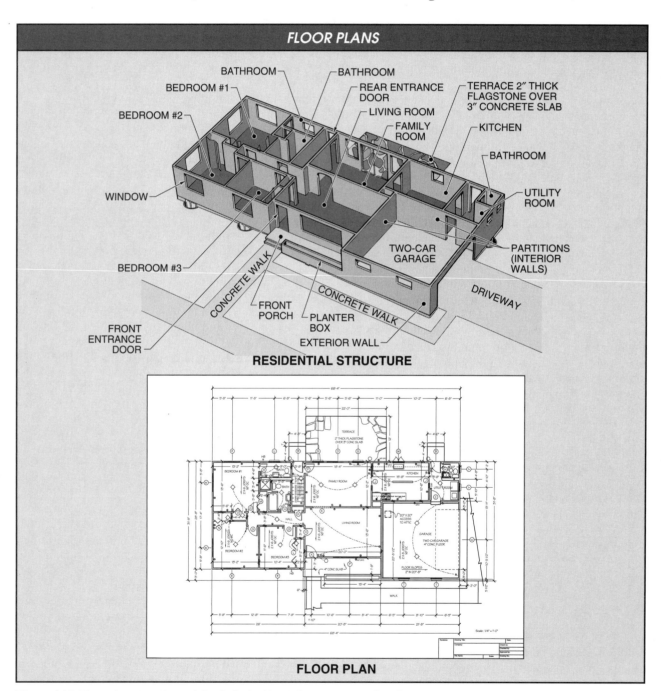

Figure 4-18. Floor plans (one for each level of a building) show exterior walls, all room partitions, doors, windows, fireplaces, stairs, bathrooms, cabinetry, and any fixtures or appliances.

Floor plans are the most common type of plan used because floor plans are used by many individuals and different trades. Floor plans are the plans most commonly recognized by the average person. Architects use floor plans to determine the best use of internal space and to meet the customer's needs, as well as meeting any building codes when designing a building. Inspectors use floor plans to issue permits. Construction trade groups such as carpenters, plumbers, masons, and electricians use floor plans to build the structure and install the required devices at the proper locations.

Realtors and customers use floor plans to determine if a building will meet expectations before actually viewing the interior. Floor plans vary in complexity depending upon the intended use by a given group. **See Figure 4-19.** For this reason, there are typically several different sets of floor plans drawn for the same building, with the most basic designed for the realtor and customer, and the detailed drawings going to the construction trades. On more complex or larger buildings, each trade will have individual floor plans showing all details required for that trade.

Southern Forest Products Association

Floor plans can be modified for electrical, HVAC, and plumbing work. No matter what the work, each floor of the structure requires a separate floor plan.

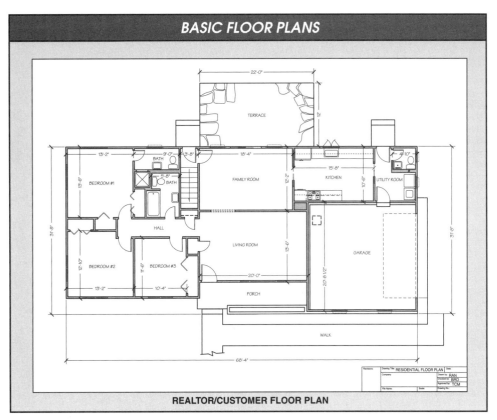

Figure 4-19. *Floor plans vary in complexity depending upon the intended use by a given group.*

Foundation Plans

A *foundation plan* is a type of drawing that indicates a building's foundation, structural supports, dimensions, and building materials. Foundation plans are used to provide excavating, construction, drainage, waterproofing, and other design information for building the foundation. Foundation plans can also indicate the type of information (internal walls, stairs, electrical, plumbing, and HVAC) found on floor plans and can serve as the basement floor plan. **See Figure 4-20.**

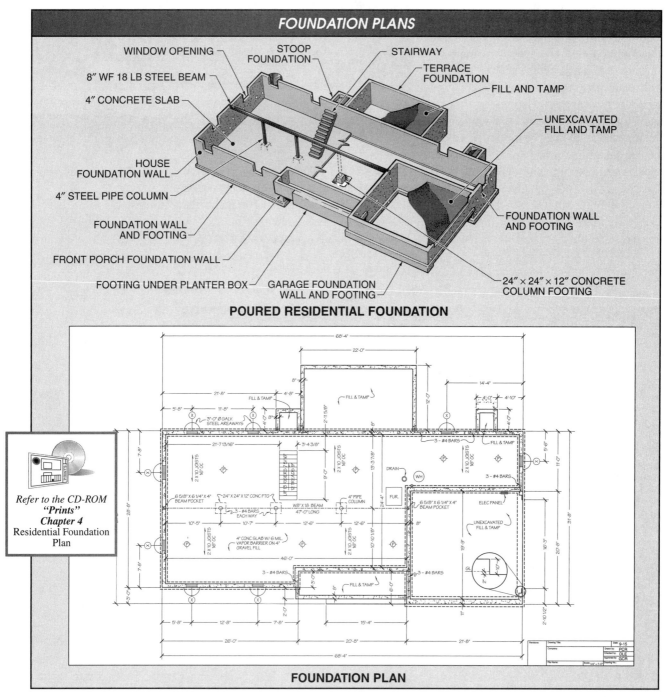

Figure 4-20. Foundation plans indicate a building's foundation, structural supports, dimensions, and materials, with excavating, construction, drainage, and waterproofing information included.

Foundation plan designs require knowledge of the landscaping that a specific piece of property has. Information concerning soil type, water-table height, frost line, and buried utilities must be known before any construction or excavating work begins. Locations of trees, rocks, ponds, and creeks and how the property drains must also be known before work begins. Other factors such as flood, earthquake, and fire potential must be considered in the design.

Structural Plans

A *structural plan* is a type of drawing that indicates the type, amount, placement, and fabrication of all materials used as structural supports of a building, bridge, or other structure. Structural plans are used to determine the building materials (steel, masonry, wood, or concrete) that are used to build the main support structure. **See Figure 4-21.**

Prefabricated and precast structural members are typically used as structural materials. On large projects, structural members are often numbered so that assembly can be completed chronologically, with the correct members fitting together. Structural plans also specify how the individual members are joined together (cemented, welded, riveted, or otherwise fastened).

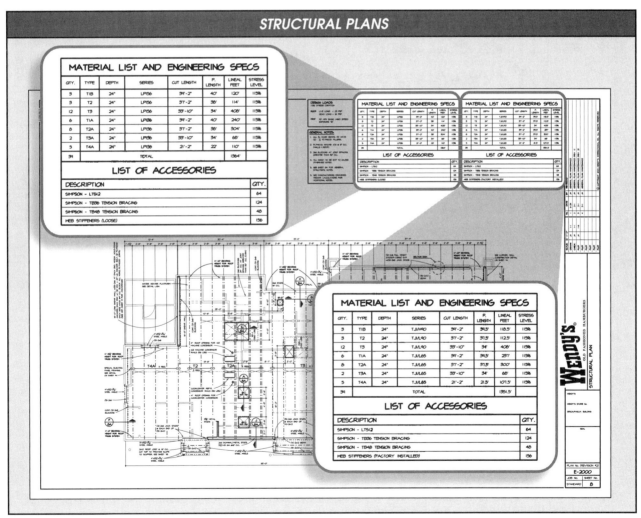

Wendy's International, Inc.

Figure 4-21. Structural plans indicate the type, amount, placement, and fabrication of all materials used as the structural supports of a building, bridge, or other structure.

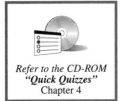

Refer to the CD-ROM "Quick Quizzes" Chapter 4

Utility Plans

A *utility plan* is a type of drawing that indicates the location and intended path of utilities such as electrical, water, sewage, gas, and communication cables. Utility plans are used along with other plans to ensure proper location and construction of a project. For example, a utility plan is used along with a foundation plan to ensure no utility lines are cut, or if need be, that lines are rerouted. **See Figure 4-22.**

All building trades use utility plans to determine the best location for bringing in the utilities. The electrical trade uses utility plans for mounting the main electrical service and any outside lighting and the plumbing trade for mounting water meters. Landscapers use utility plans to locate possible ways to camouflage some of the utility equipment. Other trade and service areas use utility plans to ensure that safety is being considered.

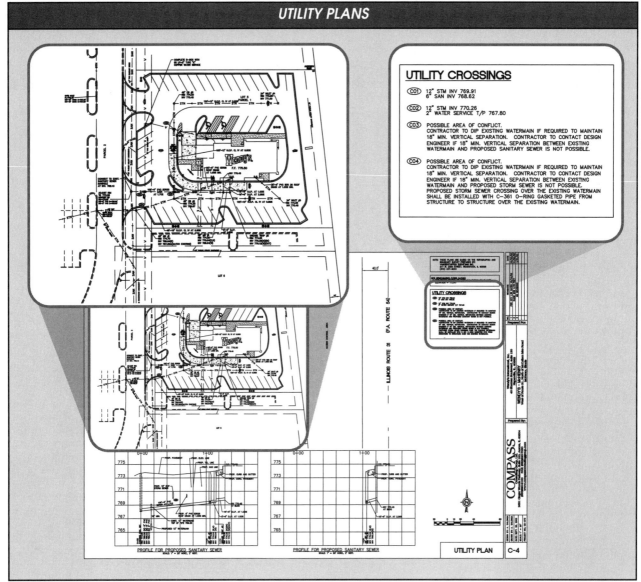

Wendy's International, Inc.

Figure 4-22. *Utility plans indicate the location and intended path of utilities such as electrical, water, sewage, gas, and communication cables.*

Chapter 4—Electrical Drawings and Plans   137

☑ Example—Installing Ceiling Lighting

4-1

Scenario:
Example 4-1 is an installation activity requiring that ceiling lighting be installed. The application is a gym and stage of a recreation center. Lo-Bay light fixtures are to be mounted above the basketball court for proper court illumination, and chain-hung fixtures will be hung above the stage for lighting effects.

Task:
Fill out a work request.
Required Work Request Information:
1. Scissor lift (8′, 12′, 16′, or 20′): _____
2. What is the center-to-center distance of the gym fixtures? _____
3. How far away (center of fixtures to outside wall) is the last row of gym light fixtures? _____
4. What is used to change the brightness of the stage lights? _____
5. How many basketball court light fixtures are used? _____

RECREATION CENTER – GYM

Reference Prints:

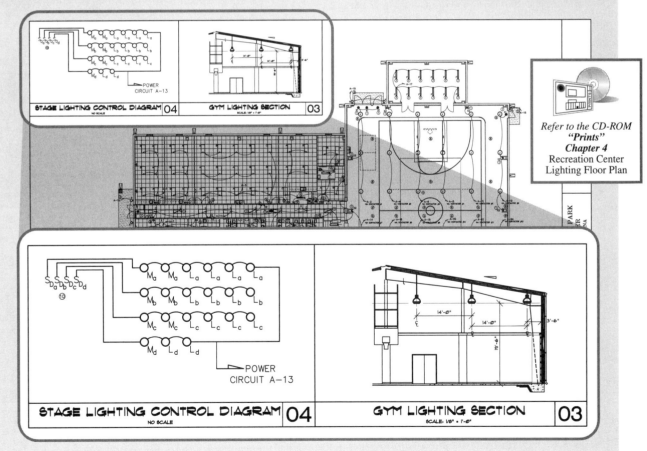

Refer to the CD-ROM
"Prints"
Chapter 4
Recreation Center
Lighting Floor Plan

138 Printreading for Installing and Troubleshooting Electrical Systems

☑ Example—Installing Ceiling Lighting

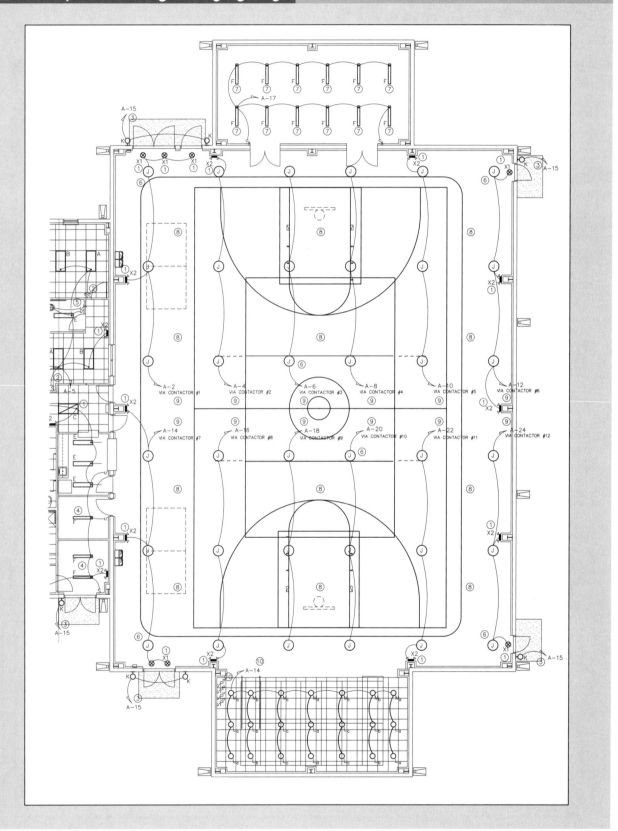

Chapter 4—Electrical Drawings and Plans   139

☑ Example—Installing Ceiling Lighting

Step 1: Get the print(s) for the Surf City Recreation Center and find the print that has the Lighting Floor Plan.

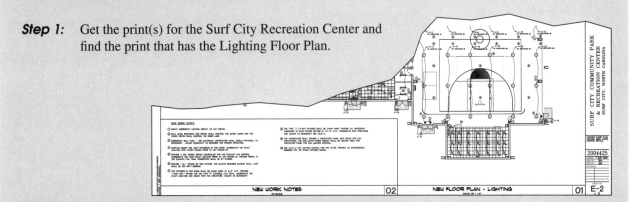

Step 2: Identify the scissor lift required. Determine the height above the floor at which the fixtures will be hung.

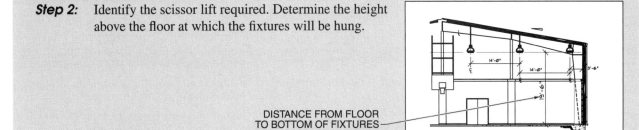

DISTANCE FROM FLOOR TO BOTTOM OF FIXTURES

Step 3: Determine the center-to-center distance of the gym light fixtures.

DISTANCE FROM CENTER OF FIXTURE TO CENTER OF NEXT FIXTURE

Step 4: Determine the distance from the center of the light fixtures (last row) to the outside wall.

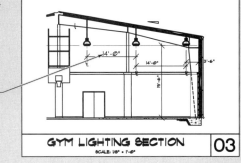

DISTANCE FROM CENTER OF FIXTURE TO CENTER OF NEXT FIXTURE

Example—Installing Ceiling Lighting

Step 5: Determine the device that is used to change the brightness of the stage lights.

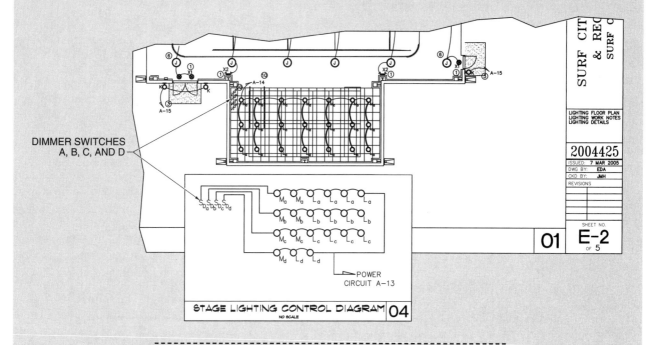

Step 6: Count the number of light fixtures above the basketball court.

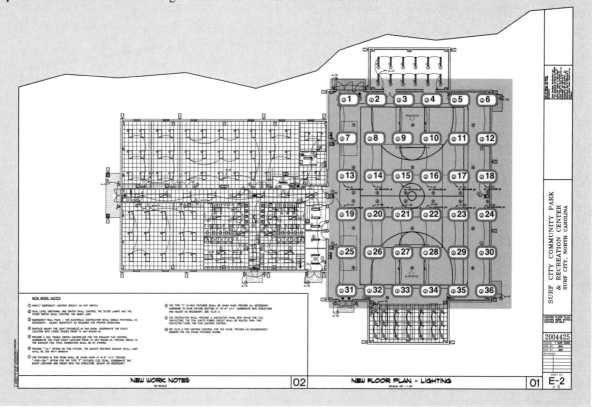

☑ Example—Troubleshooting a Room Electrical System

4-2

Scenario:
Example 4-2 is a troubleshooting activity requiring that electricians understand the correlation between electrical architectural drawings and electrical line diagrams. When problems arise, an electrician must be able to look at electrical drawings (prints) and identify the corresponding electrical line diagrams and wiring diagrams. Line diagrams allow for easier troubleshooting of a circuit or system.

Task:
Identify the corresponding line diagram for the electrical architectural drawing of the Dining Room, Living Room, and Kitchen.

Reference Prints:

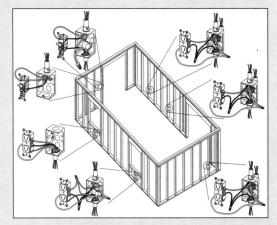

LIVING ROOM DEVICE WIRING

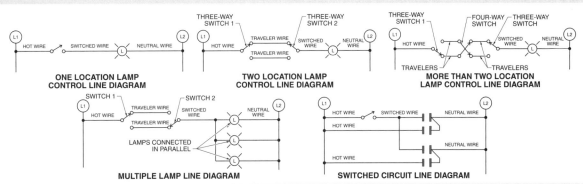

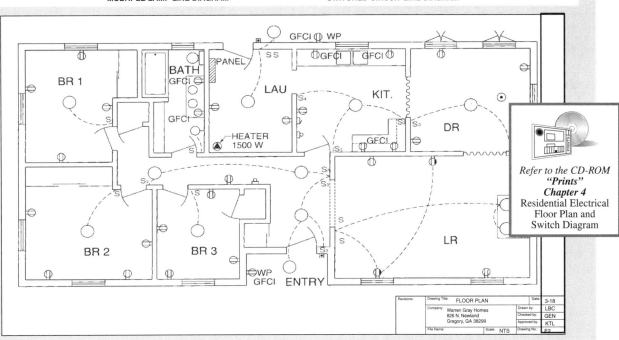

*Refer to the CD-ROM "Prints" **Chapter 4** Residential Electrical Floor Plan and Switch Diagram*

Example—Troubleshooting a Room Electrical System

Step 1: Find the electrical floor plan print for the Dining Room and identify the outlet and switch circuit.

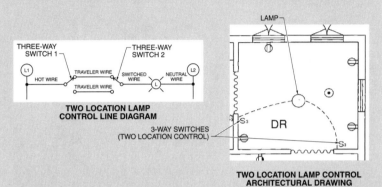
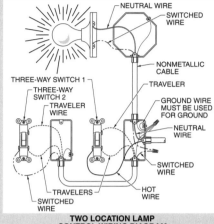

Step 2: Find the electrical floor plan print for the Living Room and identify the outlet and switch circuit.

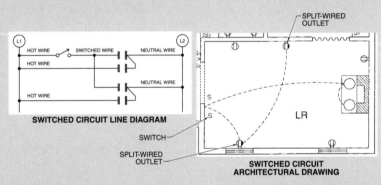
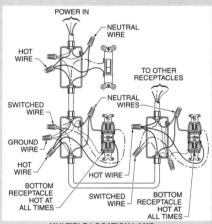

Step 3: Find the electrical floor plan print for the Kitchen and identify the lamp and switch circuit.

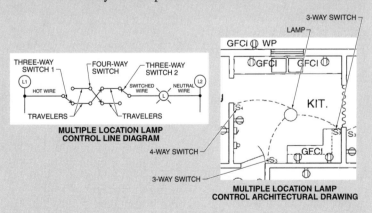
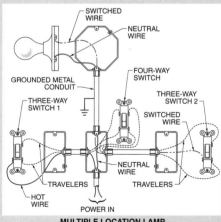

Review Questions and Activities

Electrical Drawings and Plans

Name _____ Date _____

True-False

T F **1.** Drawings, plans, and diagrams are often generically referred to as "prints".

T F **2.** A plot plan is used to show the actual layout and position of all devices and components used in an electrical circuit.

T F **3.** When an object is drawn as a two-dimensional drawing, the object is drawn using orthographic projection.

T F **4.** An assembly drawing is a type of drawing that shows as closely as possible the way individual parts or components are placed together to produce a finished piece of equipment or result.

T F **5.** Location drawings are used to define the architectural style, structural materials, and features of a building or structure.

T F **6.** The views of sectional drawings may be pictorial or orthographic.

T F **7.** Detail drawings are used by surveyors to locate the exact position of a building on a piece of property and to ensure that all easements and boundaries are met.

T F **8.** Structural plans are used to determine the building materials (steel, masonry, wood, or concrete) needed to build the main support structure.

T F **9.** A utility plan is a type of drawing that indicates the location and intended path of utilities such as electrical, water, sewage, gas, and communication cables.

T F **10.** A sectional drawing indicates the internal features of an object.

Completion

_____ **1.** A(n) ___ drawing is a type of drawing that shows the use of a specific piece of equipment or product in an application.

_____ **2.** ___ drawings are typically used during the construction and/or assembly of devices and components and are typically shown on service bulletins.

_____ **3.** A(n) ___ plan is a type of drawing that indicates land contours along with buildings and other structural information so that landscaping designs can be created.

143

_____ 4. A(n) ___ is a drawing of an object as it is viewed from above.

_____ 5. A(n) ___ is a drawing that shows exterior walls, all room partitions, doors, windows, fireplaces, stairs, cabinetry, and any fixtures or appliances.

Multiple Choice

_____ 1. ___ drawings are often used in installation and operational manuals to show where to connect external wires and position indicating lamps, switches, and displays.
    A. Detail
    B. Survey
    C. Location
    D. none of the above

_____ 2. A(n) ___ drawing is a type of drawing intended to indicate how to do work using the simplest and/or safest method.
    A. elevation
    B. instructional
    C. sectional
    D. assembly

_____ 3. A ___ plan is a type of drawing that is prepared by a licensed surveyor or civil engineer that accurately provides land contour information, dimensions, and other important feature information about a piece of property and adjacent properties.
    A. survey
    B. plot
    C. floor
    D. all of the above

_____ 4. ___ plans are used to provide excavating, construction, drainage, waterproofing, and other design information for building a foundation.
    A. Excavated
    B. Instructional
    C. Elevation
    D. Foundation

_____ 5. A drawing such as a(n) ___ drawing is used to provide as much detail as required for a clear understanding of the type and size of a device, component, or object.
    A. detail
    B. elevation
    C. sectional
    D. all of the above

Activity—Installing Commercial Light Fixtures

4-1

Scenario:
Activity 4-1 is an installation activity requiring that an electrician understands the electrical plans being used to install light fixtures and electrical equipment in the Dining Room of a Wendy's restaurant.

WENDY'S RESTAURANT — KITCHEN

Task:
Fill out a work request.

Required Work Request Information:

1. Switches will be installed 4′-6″ above the _____.
2. What are the center-to-center distances of the hanging light fixtures in the Dining Room? _____
3. How far from the wall are the hanging light fixtures mounted? _____
4. How many hanging light fixtures are there in the Dining Room? _____
5. How many lighting circuits are to be installed in the Dining Room? _____
6. How many speakers will be installed in the Dining Room? _____
7. Where will the speakers be mounted? _____

Reference Print:

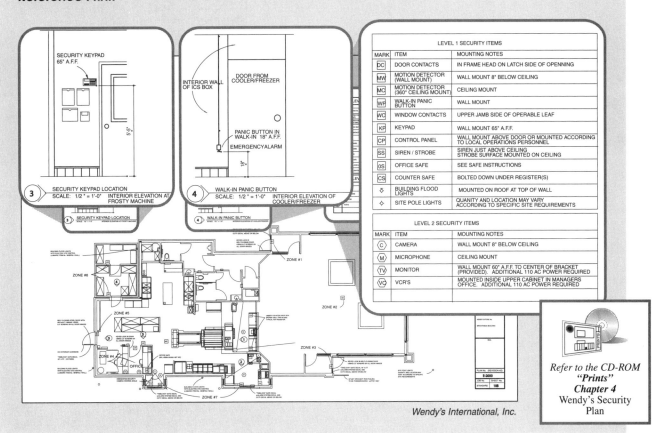

Wendy's International, Inc.

Refer to the CD-ROM "Prints" Chapter 4 Wendy's Security Plan

Activity—Installing Commercial Light Fixtures

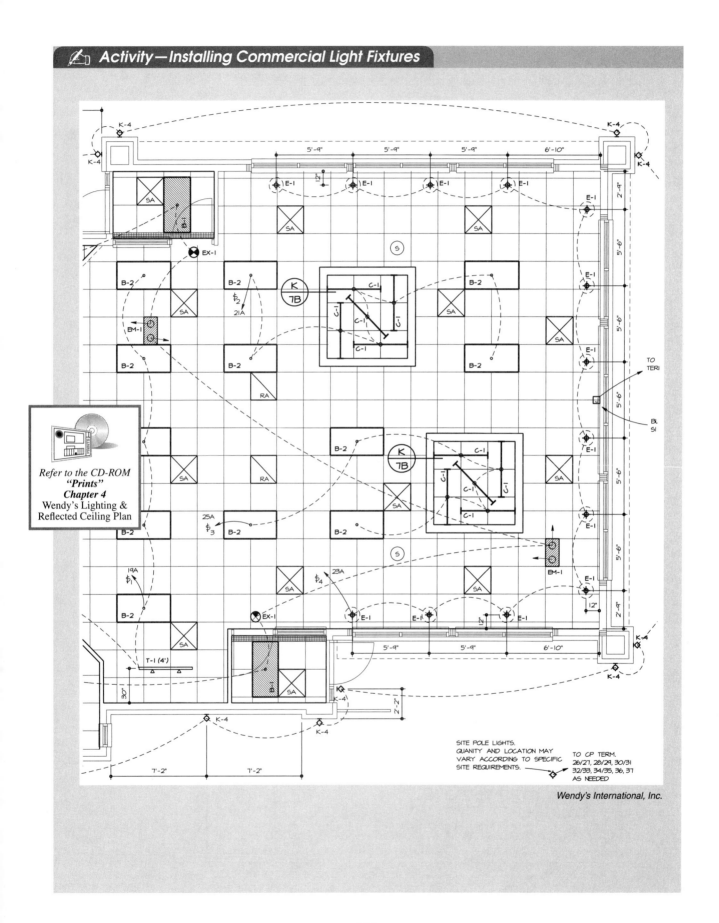

Refer to the CD-ROM "Prints" Chapter 4 Wendy's Lighting & Reflected Ceiling Plan

Wendy's International, Inc.

Activity—Troubleshooting Room Electrical Devices

4-2

Scenario:
Activity 4-2 is a troubleshooting activity requiring that electricians understand the correlation between electrical floor plans and device wiring. When problems arise, an electrician must be able to look at electrical floor plan (print) and see device positioning and equivalent wiring.

Task:
Draw the appropriate device (GFCI, outlet, and switch) per an electrical floor plan for the Kitchen and Bathroom. Indicate each device's location per the electrical floor plan using lines and arrows to point to the proper installation position on the Kitchen and Bathroom elevations.

Reference Prints:

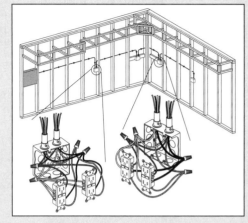

KITCHEN WIRING DEVICE

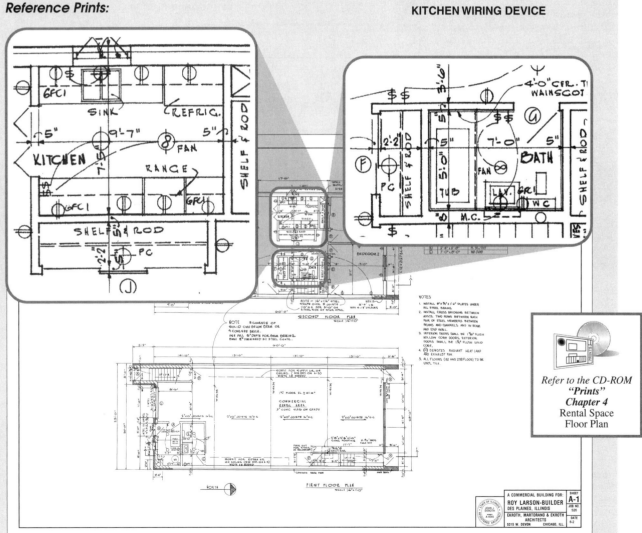

Refer to the CD-ROM
"Prints"
Chapter 4
Rental Space
Floor Plan

Activity—Troubleshooting Room Electrical Devices

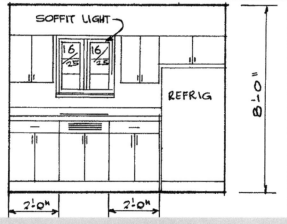
WEST KITCHEN ELEVATION

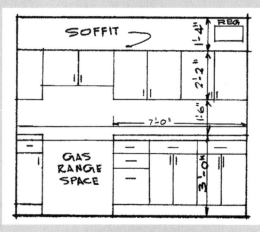
EAST KITCHEN ELEVATION

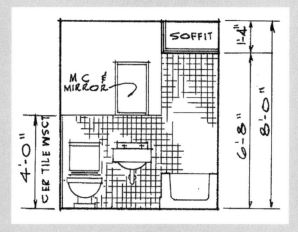
EAST BATHROOM ELEVATION

Name _____ Date _____

REFERENCE PRINT #1 (3 Phase Motor)

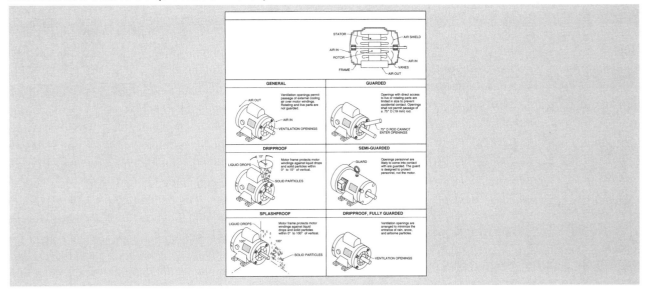

Question – Reference Print #1

_____ 1. What type of drawing is Reference Print #1 (3 phase motor)?

Reference Print #2 (PLC Mounting Bolt)

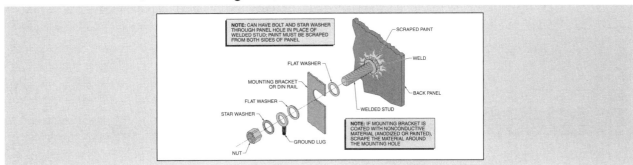

Question – Reference Print #2

_____ 1. What type of drawing is Reference Print #2 (PLC mounting bolt)?

REFERENCE PRINT #3 (Centrifugal Motor Switch)

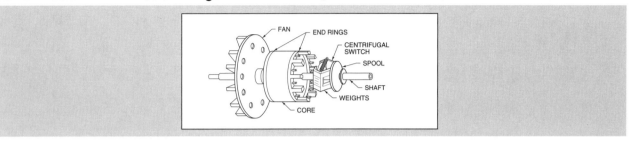

Question – Reference Print #3

_____ 1. What type of drawing is Reference Print #3 (centrifugal motor switch)?

REFERENCE PRINT #4 (PLC)

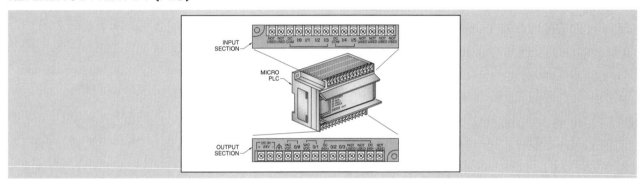

Question – Reference Print #4

_____ 1. What type of drawing is Reference Print #4 (PLC)?

REFERENCE PRINT #5 (External Gear Pump or Motor)

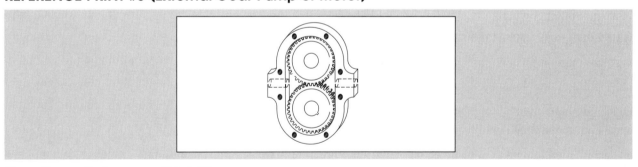

Question – Reference Print #5

_____ 1. What type of drawing Reference Print #5 (external gear pump or motor)?

 ## ADVANCED CD-ROM PRINT QUESTIONS (Research Center Elevations)

_____ 1. What elevations are shown?

_____ 2. What type of drawings are Drawing #1 and Drawing #2?

_____ 3. Are dimensions of the building provided?

Electrical and Electronic Systems

Printreading for Installing and Troubleshooting Electrical Systems

Printreading involves looking at an electrical drawing or diagram and being able to interpret the information. The type of information provided depends upon the application, equipment used, type of system, and amount of detail needed to do the work. In the electrical and electronic fields, several different types of diagrams are used to convey information.

Some types of simple diagrams, such as one-line diagrams, are used to provide the basic information required to understand a power distribution system. Others types of diagrams, such as wiring diagrams and schematic diagrams, provide detailed information on the exact connections and operation of equipment and systems.

ONE-LINE DIAGRAMS

A *one-line diagram* is an electrical drawing that uses a single line and basic symbols to show the current path, voltage values, circuit disconnect, overcurrent protection devices, transformers, and panelboards for a circuit or system. One-line diagrams are typically used with power distribution systems to provide a basic overview of the power flow, from the power entering the building, to the distribution system, to each distribution panelboard. **See Figure 5-1.**

For example, one-line diagrams typically show voltages such as 13.8 kV or 13,800 V being fed into a building and the main transformers being used for the distribution of specific voltages. High voltages are used for the distribution of large amounts of power (power = voltage × current). The high voltage is stepped down by transformers to usable voltage levels, such as 120 V, 208 V, 277 V, or 460 V, and delivered to distribution panels in a building. The distribution panelboards route power to individual loads, such as lamps, copy machines, computers, motors, and industrial machinery.

One-line diagrams are used for designing a building's power distribution system and show the required voltages, types of transformers, power panelboards, and major distribution equipment. One-line diagrams are also used after a building is operational to show the points in the system where future expansion can occur. Troubleshooting distribution-system problems, such as loss of power, low-voltages, blown fuses, or where to remove power from a section of the building during maintenance, are other uses for one-line diagrams.

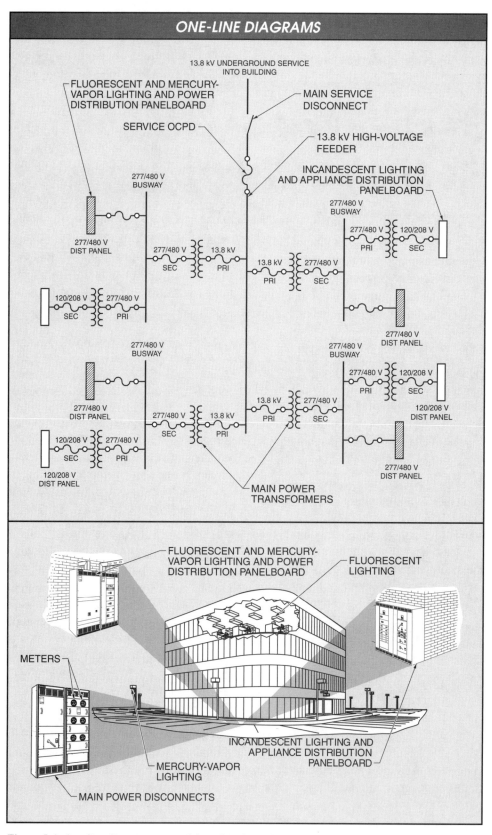

Figure 5-1. One-line diagrams are useful in identifying the major components of a power distribution system when troubleshooting power-related problems.

LADDER (LINE) DIAGRAMS

A *ladder (line) diagram* is a drawing that typically shows, using multiple rungs and graphic symbols, the logic of an electrical control circuit. Ladder diagrams show the interconnection of input devices (such as pushbuttons, limit switches, and temperature switches), output components (such as lamps, motor starters, and solenoids) and other circuit control equipment (such as timers and counters).

Ladder diagrams are one of the most important types of electrical drawings in a set of prints because ladder diagrams provide a fast and easy way to understand how individual objects are connected within a system. **See Figure 5-2.** The arrangement of a ladder diagram is simple and clear. Graphic symbols, abbreviations, and device designations are drawn per electrical standards. Lines between symbols are horizontal or vertical (never diagonal) and do not cross.

In a ladder diagram, circuit devices and components are not shown in their actual circuit positions. The electrical connections of the devices and components are what is important on a ladder diagram. For example, a ladder diagram will show that the start and stop pushbuttons are connected together by a wire but will not show whether the pushbuttons are located together in a box or panel or whether they are 100′ apart. Also, ladder diagrams do not indicate what type of pushbuttons are being used or which pushbutton is located on top or to the side of the other.

Newer ladder diagrams show the two power lines running vertically with the control circuit lines running horizontally between them. Older ladder diagrams show the two power lines running horizontally with the control circuit lines running vertically.

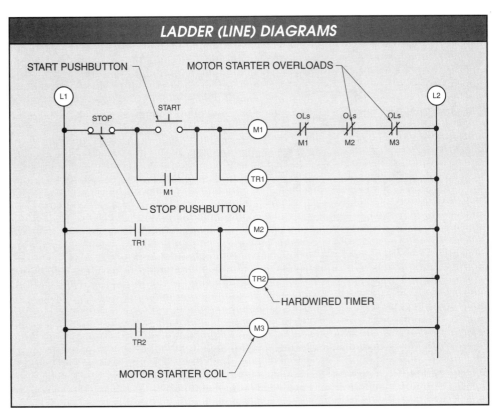

Figure 5-2. Ladder (line) diagrams use standard electrical symbols and show all circuit device and component interconnections in as simple a format as possible.

A ladder diagram of three conveyors shows how three motor starters can be sequenced so that conveyor motor 1 turns on first, conveyor motor 2 turns on 10 sec after conveyor motor 1, and conveyor motor 3 turns on 10 sec after conveyor motor 2. **See Figure 5-3.**

In the three-conveyor ladder-diagram application, the stop and start pushbuttons are the control circuit's input devices, the three motor starters are the output components, and the two timers are used to develop the required circuit logic.

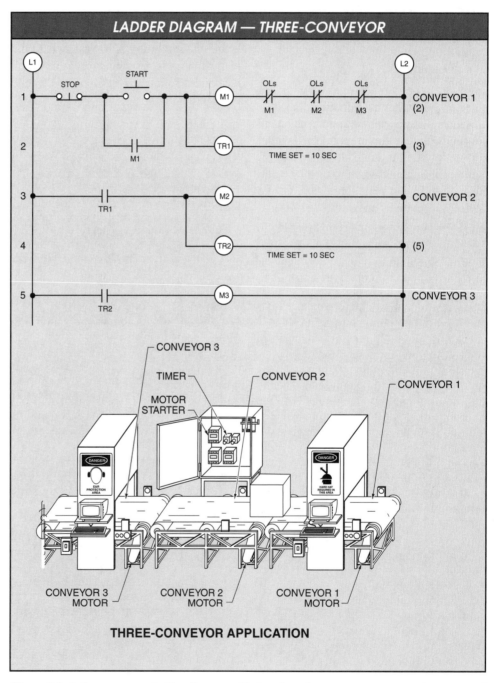

Figure 5-3. A three-conveyor ladder diagram will show how three motor starters can be sequenced so that conveyor motor 1 turns on first, conveyor motor 2 turns on 10 sec after conveyor motor 1, and conveyor motor 3 turns on 10 sec after conveyor motor 2.

PLC PROGRAMMING DIAGRAMS

A *PLC programming diagram* is a type of ladder diagram that is created on a computer and downloaded to a programmable logic controller (PLC). A *programmable logic controller (PLC)* is a solid-state control device that can be programmed and reprogrammed to control and monitor electrical circuits.

An electrical circuit designed and wired using a PLC programming diagram includes the same control circuit input devices (pushbuttons, switches, and overload contacts) and output components (lamps, solenoids, and motor starters) as an electrical circuit designed and wired using a ladder diagram. However, unlike ladder diagrams that have externally hardwired timers, relays, and counters, PLC programming diagrams have timer, relay, and counter functions as part of the PLC programming software. **See Figure 5-4.**

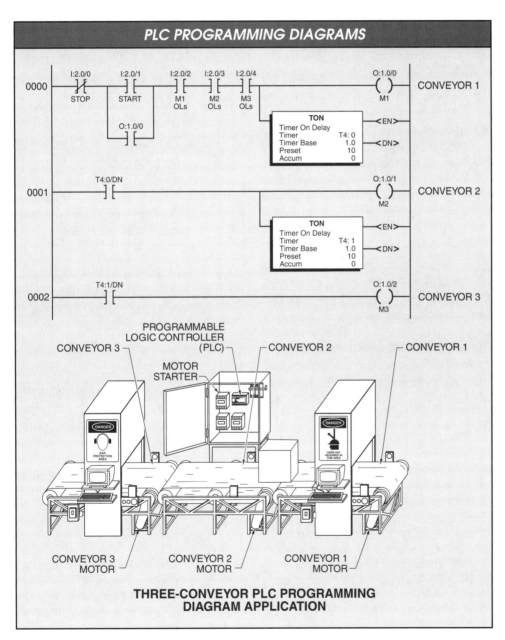

Figure 5-4. A PLC programming diagram is a control circuit that is created using PLC programming software.

For example, when using a stop/start PLC circuit to replace the hardwired timer in the three-conveyor ladder-diagram application, only the two pushbuttons, the motor starter overload contacts, and the three motor starter coils are wired to the PLC. The two timers are internally programmed using PLC programming software.

WIRING DIAGRAMS

A *wiring diagram* is a type of electrical drawing that shows the connection of input devices and output components in a circuit. Wiring diagrams are used to show how one individual device or component is wired or to show how all the devices and components of a circuit or system are wired together. Wiring diagrams are typically included with most devices and most components because most devices and components have multiple leads. For example, a wiring diagram that shows how a motor is wired for different voltages is typically included on the nameplate of a motor. **See Figure 5-5.**

Wiring diagrams showing control and power circuits connected together show as closely as possible the actual location of each device and component in the circuit. Internal and external connections are shown in sufficient detail for the electrician to be able to make or trace back connections.

Wiring diagrams are used in troubleshooting because wiring diagrams indicate the actual device and component layout with all connections. Wiring diagrams are limited when showing circuit logic because conductors are often hard to follow. When a wiring diagram is available with a corresponding ladder diagram, circuit operation and wiring can be clearly understood.

> Motor wiring diagrams use two numbering systems; power lines are numbered L1 (R), L2 (S), and L3 (T), and the leads of the motor are numbered T1 (U), T2 (V), and T3 (W). Depending on the motor design, the leads of the motor can be numbered T1 through T9 or T1 through T12.

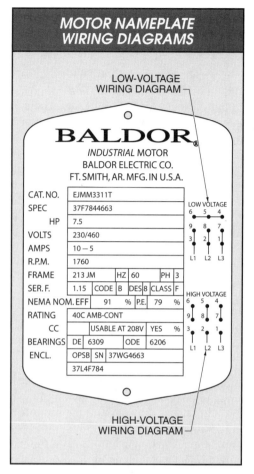

Figure 5-5. Motor nameplate wiring diagrams are used to indicate how a motor needs to be wired for various voltage levels.

Unlike ladder diagrams, wiring diagrams show the actual circuit position of each device and component. For example, a wiring diagram will show that the start and stop pushbuttons of a motor control circuit are located in the same control station. **See Figure 5-6.** This is indicated by dashed lines around the pushbuttons. The wiring diagram also shows the position of each pushbutton.

Symbols used on wiring diagrams may be the same symbols used with schematic diagrams, or they may be simple rectangles and circles. No attempt is made to show exact sizes of devices and components on wiring diagrams. Terminal connections are shown as circles even when the physical appearance of a terminal differs.

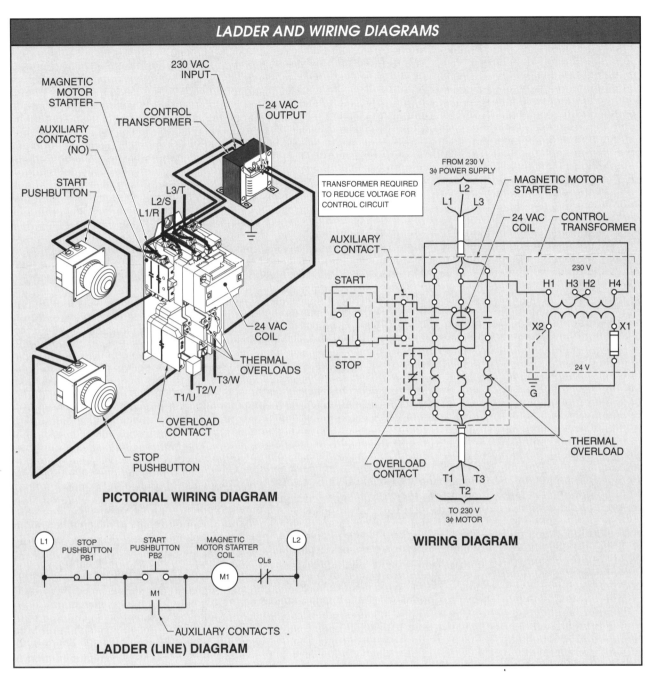

Figure 5-6. Ladder and wiring diagrams are used to indicate how individual devices and components are wired or how circuit devices and components are wired together.

SCHEMATIC DIAGRAMS

A *schematic diagram* is an electrical drawing that shows the electrical connections and functions of a specific circuit arrangement using graphic symbols. A schematic diagram does not show the physical relationship of devices and components in a circuit, but rather shows the function of each device in relationship to the other devices and components. Schematic diagrams are used to trace a circuit's operation without regard to the actual size, shape, or location of the devices and components. **See Figure 5-7.**

158 Printreading for Installing and Troubleshooting Electrical Systems

Refer to the CD-ROM
"Prints"
Chapter 5
Transistor Radio Schematic

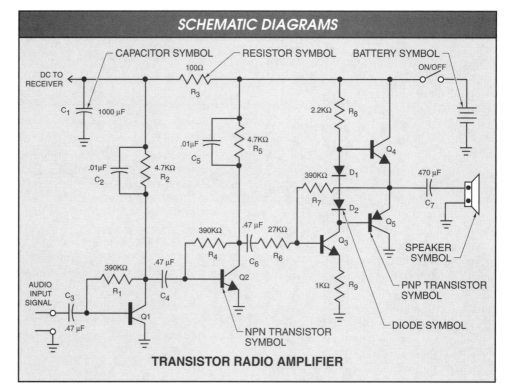

Figure 5-7. Schematic diagrams show the interconnection and function of electronic devices and components using symbols.

Schematic diagrams are typically used for designing and troubleshooting electronic circuits such as TVs or computer circuits. Schematic diagrams range from simple diagrams that show only a few devices and components, such as a timing circuit used to control a lamp or a radio amplifier circuit, to large complicated schematic diagrams that show the internal circuits of such items as a TV or computer.

Unico, Inc.

Without schematic diagrams, the troubleshooting and repair of printed circuit boards would be impossible.

INTERCONNECTING DIAGRAMS

An *interconnecting diagram* is a type of electrical drawing that shows the external connections between all system devices and components. Interconnecting diagrams are either of the wiring type or cable type. The wiring type of interconnecting diagram shows each individual wire of the system. The cable type of interconnecting diagram only shows the cables and the cable connections. For example, a cable interconnecting diagram can be used to show how all the equipment of a home entertainment center are connected together. **See Figure 5-8.**

Interconnecting diagrams must be kept simple, with as few wires or cables crossing each other as possible. Typically, interconnecting diagrams only show connections, but can include wire sizes, cable types, part numbers, reference notes, and any other information required to ensure proper connections between all equipment.

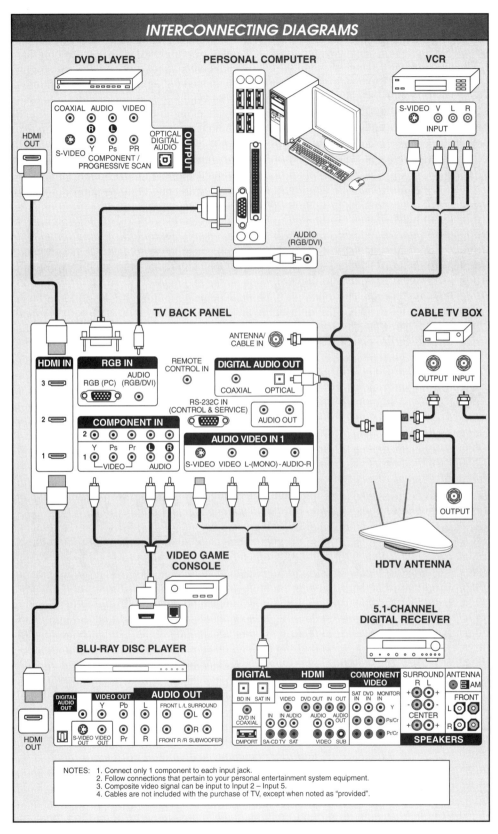

Figure 5-8. Interconnecting diagrams show how individual components of a system are connected to each other.

OPERATIONAL DIAGRAMS

An *operational diagram* is an electrical drawing that shows the operation of individual devices and components used in circuits. Operational diagrams are used to show, in simplest form, the relationship between the input and output sections of circuit devices and components.

For example, an operational diagram can show how a photoelectric sensor operates. **See Figure 5-9.** The operational diagram for a light-operated photoelectric sensor shows that the contacts of the sensor are activated (normally open are held closed and normally closed are held open) any time the light beam between the transmitter and receiver (or reflector) is interrupted. The operational diagram for a dark-operated photoelectric sensor shows that the contacts of the sensor are not activated (normally open are held open and normally closed are held closed) any time the light beam between the transmitter and receiver (or reflector) is not interrupted.

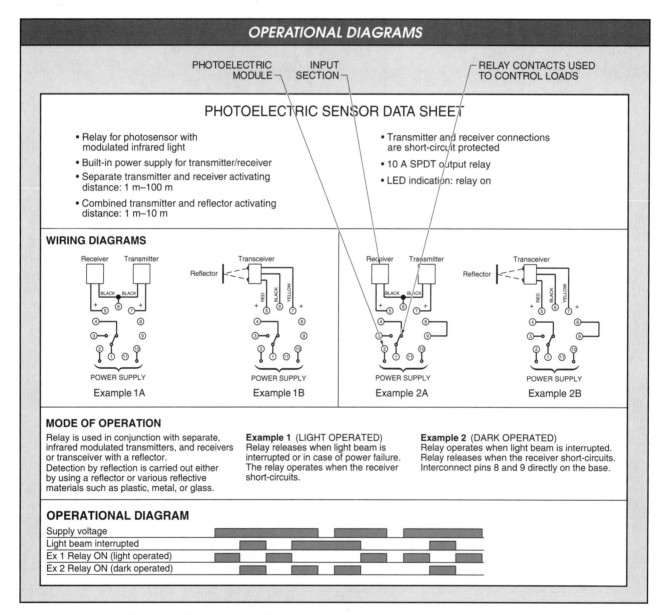

Figure 5-9. Operational diagrams show the input and output operation of devices and components.

An operational diagram and/or data sheet for a photoelectric sensor includes a wiring diagram to show the terminal connections for the transmitter, receiver, relay contacts, and power supply sections of the sensor. Also, notes are typically part of operational diagrams and data sheets.

BLOCK DIAGRAMS

A *block diagram* is a type of electrical drawing that shows the relationship between individual sections, or blocks, of a circuit or system. The circuit device or component inside each block is not shown; instead, text describes the function of the block. This solves the problem of showing complicated circuits. Block diagrams protect proprietary information (exact device or component information) that a manufacturer may prefer not to release.

Block diagrams show the contents of a circuit, device, or component. For example, a solid-state relay (SSR) block diagram can show the internal sections of a solid-state relay. **See Figure 5-10.** One block notes that a solid-state relay has a built in zero voltage sensing circuit. A zero voltage sensing circuit is used to turn a load ON only when the load voltage is at the zero crossing (no in-rush of voltage and current from the relay to the load), regardless of when the control voltage is applied. Zero voltage turn-on is recommended for resistive-type loads such as lighting and inductive-type loads such as motors. Another block notes that a solid-state relay has built-in transient protection. Transient protection prevents problems caused by stray voltages from appearing on the load side of the circuit.

Block diagrams are also used to help in troubleshooting. **See Figure 5-11.** The electrician can check the devices or components that are under suspicion of failure. Block troubleshooting diagrams can be set up with questions to check what is working and what is not. Block troubleshooting diagrams are used to troubleshoot problems by answering the questions with YES or NO until the problem is found.

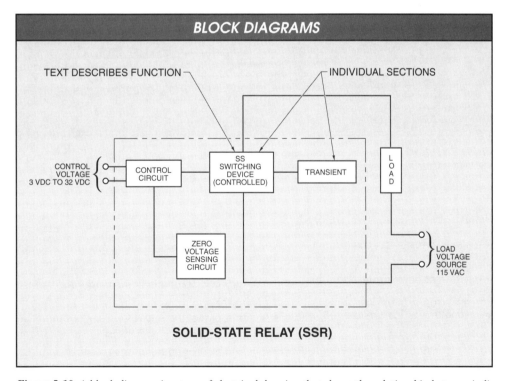

Figure 5-10. A block diagram is a type of electrical drawing that shows the relationship between individual sections, or blocks, of a circuit or system using text to describe the function of each block.

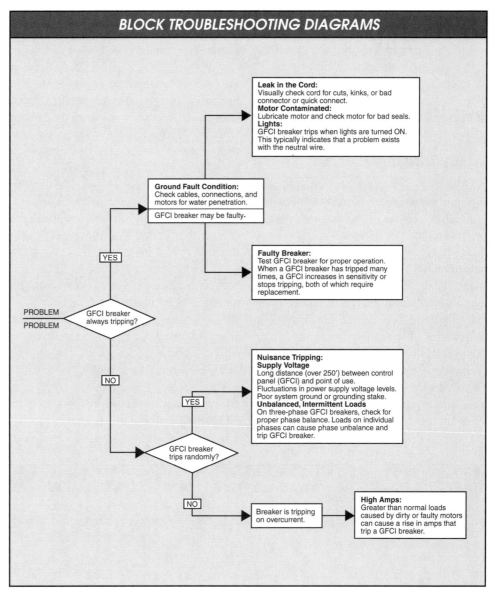

Figure 5-11. Block troubleshooting diagrams are set up with questions to check what is or what is not working in a circuit or component under test.

FUNCTION-BLOCK DIAGRAMS

All electrical circuits are designed to operate in a predetermined way. For example, a house that has a front and a back door, has two different pushbuttons that can be wired so that either pushbutton operates a doorbell when pressed. The control logic of this residential doorbell circuit is called OR logic because the front-door pushbutton "OR" the back-door pushbutton turns the circuit's output (the doorbell) ON.

Like any electrical circuit, there are several methods by which an OR circuit can be designed and connected. The two pushbuttons and doorbell can be hardwired using a ladder diagram as the reference or can be programmed using a PLC programming diagram. The two pushbuttons and doorbell can also be programmed using a function-block diagram. A function-block diagram is an alternative method for programming a PLC or programmable logic relay (PLR).

When a control circuit is programmed using function-blocks, the circuit's input devices (pushbuttons, temperature switches, pressure switches, and limit switches), are programmed as input blocks, and the circuit's output components (lamps, motors, lighting, and solenoids) are programmed as output blocks. Control function blocks are added between the input and output blocks. Available function blocks include timers (ON-delay and OFF-delay), counters, and logic blocks such as AND, OR, NOR, NAND, and XOR (exclusive OR). **See Figure 5-12.**

For example, a function block diagram can have output 1 turning ON when input 1 "AND" input 2 are activated. Output 2 will turn ON when input 1 "AND" input 2 are both activated "OR" when input 3 or input 4 is activated, but not when inputs 1 and 2 "AND" inputs 3 "OR" 4 are activated because of the exclusive OR circuit logic block. Output 3 turns ON when input 4 or input 5 is not activated (NOR logic).

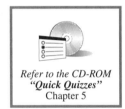

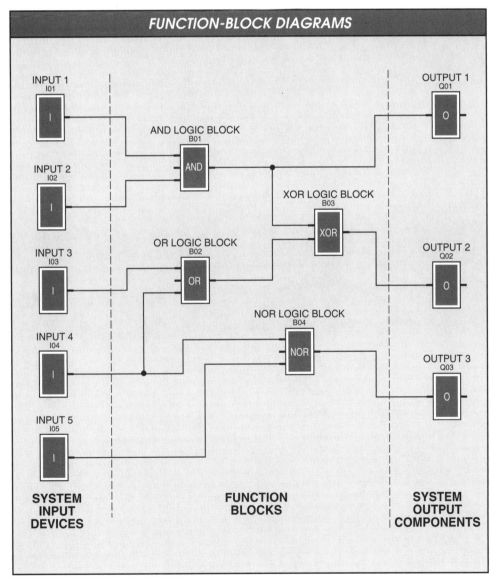

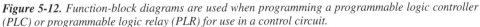

Figure 5-12. Function-block diagrams are used when programming a programmable logic controller (PLC) or programmable logic relay (PLR) for use in a control circuit.

Chapter 5—Electrical and Electronic Systems   165

Example—Replacing a Well Pump Timer

5-1

Scenario:
Example 5-1 is an installation activity requiring the replacement of a timer in a well pump-to-tank circuit. Timer 1 (TD1) needs to be replaced, but the original timer is no longer available. A timer from a different manufacturer must be used. A timer found in the maintenance store room is a "universal" timer that can be used for many applications.

Task:
Using the Well Pump to Tank print and the manufacturer wiring diagram for the timer, the operational diagram, and any technical information, determine whether the timer can be used and if yes, how it could be used.

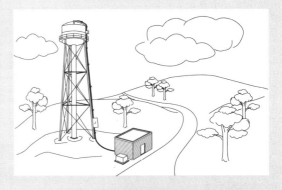

WELL PUMP HOUSE AND STORAGE TANK

Reference Prints:

STORE ROOM TIMER
(INFORMATION SUPPLIED IN BOX BY TIMER MANUFACTURER)

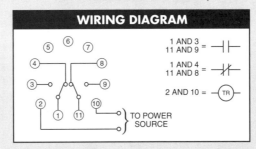

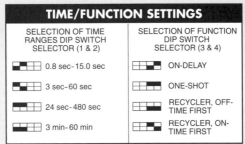

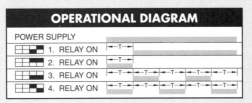

☑ Example—Replacing a Well Pump Timer

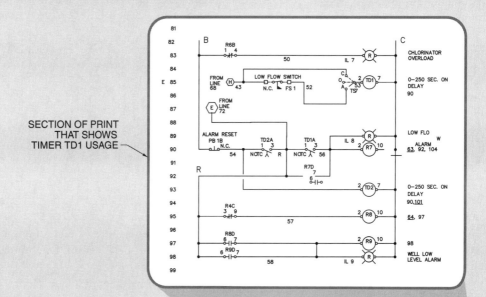

SECTION OF PRINT THAT SHOWS TIMER TD1 USAGE

Refer to the CD-ROM
"Prints"
Chapter 5
Well Pump
Ladder Diagram

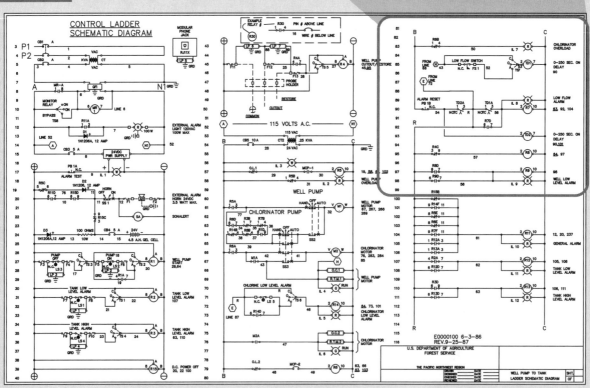

U.S. Department of Agriculture — Forest Service

☑ Example—Replacing a Well Pump Timer

Step 1: Based on the Well Pump to Tank print, determine the type of timer used in the circuit.

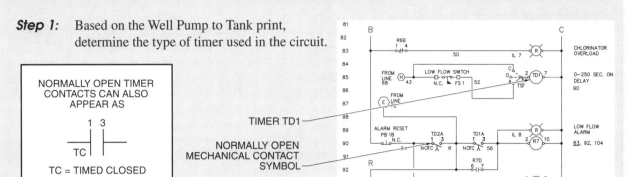

Step 2: Determine the required number and type of timer contacts.

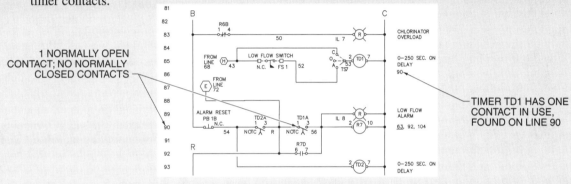

Step 3: Determine the required timing period of the timer.

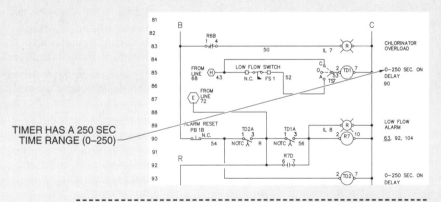

Step 4: Determine whether the store room timer is the same operational type (on-delay or off-delay) as the timer in circuit.

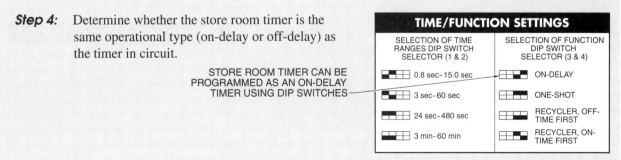

Example—Repalcing a Well Pump Timer

Step 5: Set the store room timer for the same operational function as the circuit timer.

DIP SWITCH 3 IS PLACED DOWN AND DIP SWITCH 4 IS PLACED UP

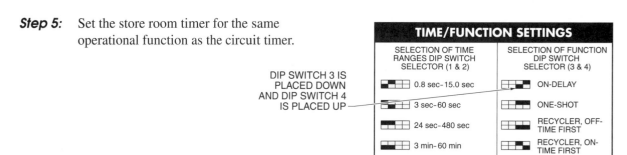

Step 6: Determine if the store room timer can be set for the same time range as the circuit timer.

STORE ROOM TIMER CAN BE PROGRAMMED FOR 4 TIMING RANGES OF 0.8 SEC TO 60 MIN

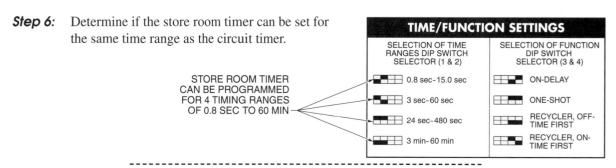

Step 7: Set the store room timer for the required timing range.

DIP SWITCH 1 IS PLACED UP AND DIP SWITCH 2 IS ALSO PLACED UP

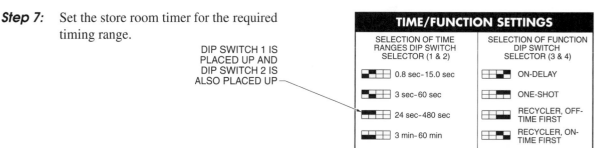

Step 8: The circuit timer used pins 2 and 7 for the coil connections. Determine which pins are used on the store room timer for connecting the coil.

STORE ROOM TIMER USES PINS 2 AND 10 TO CONNECT THE COIL

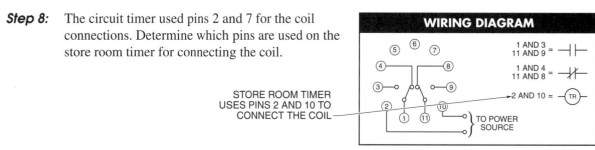

Step 9: The circuit timer used pins 1 and 3 for timer contacts. Determine which pins are used on the store room timer for the timer contacts.

STORE ROOM TIMER USES PINS 1 AND 3 OR PINS 9 AND 11 FOR NORMALLY OPEN TIMER CONTACTS

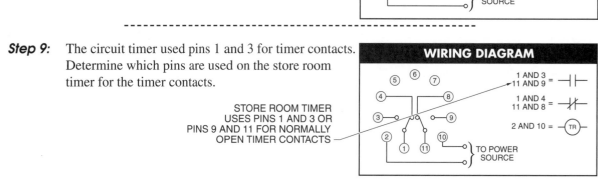

Example—Troubleshooting Waste Compacter Hydraulic Controls

5-2

Scenario:
Example 5-2 is a troubleshooting activity that requires an understanding of the operation of the electrical circuit of a waste compactor. The compactor has two hydraulic cylinders that are controlled by two solenoid-actuated directional control valves. The extend cylinder (ram) is not extending. The waste compactor control circuit uses hardwired relays.

Task:
Troubleshoot the waste compactor control circuit to determine why the extend cylinder is not extending.

WASTE COMPACTOR

Reference Print:

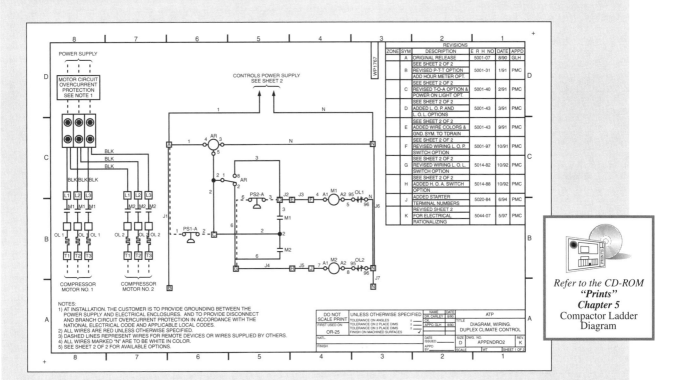

Refer to the CD-ROM "Prints" Chapter 5 Compactor Ladder Diagram

☑ Example—Troubleshooting Waste Compacter Hydraulic Controls

Step 1: Ensure an electrical test instrument(s) is available for testing electrical power and control circuits.

DIGITAL MULTIMETERS WORK WELL BECAUSE DMMS CAN MEASURE VOLTAGE, RESISTANCE, CURRENT, AND CONTINUITY AND MOST ALLOW ADDITIONAL ATTACHMENTS

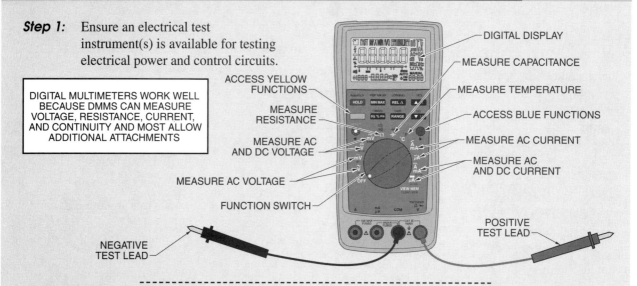

Step 2: Determine the possible problems with the control circuit that could cause the ram cylinder not to extend.

PLACE SELECTOR SWITCH (1SS) IN THE RUN POSITION ②

THE AUTOMATIC CIRCUIT IS CHECKED BY SWITCHING THE SYSTEM (SWITCH 1TGS) TO MANUAL ①

③ SHIFT SELECTOR SWITCH (2SS) TO THE EXTEND POSITION THEN TO THE RETRACT POSITION

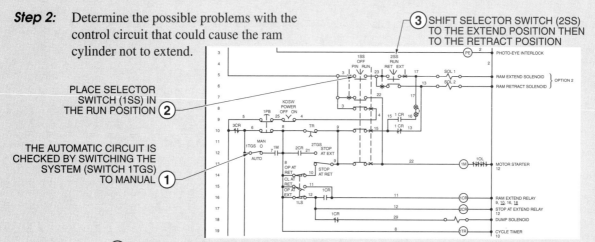

④ WHEN SYSTEM SET FOR MANUAL OPERATION THE RAM CYLINDER EXTENDS AND RETRACTS, PROBLEM MUST BE WITH AUTOMATIC CONTROL SYSTEM

Step 3: Determine control circuit test points.

MEASURE THE VOLTAGE OUT OF THE CONTROL TRANSFORMER (CHECK FOR LOW VOLTAGE CONDITION)

VOLTAGE SHOULD BE +5% TO −10% DURING ALL OPERATIONAL CONDITIONS

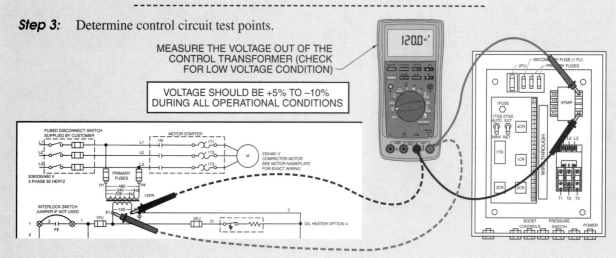

Chapter 5—Electrical and Electronic Systems 171

☑ Example—Troubleshooting Waste Compacter Hydraulic Controls

Step 4: Take voltage measurements of possible problem areas to narrow the problem down.

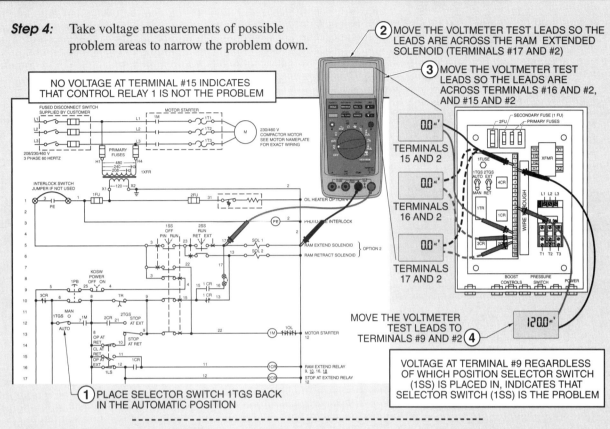

Step 5: Test the likely problem as determined by the voltage measurements.

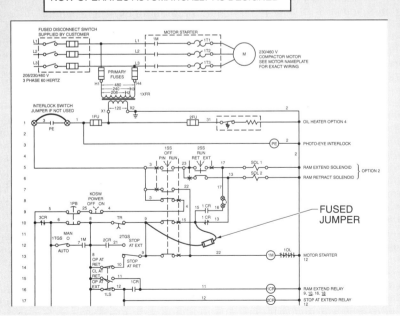

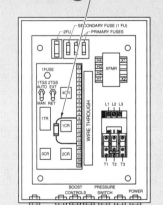

172 Printreading for Installing and Troubleshooting Electrical Systems

Example—Troubleshooting Waste Compacter Hydraulic Controls

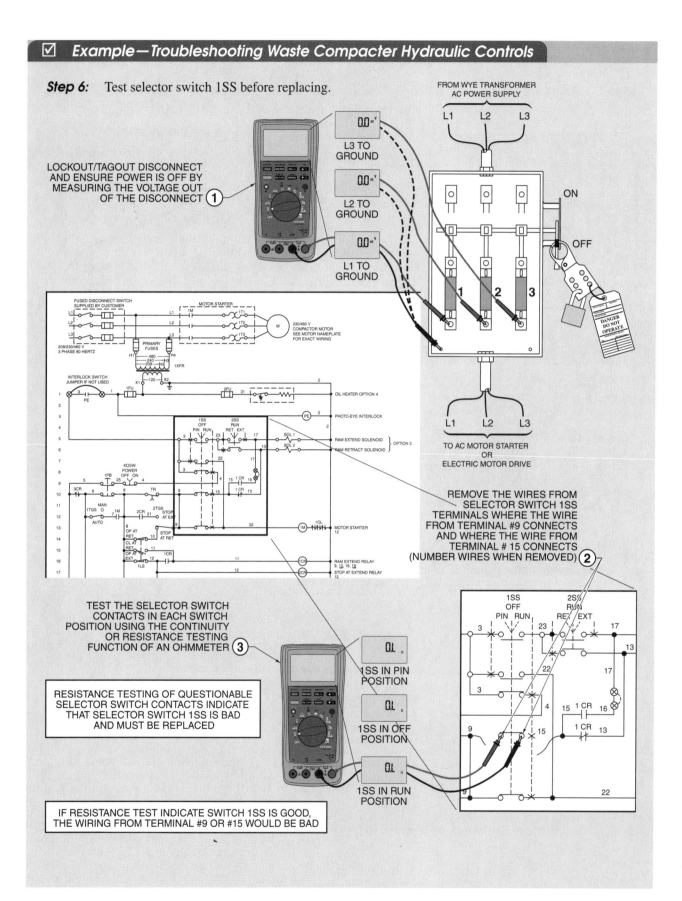

Name __ALAN RICH__ Date __11/1/11__

NO

Activity—Replacing a Photoelectric Sensor   5-1

Scenario:
Activity 5-1 is an installation activity requiring the replacement of a photoelectric sensor at the entrance to a tunnel car wash. The tunnel car wash is not activating.

Task:
Troubleshoot the photoelectric sensor and circuit to determine what type of replacement sensor to use.

TUNNEL CAR WASH
Sonny's Enterprises, Inc.

Required Information for Replacing the Photoelectric Sensor:

1. Which photoelectric sensor type (1A/2A or 1B/2B) will cover the longest distance? __1A/2A__
2. Is a light-operated or dark-operated sensor required for the application? __LIGHT__
3. In higher vibration areas, a sensor that uses a reflector works best. Which sensor example (1A, 1B, 2A, or 2B) works best for the car wash application? __1B/2B__
4. Which two pins (1, 2, 3, 4, 5, 6, etc.) would be used for the photoelectric sensor contacts shown on the print? __3+4__
5. Which two pins (1, 2, 3, 4, 5, 6, etc.) would be used for the photoelectric sensor coil shown on the data sheet? __2+10__

Reference Prints:

PHOTOELECTRIC SENSOR DATA SHEET

- Relay for photosensor with modulated infrared light
- Built-in power supply for transmitter/receiver
- Separate transmitter and receiver activating distance: 1 m–100 m
- Combined transmitter and reflector activating distance: 1 m–10 m
- Transmitter and receiver connections are short-circuit protected
- 10 A SPDT output relay
- LED indication: relay on

WIRING DIAGRAMS

Example 1A | Example 1B | Example 2A | Example 2B

MODE OF OPERATION

Relay is used in conjunction with separate, infrared modulated transmitters, and receivers or transceiver with a reflector. Detection by reflection is carried out either by using a reflector or various reflective materials such as plastic, metal, or glass.

Example 1 (LIGHT OPERATED)
Relay releases when light beam is interrupted or in case of power failure. The relay operates when the receiver short-circuits.

Example 2 (DARK OPERATED)
Relay releases when light beam is interrupted. Relay operates when the receiver short-circuits. Interconnect pins 8 and 9 directly on the base.

OPERATIONAL DIAGRAM

Supply voltage	
Light beam interrupted	
Ex 1 Relay ON (light operated)	
Ex 2 Relay ON (dark operated)	

175

Activity—Replacing a Photoelectric Sensor

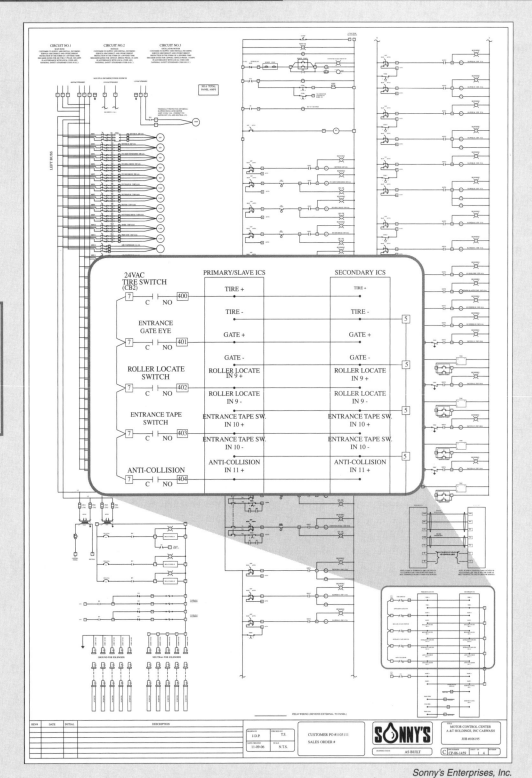

Refer to the CD-ROM
"Prints"
Chapter 5
Car Wash Motor
Control Center –
Ladder Diagram

TUNNEL CAR WASH PRINT

Sonny's Enterprises, Inc.

Name _____   Date _____

Activity—Modifying Compressor Control Circuits   5-2

Scenario:
Activity 5-2 is a troubleshooting activity that requires an understanding of a power circuit and a control circuit for a compressor in order to make additions (air dryer and automatic drain) to the circuits.

Task:
Wire the air dryer motor to the compressor control circuit.

1. What is the required voltage of the air dryer motor? _____

2. What is the additional current load to the control circuit for the 1/10 hp air dryer motor? _____

3. Which control circuit fuse (fuse number) would the current of the air dryer motor pass through? _____

4. Determine how the air dryer motor must be wired:

   a. The incoming power line from A on the power circuit print connects to which wire on the control circuit print air dryer motor add-on? _____

   b. Where would the black motor lead connect at the motor terminal block? _____

5. After a quick jog of the motor, which two motor leads must be reversed to reverse the rotation of the motor? _____

COMPRESSOR AND AIR DRYER

Reference Print

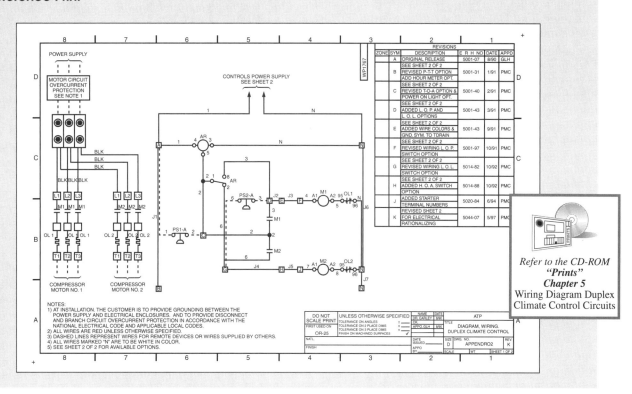

Refer to the CD-ROM
"Prints"
Chapter 5
Wiring Diagram Duplex
Climate Control Circuits

177

Activity—Modifying Compressor Control Circuits

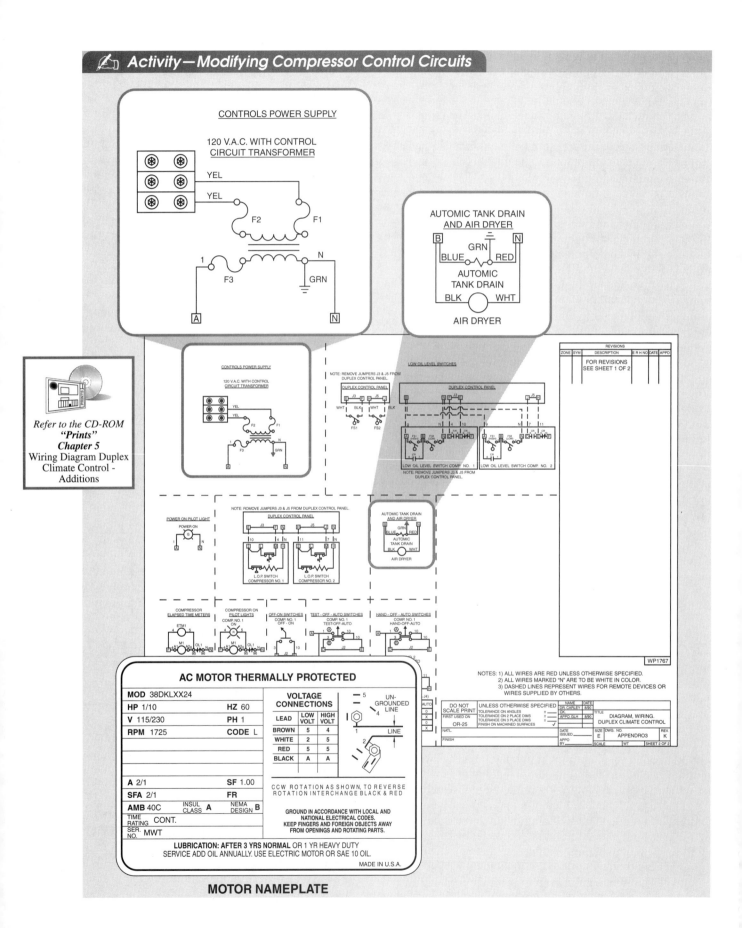

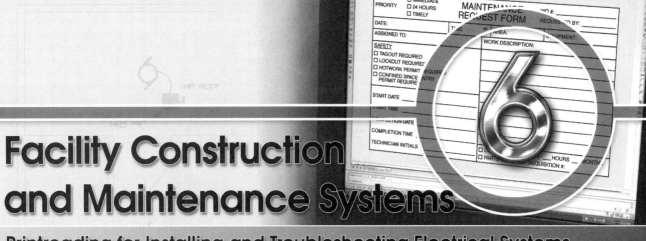

Facility Construction and Maintenance Systems

Printreading for Installing and Troubleshooting Electrical Systems

The personnel involved in the construction and maintenance of a facility have specific roles and responsibilities. The construction process and maintenance process both consist of logical and sequential sets of steps. Construction personnel and maintenance personnel use prints and written documentation to communicate information. The construction process and maintenance process must comply with federal, state, and local rules and regulations.

RESPONSIBILITIES OF CONSTRUCTION PERSONNEL

Architects, engineers, contractors, tradesworkers, and inspectors are the primary participants in the facility construction process. Each participant has a specific area or areas of expertise and a defined role. On occasion, some roles are combined, such as an electrical contractor providing the electrical engineering for a project. Some participants are only involved in a certain phase of the project, such as an electrical engineer whose only work on the project may be to calculate the distribution-system loads. Other participants are involved throughout the duration of the project, such as the architect of the facility.

Architects

The primary responsibility of an architect is to listen to the owner and accurately interpret the needs and desires of the owner in the form of a set of building prints. Architects meet with owners to develop prints and specifications. These prints and specifications are used by the owner to obtain building permits, competitive bids from contractors, and financing. **See Figure 6-1.** Architects plan and inspect a project from start to finish, ensuring the facility is constructed according to the plans.

Figure 6-1. *Architects and owners define the scope of a project and communicate the design through a set of building prints.*

181

182 Printreading for Installing and Troubleshooting Electrical Systems

An architect may be a sole practitioner or may be a member of a large architectural firm with many partners. Architects are licensed by state. Common requirements for obtaining a license include education, work experience, and passing a state exam. Prints drawn by an architect will have a state-specific architect's stamp (a seal) identifying the architect as a licensed professional. **See Figure 6-2.**

> State boards of examiners for architects have adopted rules requiring an architect to obtain an embossing seal and a rubber stamp of their private seal for imprinting documents. The seals are usually circular in shape and two inches in diameter. Also, the seal must contain the name of the state and the name and registration number of the architect.

Refer to the CD-ROM "Prints" Chapter 6 Research Center Elevations

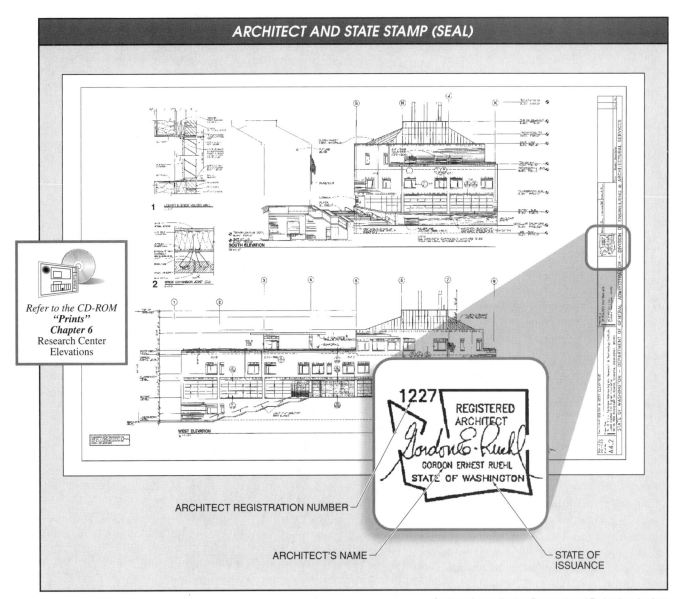

Spokane Intercollegiate Research and Technology Institute

Figure 6-2. *The stamp of an architect certifies that the architect is licensed in that state.*

Engineers

Engineers assist architects in planning a project. Electrical, mechanical, structural, and civil engineers are typically involved in the construction process. Engineers assist the architect with specific design issues to guarantee the building complies with national and local building codes. **See Figure 6-3.** Each type of engineer provides design expertise in that particular field, such as an electrical engineer calculating the size of the electrical service required; a mechanical engineer designing HVAC systems for occupant comfort; or a structural engineer designing the columns and beams to safely support and withstand the forces on the building.

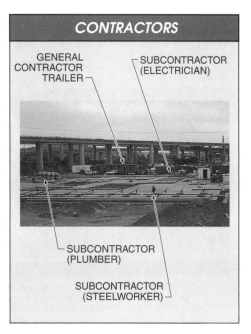

Figure 6-4. The general contractor provides the labor (subcontractors), work coordination, and material to build the frame (core) or shell of a building and the interior spaces.

Figure 6-3. Engineers assist the architect with specific design issues to guarantee a building complies with national and local building codes.

Contractors

Contractors fall into two categories, general contractors and subcontractors. A general contractor is responsible for coordinating all work done on a construction project including work done by subcontractors. The general contractor provides the labor and material to build the frame or shell of a building and the interior spaces. **See Figure 6-4.**

The subcontractors are under contract to the general contractor and are responsible for trade-specific portions of the construction or maintenance project, such as electrical, mechanical, roofing, and plumbing.

Tradesworkers

The actual construction of any facility is accomplished by tradesworkers. **See Figure 6-5.** Apprentices and journeymen that are supervised by foremen perform all work. A very large project or a particularly complex project may require a general foreman and/or a project superintendent.

Typically, journeymen have completed a three- to five-year apprenticeship program for a particular trade, such as electrical, plumbing, sheet metal, or carpentry. Many states require journeymen to be licensed through state testing in order to practice a trade. Apprentices work under the supervision of a journeyman. Typical apprenticeship programs consist of on-the-job training and classroom instruction.

CONSTRUCTION TRADEWORKERS		
STRUCTURAL	**MECHANICAL**	**FINISHING**
• Carpenters • Construction equipment operator • Brick masons • Block masons • Stonemasons • Cement masons • Concrete finshers • Iron and metal workers	• Pipe layers • Plumbers • Pipefitters • Steamfitters • Electricians • Sheetmetal workers • Heating and air conditioning • Refrigeration • Communication	• Carpenters • Drywall installers • Ceiling tile installers • Tapers • Plasterers and stucco masons • Segmental pavers • Terrazzo workers • Painters and paperhangers • Glaziers • Roofers • Carpet layers • Floor and tile installers and finishers • Insulation workers

Figure 6-5. Tradeworkers on construction projects are typically classified as structural, mechanical, or finishing workers.

A *foreman* is a worker who manages labor and material for the portion of the project relevant to the foreman's trade. Foremen are the contractor's representatives on the job site. Foremen are responsible for the successful completion of a portion of a project and are required to interact with architects, engineers, foremen from other trades, and inspectors. A *general foreman* is a foreman who supervises a group of project foremen.

Building Inspectors

Cities, counties, states, and the federal government all employ building inspectors. The job of the third party building inspector is to inspect a construction project and ensure the project is built to code. An inspector makes visits during the course of a project and checks on specific aspects of the job, such as building framing. **See Figure 6-6.**

Some building inspectors are responsible for all aspects of a construction project, while other inspectors specialize in a specific trade. In addition to building inspectors, fire-department officials will inspect certain aspects of a job, such as the placement of exit signs, the number and type of fire sprinklers, and the proper functioning of fire alarm and life-safety systems.

Figure 6-6. Building inspectors conduct many inspections of a building before the walls are covered and once or twice after the walls are finished.

When working on old structures, historic building inspectors can be very helpful. Historic building inspectors have met specific experience requirements and have proven expertise in evaluating the conditions of historic structures.

OVERVIEW OF CONSTRUCTION PROCESS

The process of constructing a building or structure involves a number of various contractors and tradesworkers. The construction process includes site preparation, construction of the building core, electrical construction, and mechanical construction. A contractor and/or tradesworker may be present for the entire length of a construction project or for only a small portion of the project.

Work performed by the various trades often occurs at the same time and can overlap. The general contractor must always provide coordination between the various subcontractors on a construction project. Written documentation is used to facilitate communication, enhance coordination between the trades, and ensure the construction process proceeds with a minimum of delays or cost overruns.

Site Preparation

Before a new facility can be constructed, the building site must be prepared. In some cases, the first step in site preparation is the demolition of old structures. When no demolition is required, earthmoving equipment is used to grade, move, and add soil as needed. After the earthmoving work is completed, foundation piles (caisons) may need to be driven into the ground to provide a solid base for the building if the soil is unstable. **See Figure 6-7.**

During the site preparation phase of the project, conduits and pipes for electricity, water, gas, sewage, and drainage are put into the earth. The conduits and pipes bring the utilities into the building and distribute the utilities to the appropriate areas throughout the ground floor. After all earthmoving work is done, piles are driven, the correct number and type of pipes are installed in the earth, and a concrete foundation is poured. The foundation is the base upon which the new building will be constructed.

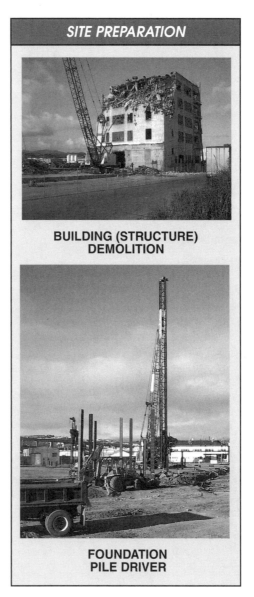

Figure 6-7. Site preparation begins with the demolition of old structures if necessary and bringing earthmoving equipment to the construction site.

Building Core

The building core consists of the frame, the exterior walls, the roof that encloses the building, and the mechanical elements of the building. After the foundation is poured, the exterior walls are built and the roof is installed. **See Figure 6-8.** While the building exterior is under construction, various tradesworkers are on the job installing specific portions of the project, such as sheet-metal workers installing ductwork for the HVAC system.

Figure 6-8. The foundation and frame of a building are the main sections of building's core.

United States Gypsum Company

Interior work on a structure must start as soon as possible and continue, even through winter, if the structure's planned completion date is to be met.

Depending on the size and urgency of a project, construction can be planned so that a building is totally enclosed (having the exterior walls and roof in place) before the start of winter. Having a building enclosed allows tradesworkers to work inside during bad weather. After the building is enclosed, the interior rooms and spaces can be built. Some of the tradesworkers who worked on the exterior of the building will work on the interior. Other tradesworkers only work on the interior portion of the building, such as drywallers, carpet installers, and painters. Whether the exterior or interior portion of a construction project is being worked on, it is common for multiple trades to be working on the project at the same time. Cost overruns, delays, and legal disputes are avoided by maintaining constant communication and coordination between individual contractors and tradeworkers.

Electrical Construction

Electrical construction typically takes place during the entire construction project. **See Figure 6-9.** Electrical contractors must work closely with the general contractor and other subcontractors throughout the project to ensure the job is done on time and on budget. Electricians must do the following work as the project progresses:

1. Install temporary electrical service before site work begins.
2. Install conduit in the earth before the foundation is poured.
3. Install power distribution panels during and after the construction of the building exterior.
4. Install power and lighting circuits for the interior rooms and spaces.

> Construction permits are issued by local building departments. Large projects typically have several permits that are required: electrical, elevator, fire sprinkler, plumbing, and general building. Each contractor must obtain a permit for their portion of the project.

As the use of automation and networking has grown in homes, offices, and factories, so has the scope of electrical construction. In addition to power and lighting circuits, electricians now install access control systems, fire alarm systems, HVAC temperature control systems, and low-voltage wiring for voice, data, and video (VDV) systems. Some electrical contractors have separate divisions that specialize in different systems, such as fire alarm, HVAC controls, or VDV systems. **See Figure 6-10.**

ELECTRICAL CONSTRUCTION

Step 1. Install temporary electrical service before site work begins.

Step 2. Install conduit in the earth before the foundation is poured.

Steel Tube Institute

Step 3. Install power distribution panels during and after the construction of the building exterior.

Step 4. Install power and lighting circuits for the interior rooms and spaces.

Figure 6-9. Electricians on a construction site must install temporary electrical service, conduit banks in trenches prior to concrete pouring, power panels, and interior power and lighting circuits.

Mechanical Construction

Mechanical construction takes place throughout the duration of a construction project. Mechanical construction requires more than one trade. **See Figure 6-11.** Plumbers install piping systems for cold water, hot water, and waste. Pipefitters install piping systems for HVAC and other systems in buildings. Pipefitters also install piping systems for processes in industrial facilities, for medical gases in hospitals, and for instrumentation in biotechnology research facilities.

VDV SYSTEM TESTING

Figure 6-10. Some electrical contractors have separate divisions that specialize in different systems, such as fire alarm, HVAC control, and VDV systems.

ISOMETRIC PIPE DIAGRAMS

Mechanical system piping includes sanitary and stormwater drainage piping, hot and cold water supply piping, natural gas piping, HVAC ductwork, boiler piping, and other piping systems. Isometric piping diagrams provide information on pipe diameter, material, type of fittings, purpose, and quantity required. Isometric piping diagrams are not typically provided with a set of prints for small- or medium-sized projects, leaving the plumbing contractor to develop the diagram if needed.

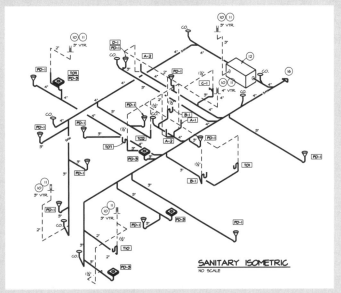

Isometric piping diagrams are, however, provided for industrial and large commercial projects. Industrial and large commercial projects typically require many isometric piping diagrams.

MECHANICAL CONSTRUCTION

Figure 6-11. Pipefitters, sprinkler fitters, plumbers, and sheet-metal workers perform the majority of mechanical-system work on construction projects.

Sprinkler fitters install piping systems for fire suppression, fire sprinklers, fire pumps, and specialized fire-suppression systems for computer rooms. Sheet-metal workers install ductwork to carry conditioned air from furnaces, air purifiers, and air conditioning units to occupied spaces.

Coordination is required between each of the trades when working on the mechanical systems of a facility. It is common for all the trades to be employed by a single mechanical contractor whose crew performs all of the mechanical system work. In addition to coordination between the trades, the mechanical contractor must also coordinate work with all the trades that are not part of the mechanical contractor's crew, especially electricians.

A great deal of coordination is required between the mechanical and electrical system contractors because many mechanical systems have electrical components, such as electric motors and solenoids. **See Figure 6-12.** Also, the placement of electrical devices and components is dependent upon where mechanical components are installed, such as with the location of duct smoke detectors.

Duct smoke detectors must be mounted at specific locations within the sheet-metal ductwork. Finally, there are many locations in a building where mechanical and electrical system components compete for space. For example, a crowded mechanical room can have sheet-metal ducts, electrical conduit, and piping that all need to be installed.

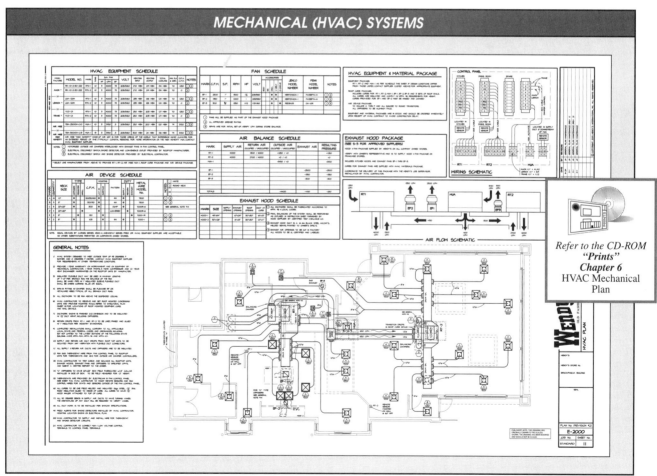

Wendy's International, Inc.

Figure 6-12. The tradeworkers working on mechanical systems must be aware of the specific locations of electrical devices and components and must know how to allocate space for various mechanical, electrical, and instrumentation equipment.

Construction Documentation

In addition to prints and specifications, there are other types of documentation used in the construction of a building. Regardless of the size of a construction project, conflicts and issues will arise. For example, information on a print may be missing or unclear, a contractor may not complete a portion of work by a specified time, or two prints for the same project may contain conflicting information, such as an HVAC print and an electrical print.

Construction documentation is used to identify problems and issues and provide a means of communicating and resolving those problems and issues. Requests for information (RFIs) and punch lists are the primary forms of construction documentation. Print/project conflict resolution uses construction documentation to solve print or project conflicts.

Request for Information (RFI). A *request for information (RFI)* is a formal document generated by a contractor requesting information about items that are missing or unclear on a print or specification. An RFI is sent to the person who can provide the required information, such as the project architect or engineer. Copies of RFIs are sent to various other parties as well to inform them of the specific issue. Typically, contractors have multiple copies of preprinted RFI forms that foremen can fill out. **See Appendix.**

The format of an RFI varies from contractor to contractor. However, most RFIs contain certain elements. **See Figure 6-13.** Typically RFIs include the following information:

- sender and recipient names
- date of RFI, subject, RFI #, and date of reply
- category or type of RFI
- reference to contract drawing
- area for the request
- area for the reply
- persons receiving copy of the RFI

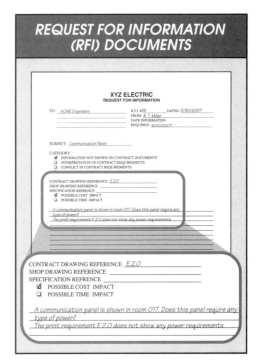

Figure 6-13. Request for information (RFI) documents include the names of the sender and recipient, date, subject of the RFI, category or type of RFI, drawing reference number, request space, reply space, and a list of people who will receive a copy of the RFI.

Punch Lists. A *punch list* is a formal document generated by an architect listing items that a contractor missed, partially completed, or did not complete per the prints and specifications. Typically, a punch list contains sections that include multiple items for each contractor. The punch list is generated at the end of a construction project. The architect uses the punch list to ensure each contractor completes the part of the project per the prints and specifications. **See Appendix.**

The punch-list format varies from architect to architect. However, most punch lists contain certain elements. **See Figure 6-14.** Punch lists commonly include the following:

- project name, date, and person generating the punch list
- location and description of the punch-list item
- party responsible for correcting the item and the required completion date
- cost impact of the item

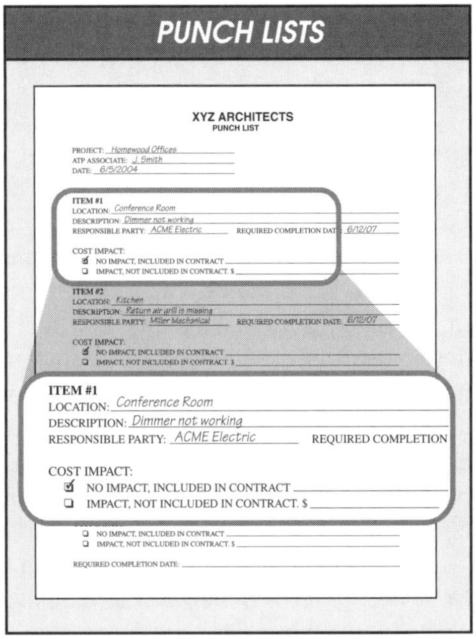

Print/Project Conflict Resolution. Print conflicts occur for many different reasons. It is important that the conflicts are resolved quickly to avoid delays or additional project costs. Common conflicts can include an item that cannot be installed per the print, such as a light fixture that does not fit in a specific location; a difference between the written dimension on a print and the actual dimension in the field; or two different prints with conflicting information, such as an electrical print and a mechanical print that indicate different horsepower ratings for the same chilled water pump drive motor.

The specifications typically direct the contractor to consult the architect or engineer. **See Figure 6-15.** The contractor sends an RFI to the architect or engineer to obtain a decision. Quite often, the specifications also require the contractor to make necessary changes without any additional cost to the owner.

Figure 6-14. Punch lists contain the project's name, name of the person who created the punch list, location and description of the specific punch-list items, name of the contractor that will correct the punch-list item, completion date, and a cost-impact statement.

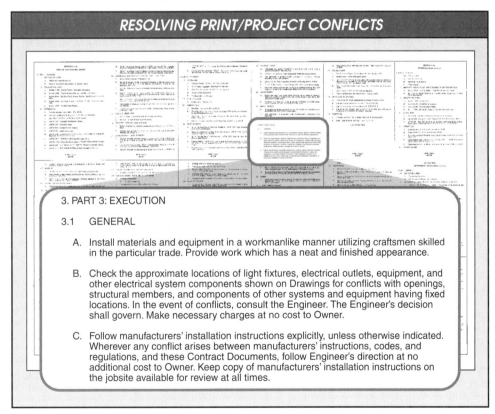

Figure 6-15. Specifications typically require that a contractor consult the architect or project engineer in the event of print or project conflict.

ELECTRICAL MAINTENANCE SAFETY

As an electrician or maintenance person, electrical safety is a concern shared by all in facility and industrial maintenance. Each year, electrical accidents cause thousands of people to be shocked and/or burned, with electrical failures causing at least a billion dollars in damage.

With building systems becoming more integrated and facility and industry owners looking towards "green" for new designs, the importance of continued facility operation is imperative. New technologies, such as renewable energy systems and on-site power generation, are becoming an integral part of many projects, and electrical safety is an essential element.

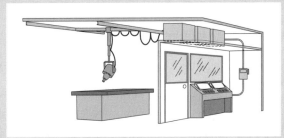

Each city, county, and/or state must establish an electrical safety committee and designate an AHJ (authority having jurisdiction) for interpreting the electrical requirements of OSHA, the NEC®, and any standards set by local governments. All electrical personnel must be trained in safe electrical work practices.

The ten principles of electrical safety are as follows: plan every job, anticipate unexpected events, use the right tool for the job, use procedures as tools, isolate equipment, identify hazard areas, minimize hazards, protect personnel, assess people's abilities, and audit the safety principles.

RESPONSIBILITIES OF MAINTENANCE PERSONNEL

Companies or institutions typically employ maintenance technicians to perform maintenance tasks in their own facilities. The duties and responsibilities of maintenance departments are determined by their structure. Maintenance departments range in size from one person in a small facility to hundreds of people in a large facility. Large maintenance departments will have maintenance technicians responsible for specific trades, such as electrical, millwright, instrumentation, sheet metal, welding, plumbing, and pipefitting.

Maintenance personnel are either trained and hired, or hired and then trained, to perform maintenance in a specific trade. For example, electrical repairs may only be performed by electricians, while machine repairs are performed by machinists and millwrights. Stationary engineers operate boilers. As the reliance on steam for heating and powering industrial processes diminishes in some facilities, the stationary engineer has become responsible for heating, cooling, and industrial process equipment.

The International Union of Operating Engineers (IUOE) is active in developing apprentices to become stationary engineers as operator duties expand into areas traditionally performed by maintenance personnel. The result is that IUOE-trained apprentices are sometimes referred to as operating engineers.

Regardless of the title, some maintenance personnel are broadly classified as specialists or multiskilled technicians. Specialists have a great amount of expertise in one trade. Multiskilled technicians have experience and training in several trades. The structure of the maintenance department is determined by the classification of maintenance personnel. **See Figure 6-16.**

Maintenance work is typically divided into the two general categories of facility maintenance and industrial maintenance. **See Figure 6-17.** A *facility maintenance technician* is a maintenance person who operates, maintains, and repairs building systems and equipment in hotels, schools, office buildings, and hospitals. Building systems include HVAC systems, fire-protection systems, and security systems. An *industrial maintenance technician* is a maintenance person who operates, maintains, and repairs production systems and equipment in industrial settings. Industrial systems include processing machinery, casing machinery, labeling machinery, and packaging machinery.

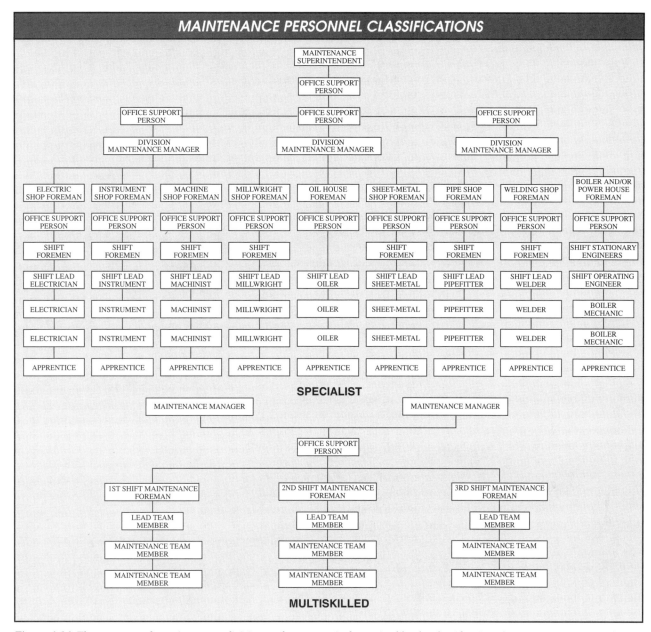

Figure 6-16. The structure of a maintenance division or department is determined by the classification of maintenance personnel.

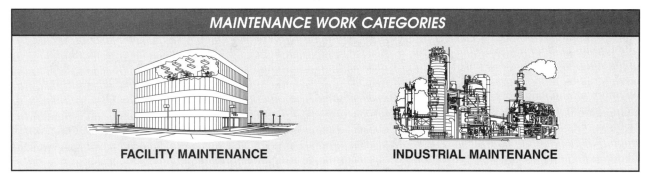

Figure 6-17. Facility maintenance is the maintenance of building systems and equipment in hotels, schools, office buildings, and hospitals, while industrial maintenance is the maintenance of production systems and equipment in industrial settings.

OVERVIEW OF MAINTENANCE PROCESS

The maintenance process involves maintenance technicians and a wide variety of facility and equipment tasks. A multiskilled maintenance technician may perform a routine maintenance task, such as changing the filters on an air conditioning unit; while a specialist technician typically would perform a more complex task, such as calibrating a pressure switch. Some maintenance tasks require more than one maintenance technician and/or more than one type of specialist technician.

The two main types of maintenance are preventive maintenance and predictive maintenance. Most maintenance departments use a combination of preventive and predictive maintenance to keep equipment and infrastructure running trouble free. Written documentation is used to communicate within the maintenance department, communicate with other departments in the company, and keep track of the maintenance process.

Preventive Maintenance

Preventive maintenance (PM) is a combination of unscheduled and scheduled work required to maintain equipment in peak operating condition. PM minimizes equipment malfunctions and failures and maintains optimum production efficiency and safety conditions in a facility. **See Figure 6-18.** Properly performed PM results in increased service life, reduced downtime, and greater overall facility or plant efficiency.

The PM tasks and frequency of the tasks associated with each piece of equipment are determined by manufacturer's specifications, equipment manuals, experience of maintenance personnel, and trade publications. Examples of preventive maintenance include repairing a piece of equipment after an unexpected breakdown; inspecting equipment for conditions, such as unusual noises, leaks, or excessive heat, that may indicate potential problems; and adjusting and replacing parts to maintain the proper operating condition of the equipment.

Figure 6-18. Preventive maintenance (PM) is the work required to keep equipment in peak operating condition.

Predictive Maintenance

Predictive maintenance (PDM) is the monitoring of wear conditions and equipment characteristics against a predetermined tolerance to predict possible malfunctions or failures. Equipment-operation data is gathered and analyzed to find trends in performance and component characteristics. Corrective repairs are made as required. **See Figure 6-19.**

PDM requires a substantial investment in training and equipment, and is typically used on expensive or critical equipment. Data collected through equipment monitoring is analyzed to check if values are within acceptable tolerances. Common PDM procedures include visual and auditory inspection, vibration analysis, lubricating-oil analysis, and thermography.

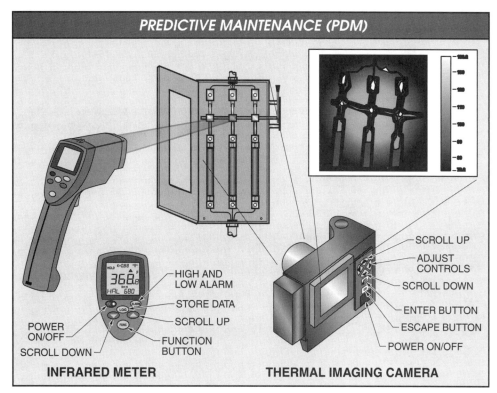

Figure 6-19. Predictive maintenance (PDM) involves the use of special equipment to monitor and analyze machine performance.

An important part of a preventive maintenance program is the recording and interpretation of data relevant to each piece of equipment. Maintenance documentation is the only way to preserve data accumulated over the life of a facility. Circuit modifications must also be recorded in the maintenance documentation.

Maintenance Documentation

In addition to prints and equipment manuals, maintenance personnel use a variety of documentation. This documentation provides a means of communication and a method of tracking the maintenance process. The primary types of documentation are work orders and logbooks.

The size of the maintenance department and the amount of equipment data collected determine the method of documentation. A paper-based system is adequate for small maintenance departments. Large departments, however, typically require a computerized maintenance management system (CMMS). In addition to the work-order and logbook functions, a CMMS can provide equipment analysis and inventory control functions.

Work Orders. A *work order* is a document that describes the work a maintenance person is required to perform. Work orders are also used to organize, schedule, and monitor various maintenance tasks. Work orders can be generated using paper forms or a CMMS. **See Figure 6-20.** Work orders typically include the following:
- job number
- date and time
- name of equipment
- location of equipment
- work description
- safety requirements
- parts ordered to complete repair

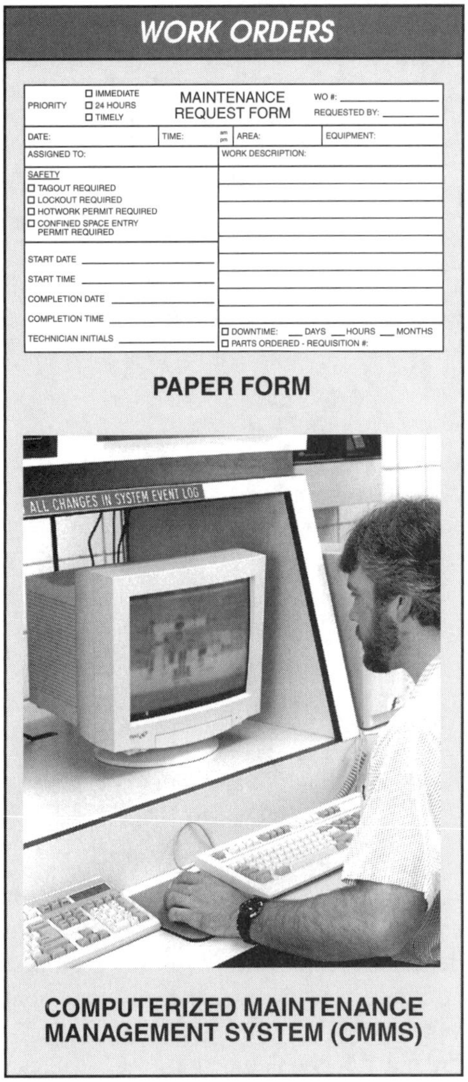

The logbook serves as the primary means of communication between different shifts and personnel about a piece of equipment.

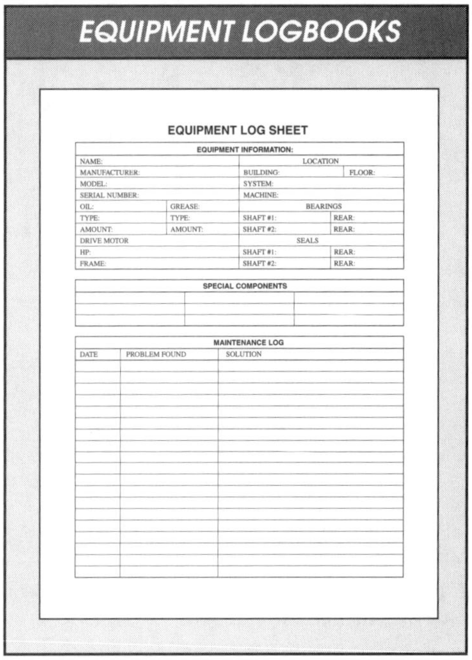

Figure 6-20. Work orders are generated using paper forms or a computerized maintenance management system (CMMS) and are used to notify maintenance personnel of tasks that need to be completed.

Figure 6-21. An equipment logbook is a binder or electronic file that documents all work performed on a piece of equipment (back to installation) by maintenance personnel.

Efficient maintenance departments have time logs available for their staff. Records, including time logs, must be maintained in whatever format is suitable for comparison and evaluation.

Equipment Logbooks. An *equipment logbook* is a binder or electronic file that documents all work performed by maintenance personnel on a piece of equipment, beginning with installation. **See Figure 6-21.** When a work order is received for a piece of equipment, maintenance technicians review the logbook for that piece of equipment. Based on the information in the logbook, maintenance personnel might find the problem occurred in the past and the action that was taken to solve the problem.

Time Logs. A *time log* is a binder or electronic file that documents all work using job numbers and the time taken to perform the work by maintenance personnel. **See Figure 6-22.** Maintenance personnel begin each day by reviewing the time logs from the previous shifts. Based on the information in the time logs and the type of work orders received, maintenance personnel are able to create a list of tasks to be completed.

TIME LOGS

	TIME LOG		
Name: J. Smith		Shift: 1	Date: 5/11/07

Time	Job#	Task	Comment
1. 2 hrs	31236	PM #3 oven, PM conveyors & conveyor drives #5, #6, #7, & #8	No unusual conditions
2. 1 hr	63881	Replaced Mixer #6 drive motor (with Gene)	Motor overheated and burned out when mixer locked up
3. 0.5 hr	63888	Replaced limit switch on #2 palletizer skid table	Skid table would not move
4. 2 hrs	63889	Repaired drive chain guard on #3 conveyor	Used portable arc welder #2 *Reordered 6010 welding rod
5. 5 hrs	63891	Cleaned #1 alcohol centrifugal-replaced two nozzles (with Chris)	Not shooting, plugged solid, found two washed out nozzles
6.			
7.			

Figure 6-22. A time log is a binder or electronic file that documents all work and the time taken to perform the work.

RULES AND REGULATIONS

The construction process and maintenance process are governed by rules established by regulatory agencies. A *regulatory agency* is a federal, state, or local government organization that establishes rules and regulations related to safety, equipment installation, equipment operation, and health. Regulatory agencies employ inspectors to ensure that companies comply with the rules and regulations. Failure to follow the rules generally results in fines.

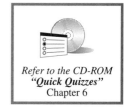

Refer to the CD-ROM "Quick Quizzes" Chapter 6

Example—Filling Out a Request for Information

6-1

Scenario:
Example 6-1 is an installation activity where the Communications Room (317) of an office building contains only one duplex outlet and is not on a dedicated circuit. The communications room is likely to contain telephone and data communications equipment.

Task:
Fill out an RFI to send to Acme Engineering. Copy the architect and project manager and provide the complete panel designation and circuit number for the outlet. Tell the engineer whether spare circuits are available in the panel that presently feeds the communications room outlet. Communications equipment typically requires multiple duplex outlets on dedicated circuits. Also, in order to avoid a conflict with the communications equipment, ask for the exact location of the outlet(s).

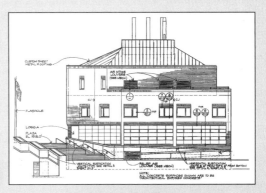

OFFICE BUILDING

Reference Prints:

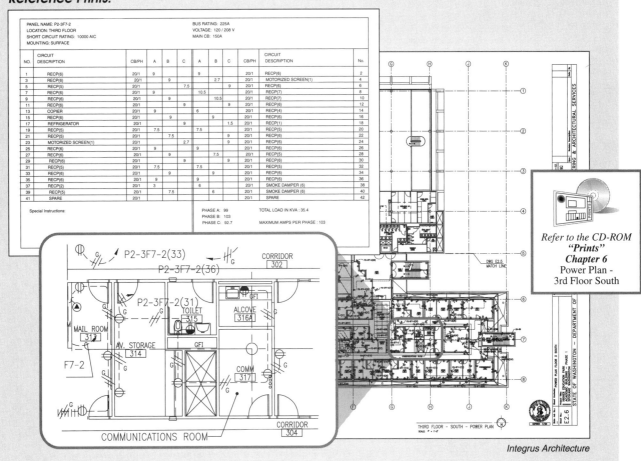

Refer to the CD-ROM "Prints" Chapter 6 Power Plan - 3rd Floor South

Integrus Architecture

✓ Example—Filling Out a Request for Information

Step 1: Fill in name of RFI recipient and date.

XYZ ELECTRIC
REQUEST FOR INFORMATION

ACME Engineering   R.F.I. #: _1_   DATED: _8/16_
FROM: _John Miller_
DATE INFORMATION
REQUIRED: _8/23_

Step 2: Fill in RFI subject.

SUBJECT: _Communications Room (317)_

Step 3: Check appropriate RFI category.

CATEGORY:
- ☑ INFORMATION NOT SHOWN ON CONTRACT DOCUMENTS
- ☑ INTERPRETATION OF CONTRACT REQUIREMENTS
- ☐ CONFLICT IN CONTRACT REQUIREMENTS
- ☐ COORDINATION ISSUES

Step 4: Fill in drawing references (prints).

CONTRACT DRAWING REFERENCE _E2.6 & E5.5_
SHOP DRAWING REFERENCE _____
SPECIFICATION REFRENCE _____

Step 5: Check impact category.

- ☑ POSSIBLE COST IMPACT
- ☑ POSSIBLE TIME IMPACT

Step 6: List panel designation and circuit number under inquiry.

The duplex outlet is in Communications Room (317) and is on circuit P2-3F7-2 (31). This circuit is shared with other duplex outlets.

Step 7: List RFI questions.

- Should the duplex outlet in Room 317 be on a dedicated circuit?
- Should there be more than one duplex outlet in Room 317?
- What is the exact mounting location for the duplex outlet(s) in Room 317?

Step 8: List available spare circuits in designated panel.

Note: There are no spare circuits in local electrical panels.

Step 9: Check who to copy.

CC:
- ☐ GENERAL CONTRACTOR   ☐ ATP SUPERINTENDENT   ☑ ATP PROJECT MNGR
- ☑ ARCHITECT   ☐ ATP ENGINEER   ☐ ATP PURCHING

☑ Example—Filling Out a Maintenance Work Order

6-2

Scenario:
Example 6-2 is a troubleshooting activity where work orders contain specific maintenance task information. To avoid delays, the maintenance person must have all of the necessary equipment and tools to complete the work on the first trip to the task location.

Task:
Fill out a work order. Work orders contain the time and date of issue, the area where the task is located, and a description of the work to be performed.

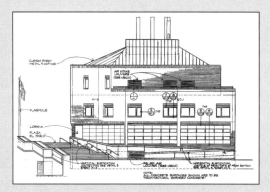

OFFICE BUILDING

Required Information to Fill Out Work Order:
1. What is the room number of the Dean's office? _____
2. What is the Lum number of the fixtures used in the Dean's office? _____
3. How many lamps are required to relamp the Dean's office? _____
4. What is the panel designation and circuit number for the circuit feeding the lights in the Dean's office? _____

5. Light fixture information:
   a. fixture type: _____
   b. fixture voltage: _____
   c. fixture description: _____

Reference Prints:

LUM #	TYPE	VOLTAGE	LUMINAIRE DESCRIPTION	MANUFACTURER/ SERIES #	LAMPS & COLOR TEMP	FINISH	MOUNTING
F1	FLUOR	277V	2'x4' STATIC TROFFER FIXTURE W/AL DOOR FRAME AND SPEC GRADE ACRYLIC #12 PATTERN LENS STEEL HOUSING, POLYESTER ENAMEL FINISH	COLUMBIA #5PS	(2) FO32/31K (32 WATTS)	WHITE	RECESS
F1A	FLUOR	277V	SAME AS TYPE F1 EXCEPT W/EMERGENCY BATTERY BACKUP	SEE TYPE F1	SEE TYPE F1	WHITE	SEE TYPE F1
F2	FLUOR	277V	2'x4' STATIC TROFFER FIXTURE W/AL DOOR FRAME AND SPEC GRADE ACRYLIC #12 PATTERN LENS STEEL HOUSING, POLYESTER ENAMEL FINISH	COLUMBIA #5PS	(3) FO32/31K (32 WATTS)	WHITE	RECESS
F2A	FLUOR	277V	SAME AS TYPE F2 EXCEPT W/EMERGENCY BATTERY BACKUP	SEE TYPE F2	SEE TYPE F2	WHITE	SEE TYPE F2
F3	FLUOR	277V	2'x4' STATIC TROFFER FIXTURE W/AL DOOR FRAME AND SPEC GRADE ACRYLIC #12 PATTERN LENS STEEL HOUSING, POLYESTER ENAMEL FINISH	COLUMBIA #5PS	(4) FO32/31K (32 WATTS)	WHITE	RECESS
F4	FLUOR	277V	2'x4' STATIC TROFFER FIXTURE W/3" DEEP 16-CELL (2x8 CELLS) AL SEMI-SPECULAR LOW IRIDESCENT PARABOLIC LOUVER, STEEL HOUSING	COLUMBIA #P4-242	(2) FO32/31K (32 WATTS)	WHITE	RECESS
F4D	FLUOR	277V	SAME AS TYPE F4 EXCEPT W/DIMMABLE ELECTRONIC BALLAST	SEE TYPE F4	SEE TYPE F4	WHITE	SEE TYPE F4
F5	FLUOR	277V	2'x4' STATIC TROFFER FIXTURE W/3" DEEP 24-CELL (3x8 CELLS) AL SEMI-SPECULAR LOW IRIDESCENT PARABOLIC LOUVER, STEEL HOUSING	COLUMBIA #P4-243	(3) FO32/31K (32 WATTS)	WHITE	RECESS
F5A	FLUOR	277V	SAME AS TYPE F5 EXCEPT W/EMERGENCY BATTERY BACKUP	SEE TYPE F5	SEE TYPE F5	WHITE	SEE TYPE F5
F6	FLUOR	277V	2'x4' STATIC TROFFER FIXTURE W/3" DEEP 32-CELL (4x8 CELLS) AL SEMI-SPECULAR LOW IRIDESCENT PARABOLIC LOUVER, STEEL HOUSING	COLUMBIA #P4-244	(4) FO32/31K (32 WATTS)	WHITE	RECESS
F6A	FLUOR	277V	SAME AS TYPE F6 EXCEPT W/EMERGENCY BATTERY BACKUP	SEE TYPE F6	SEE TYPE F6	WHITE	SEE TYPE F6

Integrus Architecture

Example—Filling Out a Maintenance Work Order

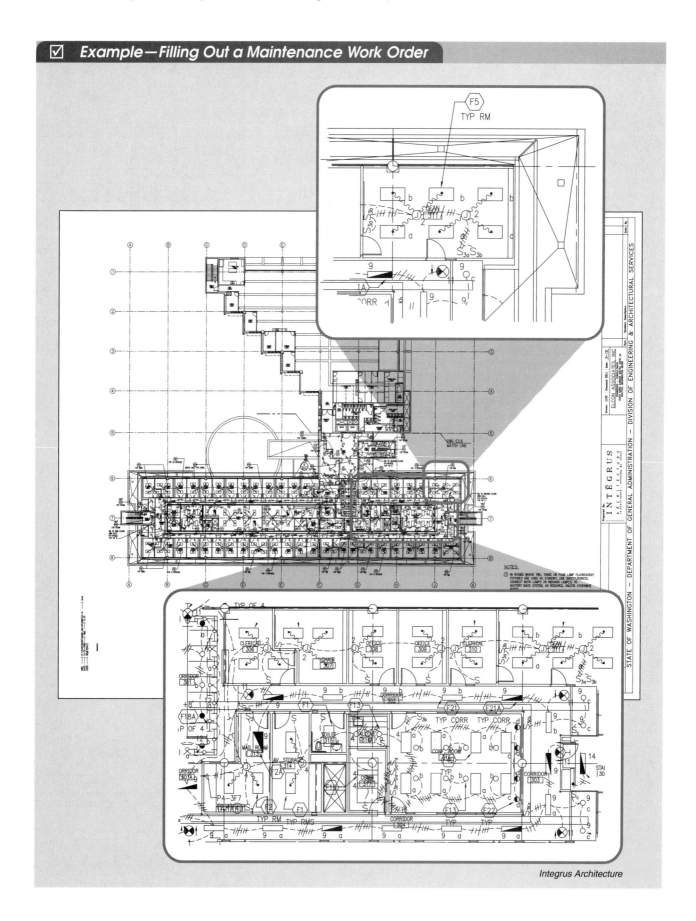

Integrus Architecture

☑ Example—Filling Out a Maintenance Work Order

Step 1: Check work order request priority.

MAINTENANCE WORK REQUEST

PRIORITY:
- ❏ IMMEDIATE
- ☑ 24 HOURS
- ❏ TIMELY

Step 2: Fill in work order number and date.

W.O. #: _167_
REQUESTED BY: _Dean Miller_
DATE: _6/17_   TIME: _4:00 P.M._

Step 3: Fill in task location and equipment.

LOCATION: _3rd Floor, Dean's Office_
EQUIPMENT: _Light Fixture_
ASSIGNED TO: _____
TECHNICIANS INITIALS: _____

MAINTENANCE WORK REQUEST

PRIORITY:
❏ IMMEDIATE   W.O. #: _____
❏ 24 HOURS   REQUESTED BY: _____
❏ TIMELY   DATE: _____ TIME: _____

LOCATION: _____   START DATE: _____
EQUIPMENT: _____   START TIME: _____
ASSIGNED TO: _____   COMPLETION DATE: _____
TECHNICIANS INITIALS: _____   COMPLETION TIME: _____

WORK DESCRIPTION:
_____
_____
_____
_____
_____
_____
_____
_____
_____
_____

❏ LOCKOUT/TAGOUT REQUIRED
❏ PARTS ORDERED   REQUISITION #: _____

Step 4: Fill in work description.

Several lamps are burnt out.
Relamp all light fixtures.
Coordinate relamping with Dean Miller's
administrative assistant. (Ext. 7469)

Step 5: Identify the Dean's office room number and fixture Lum number.

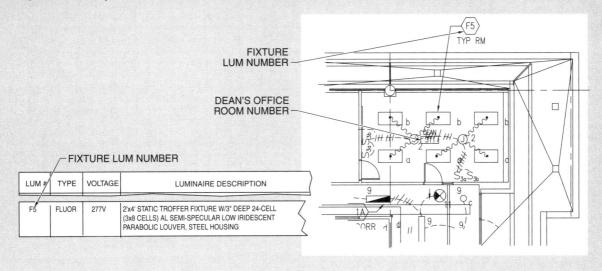

FIXTURE LUM NUMBER

LUM #	TYPE	VOLTAGE	LUMINAIRE DESCRIPTION
F5	FLUOR	277V	2'x4' STATIC TROFFER FIXTURE W/3" DEEP 24-CELL (3x8 CELLS) AL SEMI-SPECULAR LOW IRIDESCENT PARABOLIC LOUVER, STEEL HOUSING

Example—Filling Out a Maintenance Work Order

Step 6: Determine the number of lamps required to relamp the Dean's office.

LUM #	TYPE	VOLTAGE	LUMINAIRE DESCRIPTION	MANUFACTURER/ SERIES #	LAMPS & COLOR TEMP	FINISH	MOUNTING
F5	FLUOR	277V	2'x4' STATIC TROFFER FIXTURE W/3" DEEP 24-CELL (3x8 CELLS) AL SEMI-SPECULAR LOW IRIDESCENT PARABOLIC LOUVER, STEEL HOUSING	COLUMBIA #P4-243	(3) FO32/31K (32 WATTS)	WHITE	RECESS

— LAMPS PER FIXTURE

```
LAMPS/FIXTURE        3
NO. OF FIXTURES     ×6
TOTAL NO. OF LAMPS  18
```

NUMBER OF FIXTURES IN DEAN'S OFFICE

Step 7: Identify the type, voltage, and description of the fixtures used in the Dean's office.

LUM #	TYPE	VOLTAGE	LUMINAIRE DESCRIPTION	MANUFACTURER/ SERIES #	LAMPS & COLOR TEMP	FINISH	MOUNTING
F5	FLUOR	277V	2'x4' STATIC TROFFER FIXTURE W/3" DEEP 24-CELL (3x8 CELLS) AL SEMI-SPECULAR LOW IRIDESCENT PARABOLIC LOUVER, STEEL HOUSING	COLUMBIA #P4-243	(3) FO32/31K (32 WATTS)	WHITE	RECESS

TYPE OF FIXTURE — VOLTAGE OF FIXTURES — FIXTURE DESCRIPTION

Step 8: Determine the panel designation and the circuit number for the circuit feeding the lights in the Dean's office.

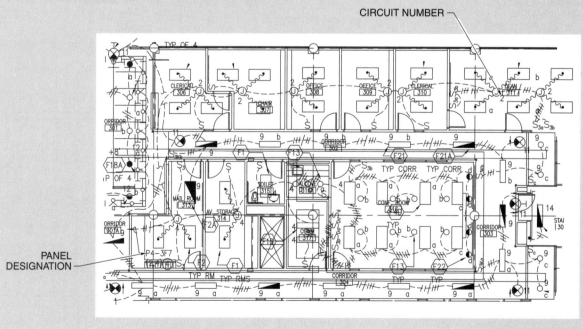

CIRCUIT NUMBER

PANEL DESIGNATION

Review Questions and Activities
Facility Construction and Maintenance Systems

Name _____ Date _____

True-False

T F 1. The primary responsibility of an architect is to listen to the owner and accurately interpret the needs and desires of the owner in the form of a set of building prints.

T F 2. A foreman is a worker who manages personnel and material for the portion of the project relevant to the foreman's trade.

T F 3. Mechanical construction consists of the frame, the exterior walls, the roof that encloses the building, and the mechanical elements of the building.

T F 4. The two main types of maintenance are preventive maintenance and predictive maintenance.

T F 5. Preventive maintenance is the monitoring of wear conditions and equipment characteristics against a predetermined tolerance to predict possible malfunctions or failures.

T F 6. The primary types of documentation are time logs and equipment logbooks.

T F 7. A work order is a document that describes the work a maintenance person is required to perform.

T F 8. A time log is a binder or electronic file that documents all work using job numbers and the time taken to perform the work by maintenance personnel.

T F 9. The construction process and maintenance process are governed by rules established by regulatory agencies.

T F 10. Building inspectors inspect all aspects of a construction project, including the placement of exit signs, the number and type of fire sprinklers, and the proper functioning of fire alarm and life safety systems.

Completion

_____ 1. ___ assist the architect with specific design issues to guarantee the building complies with national and local building codes.

_____ 2. The job of a(n) ___ is to inspect a construction project and ensure the project is built to code.

_____ 3. During the ___ phase of a project, conduits and pipes for electricity, water, gas, sewage, and drainage are put into the earth.

_____ 4. A(n) ___ is a formal document generated by a contractor requesting information about items that are missing or unclear on a print or specification.

_____ 5. A(n) ___ is a binder or electronic file that documents all work performed by maintenance personnel on a piece of equipment.

Multiple Choice

_____ 1. A ___ is responsible for coordinating all work done on a construction project and for providing labor and materials.
    A. tradesworker
    B. building inspector
    C. general contractor
    D. foreman

_____ 2. The construction process includes site preparation and ___.
    A. construction of the building core
    B. electrical construction
    C. mechanical construction
    D. all of the above

_____ 3. Plumbers install piping systems for ___.
    A. hot water, cold water, and waste
    B. HVAC systems
    C. instrumentation
    D. fire sprinklers

_____ 4. A ___ contains items that a contractor missed or only partially completed.
    A. time log
    B. punch list
    C. request for information (RFI)
    D. none of the above

_____ 5. During the construction of a building or facility, ___ typically takes place during the entire construction process.
    A. mechanical construction
    B. electrical construction
    C. predictive maintenance
    D. punch list resolution

Name _____ Date _____

Activity—Filling Out an RFI (Device Location)   6-1

Scenario:
Activity 6-1 is an installation activity requiring that the prints for a large office building be reviewed. Typically this requires looking at multiple electrical, mechanical, and/or other prints.

A common problem is that the location and/or number of electrical devices on one print does not agree with the location and/or number on another print. When a conflict or problem is discovered, an RFI must be sent to the engineer.

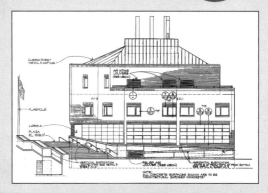

OFFICE BUILDING

Task:
Fill out a second RFI for the engineer of the project at ACME Engineering and copy the mechanical contractor, architect, ATP Project manager, and Miller Mechanical. Describe the locations of any flow switches and any tamper switches. Ask how many flow and tamper switches are in the basement and what their correct locations are, and request that an answer be given within a week.

1. Are there any notes about flow switches and tamper switches on the prints? _____
   If so, what do the notes say? _____
2. Highlight any tamper switches and any flow switches on the prints. Does the number of flow and tamper switches in the basement agree with the Fire Alarm Riser Diagram? _____

Reference Prints:

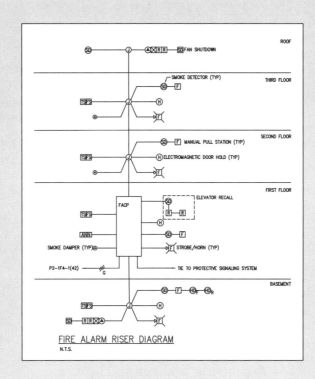

Refer to the CD-ROM
"Prints"
Chapter 6
Fire Alarm
Riser Diagram

207

Activity—Filling Out an RFI (Device Location)

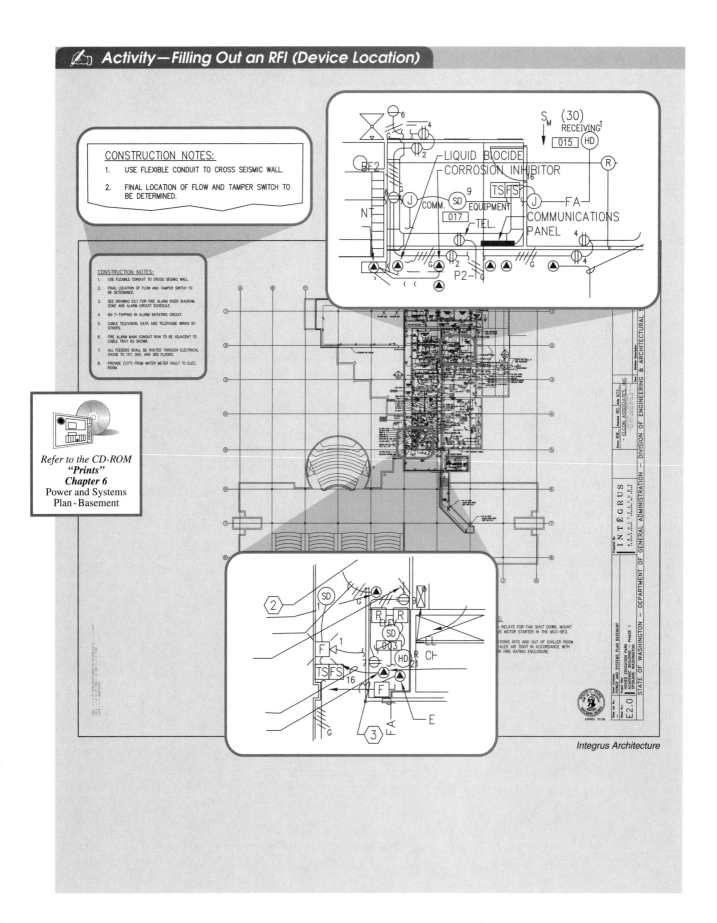

Integrus Architecture

Name _____ Date _____

Activity—Filling Out a Work Order (Pump)

6-2

Scenario:
Activity 6-2 is a troubleshooting activity where work orders must contain specific maintenance and safety information about a pump in order for work to be done.

Task:
Write a work order that includes lockout/tagout. Find the location of the pump and the source of power for the pump. Lockout/tagout the disconnect for the pump with a properly filled-out tag.

Required Information to Fill Out Work Order:
1. Describe the location of P-4. _____
2. What is the designation of the panel that is the power source for P-4? _____
3. In which room is the power source for P-4 located? _____
4. Is a local disconnect available at P-4 for a lockout/tagout? _____
5. Fill out work order.
6. Fill out lockout/tagout tag.

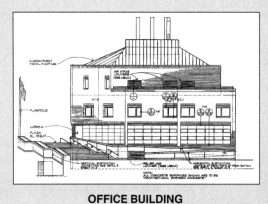

OFFICE BUILDING

Reference Prints

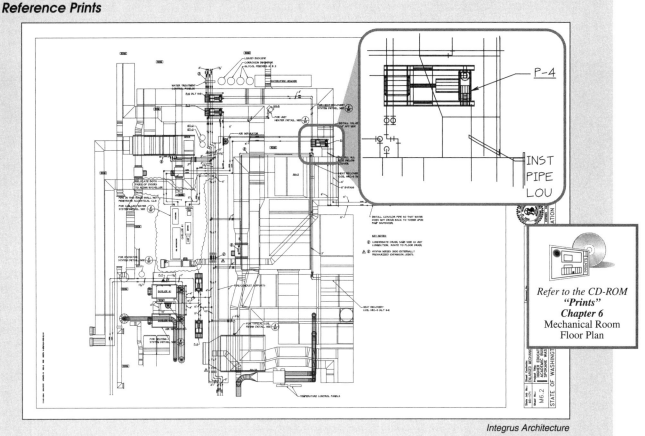

Refer to the CD-ROM
"Prints"
Chapter 6
Mechanical Room
Floor Plan

Integrus Architecture

209

Activity—Filling Out a Work Order (Pump)

Integrus Architecture

Trade Competency Test
Facility Construction and Maintenance Systems

Name _____ Date _____

REFERENCE PRINT #1 (HVAC Equipment Schedule)

	MANU-FACTURE	MODEL NO.	MARK	TONS	S/A FAN HP	S/A FAN CFM	S/A FAN SP	VOLT	HEATING INPUT	HEATING OUTPUT	TOTAL COOLING	MAX FLA @ 208V.	O.F.A. C.F.M.	NOTES
RT1 AND RT2	AAON	RK-10-2-EO-222	RTU-1	10	3	4000	.75	208/3/60	270 MBH	219 MBH	134 MBH	70	1250	1,2
		RK-10-2-EO-222	RTU-2	10	3	4000	.75	208/3/60	270 MBH	219 MBH	134 MBH	70	0	2
	OR LENNOX	LGA-120H	RTU-1	10	3	4000	.75	208/3/60	235 MBH	188 MBH	124 MBH	70	1250	1,2
		LGA-120H	RTU-2	10	3	4000	.75	208/3/60	235 MBH	188 MBH	124 MBH	70	0	2
	OR TRANE	YCD-121	RTU-1	10	3	4000	.75	208/3/60	250 MBH	203 MBH	127 MBH	70	1250	1,2
		YCD-121	RTU-2	10	3	4000	.75	208/3/60	250 MBH	203 MBH	127 MBH	70	0	2
MUA	AIR WISE	TBA-250/DX-C10	MUA-1	10	3	3150	.5	208/3/60	250 MBH	200 MBH	120 MBH	79	3150	1,3
	OR AIR WISE HIGH HUMIDITY	TBA-300/DX-C13	MUA-1	13	3	3150	.5	208/3/60	300 MBH	24 MBH	156 MBH	79	3150	1,3

AIR WISE "HIGH HUMIDITY" MAKE-UP AIR UNIT IS FOR THOSE AREAS OF THE WORLD THAT EXPERIENCE HUMID CLIMATES FOR AN EXTENDED PERIOD OF TIME (TYPICALLY COASTAL REGIONS). FOR MORE INFORMATION ON THE HIGH HUMIDITY MUA, CONTACT HVAC EQUIPMENT SUPPLIER.

NOTES:
1. MOTORIZED OUTSIDE AIR DAMPERS INTERLOCKED WITH EXHAUST FANS IN FAN CONTROL PANEL.
2. ELECTRICAL DISCONNECT SWITCH, SMOKE DETECTOR, AND CONVENIENCE OUTLET PROVIDED BY ROOFTOP MANUFACTURER.
3. ELECTRICAL DISCONNECT SWITCH AND SMOKE DETECTOR PROVIDED BY ELECTRICAL CONTRACTOR.

* SELECT ONE MANUFACTURER FROM ABOVE TO PROVIDE RT-1, RT-2, AIR WISE MUA-1, ROOF CURB PACKAGE AND AIR DEVICE PACKAGE.

Questions – Reference Print #1

_____ 1. ___ are acceptable manufacturers for RT1 and RT2.

_____ 2. The model number for a make-up air unit used in high humidity locations is ___.

  T   F   3. Motorized outside air dampers are interlocked between RTU-1 and exhaust fans.

_____ 4. The input voltage for a Lennox LGA-120H is ___.

_____ 5. The maximum full load amps for all models of rooftop (RT) units is ___.

_____ 6. The ___ provides an electrical disconnect switch and smoke detector for MUA-1.

_____ 7. The supply fan for the Trane RTU-1 moves ___ cubic feet per minute (cfm) of air.

_____ 8. The maximum full load amps for either model of MUA-1 is ___.

REFERENCE PRINT #2 (Office Building – South Elevation)

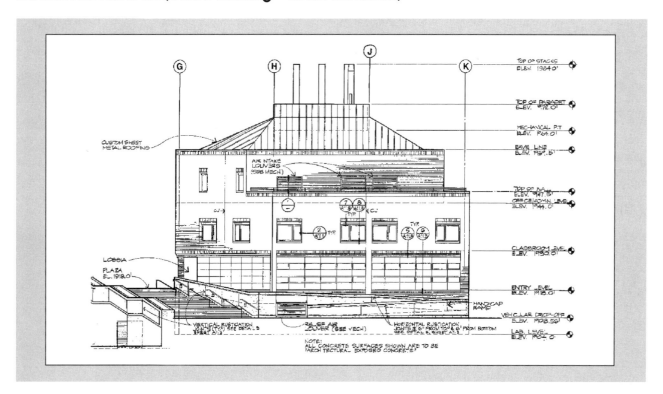

Questions – Reference Print #2

_____ 1. The elevation at the top of the stacks is ___.

_____ _____ 2. The stacks are located between column line ___ and ___.

_____ 3. The ___ louver is located near the base of the handicap ramp.

_____ _____ 4. Details on the vertical rustication joints are found on sheet ___ detail ___.

_____ 5. There are ___ air intake louvers shown.

_____ 6. Additional information about the relief air louver can be found in the ___ section of prints.

_____ 7. The classroom elevation level is ___.

 ADVANCED QUESTIONS (CD-ROM Reference Print – Office Building South and West Elevations)

_____ 1. The print was prepared by ___.

_____ 2. The scale for the south elevation is ___.

_____ 3. The exterior stairs shown in the west elevation have ___ light fixtures.

_____ 4. The scale for the brick expansion joint detail is ___.

_____ 5. Office/administration level windows are ___ tall.

212

Residential and Commercial Power and Lighting Systems

Printreading for Installing and Troubleshooting Electrical Systems

Both residential and commercial construction projects use power and lighting prints. The most common types of power and lighting prints are the power floor plan and the lighting floor plan. Commercial construction projects use separate floor plans of each floor for power and lighting circuits. Residential construction projects typically combine the power and lighting circuits on a single power floor plan.

Other types of power and lighting prints include site plans, single-line diagrams, details, and schedules. Electricians install raceways, mount equipment, pull wire, and terminate conductors based on the information contained in the power and lighting floor plans. Power and lighting floor plans are the primary prints electricians use to complete a project.

SITE PLANS

A *site plan* is a drawing that depicts a complete building site, the layout of the planned buildings, and all of the utility items installed below ground. **See Figure 7-1.** Site plans are drawn at a small scale because site plans typically depict large areas of property. Scales used on site plans are typically $1'' = 30'$ and $1'' = 40'$.

Site plans can contain information on many different types of electrical circuits. Site plans depict electrical power circuits, lighting circuits, grounding grids, low-voltage circuits (closed circuit TV and intercom), and utility feeds from power and telephone companies. Many site plans also contain numerous sheet notes pertaining to electrical installations, detail drawings of underground conduit installations, and detail drawings of pole bases for light fixtures.

Site plans are used early in a construction project. Electricians must carefully review the site plan and notes to ensure all of the required raceways (conduits), junction boxes, and pole bases are properly installed below ground. **See Figure 7-2.** A mistake in the installation of below ground items can be very expensive to correct.

> Site plans show electrical easements. Easements allow utility companies to use small portions of a property to route cable, conduit, and pipes.

214 Printreading for Installing and Troubleshooting Electrical Systems

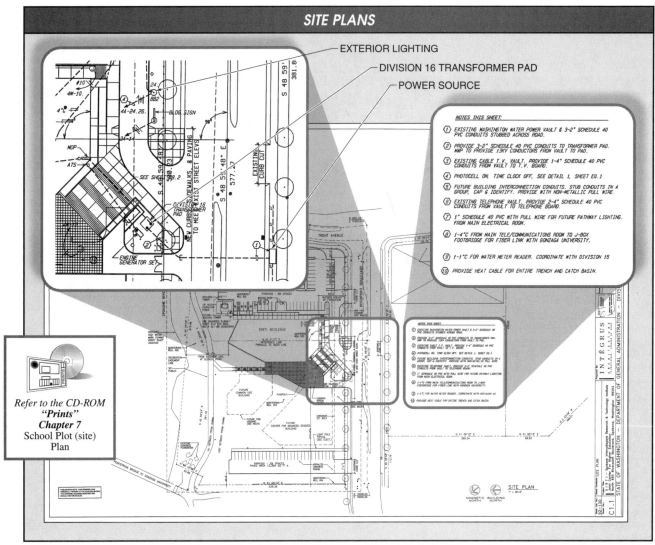

Integrus Architecture

Figure 7-1. *A site plan is a drawing of the complete building site. The plan shows electrical equipment to be installed below ground, such as power circuits, lighting circuits, grounding grids, low-voltage circuits, and utility feeds from power and telephone companies.*

The installation of below ground electrical items requires coordination between the general contractor and all subcontractors, such as plumbers installing underground sewer and water lines. Because site work is typically performed in phases, detailed notes must be kept by contractors on the progress of all items installed below ground.

POWER PRINTS

Power floor plans and single-line diagrams are the two power prints most commonly used by electricians. Power floor plans show the position of outlets, permanently connected loads, panels, and transformers for a specific floor or section of a building. Locations, circuit numbers, and special installation information are shown on power floor plans.

Single-line diagrams provide an overview of the power backbone of a building. Single-line diagrams show the electric service, feeders, distribution panels, branch-circuit panels, transformers, and motor control centers.

Chapter 7—Residential and Commercial Power and Lighting Systems   **215**

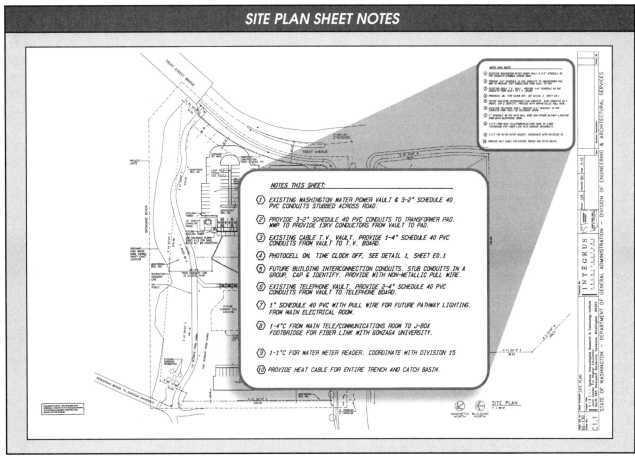

Integrus Architecture

Figure 7-2. *The sheet notes on site plans include electrical information on underground material usage, existing electrical and utility items, future expansion, and property and climate.*

> The NEC® has minimum depth of cover requirements for underground installations of direct buried cables, conduit, and raceways. The requirements vary based on the type of cable or raceway, location, and type of soil.

Power Floor Plans

A *floor plan* is a drawing that provides a plan view of each floor of a building. A floor plan shows the elevator shafts, equipment locations, doors, plumbing fixtures, rooms, stairways, walls, and windows on a given floor. Typically, a description and/or room number is given to each room, including closets and corridors.

A *power floor plan* is a print that shows all the power circuits for a specific floor of a building. A power floor plan shows electrical receptacles, panels, directly connected loads, and transformers in addition to the other features typically found on a floor plan. Both high-voltage (277/480 VAC) and common-voltage (120/208 VAC) loads are shown on power floor plans. Electrical receptacles location, required wiring, and installation information are also given on power floor plans. **See Figure 7-3.**

A power floor plan may also show telephone and data outlets, fire alarm devices, temperature control, and building automation components. Although power floor plans for commercial and residential projects contain much of the same types of information, there are significant differences between the two.

POWER FLOOR PLANS

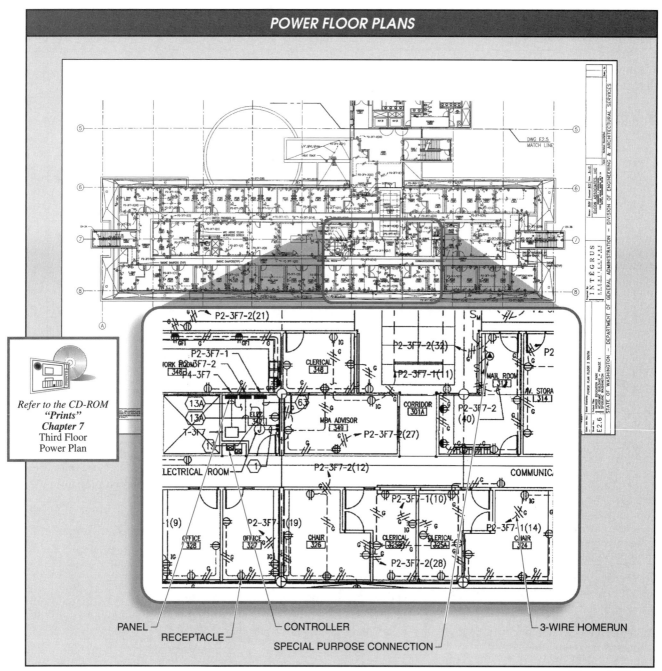

Refer to the CD-ROM "Prints" Chapter 7 Third Floor Power Plan

Integrus Architecture

Figure 7-3. *A power floor plan shows the location of all features typically found on floor plans plus receptacles, panels, directly connected loads, and transformers used in the distribution of power on a specific floor.*

Commercial Power Floor Plans. Commercial power floor plans are used for banks, hospitals, office buildings, schools, car washes, retail stores and other commercial buildings. Typically, receptacles in a commercial building are mounted at the same height, with the locations of the receptacles identified on the power floor plan. Most power floor plans or specifications provide the mounting height and whether the outlets are to be installed exactly per provided dimensions or to the nearest stud.

Some power floor plans provide exact dimensions for every receptacle. Typically, architects and engineers only provide dimensions for receptacles that are not considered standard installations.

In some installations only a general location is provided for certain receptacles and electrical devices. General location installations include mechanical rooms in commercial buildings and manufacturing areas in industrial facilities. A general location is provided for items such as pumps, fans, chillers, and manufacturing equipment because the equipment is installed by various trades and certain field conditions may require some minor location changes. **See Figure 7-4.**

When a general location is provided, electricians must consult the mechanical drawings, detail drawings, and/or equipment drawings for the exact location of equipment power connections. For example, the mechanical drawings of a project must be consulted to determine where the power connections for a chiller will be installed.

> Schedules provide reference numbers for circuit breaker locations. The actual location of a circuit breaker is found on the corresponding elevation drawing, which uses the same circuit breaker reference number as the schedule.

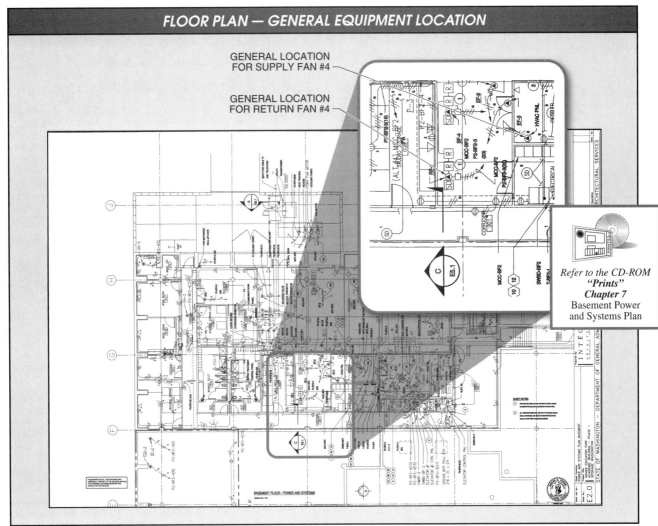

Integrus Architecture

Figure 7-4. *Only a general location is provided on floor plans for items such as pumps, fans, chillers, and manufacturing equipment. Electricians must consult other drawings or specification sheets for the exact location of equipment power connections.*

In addition to the location of receptacles and electrical devices, circuit numbers and panel designations are provided on commercial power floor plans. Each individual circuit is indicated using dashed or solid lines. The lines indicate circuit connections only, not the actual route of a cable or raceway. Hash marks drawn almost perpendicular to the dashed or solid lines denote the number of conductors required. All hash marks are equal in length with the exception of the neutral conductor, which has longer hash marks. The type of arrow on the end of a line can indicate whether a circuit returns to a motor control panel or other panel such as a service panel (homerun). **See Figure 7-5.**

Power floor plans are drawn for each floor of a commercial building. More than one power floor plan may be necessary if a specific floor is extremely large. Power floor plans contain sheet notes and details that provide extra information about special outlets, electrical rooms, and installation. Electricians must always consult a project's prints and specifications to determine the proper location for outlets and electrical devices. Although a commercial power floor plan often shows other devices and components, such as fire alarm devices or telephone outlets, some projects have a separate floor plan for these types of devices. Commercial power floor plans do not show light fixtures or switches.

Refer to the CD-ROM
"Prints"
Chapter 7
Equipment
Layout

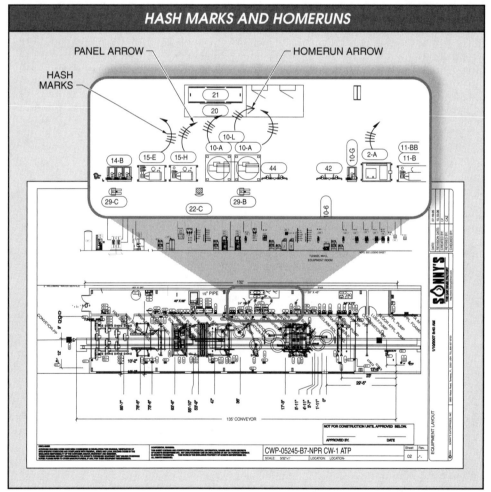

Sonny's Enterprises, Inc.

Figure 7-5. *Hash marks are drawn perpendicular to dashed or solid lines and denote the number of conductors being used in the run. The type of arrow on the end of the dashed or solid line indicates whether the circuit returns to a motor control panel or other panel such as a service panel (homerun).*

Residential Power Floor Plans. Typically, there is only one floor plan used for residential construction projects. A residential power floor plan is most often shown using the project floor plan. Like a commercial power floor plan, residential power floor plans provide the locations of receptacles, electrical devices, telephone and data outlets, and thermostats. **See Figure 7-6.**

Because it is the project floor plan, a residential power floor plan also shows interior and exterior building dimensions, room dimensions, and locations of light fixtures, switches, and HVAC registers. A residential power floor plan also gives information on plumbing fixtures, doors and windows, floor and wall coverings, and floor and roof framing.

Like a commercial project, most outlets of a residential project are mounted at the same height, with the location of each identified on the residential power floor plan. Information on receptacle and switch mounting heights, and whether mounting locations are per Article 210 at the NEC®, or are found on power floor plans or in the specifications. Ceiling light fixtures are typically centered in a room. Dimensions are only shown for light fixtures, outlets, and switches that are not typical installations.

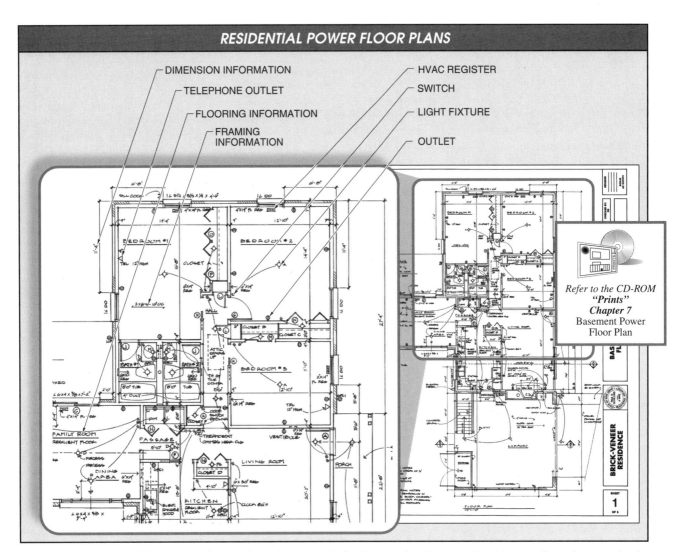

Figure 7-6. A residential floor plan and power floor plan are typically the same drawing. A residential power floor plan provides the location of electrical outlets, electrical devices, telephone and data outlets, thermostats, light fixtures, and switches.

The circuits for light fixtures, switches, and switched outlets are indicated by dashed or solid lines. The lines only indicate circuit connections, not the actual route of a cable or raceway. An arrow on the end of a line indicates a home run. Unlike commercial power floor plans, circuit numbers, hash marks, and panel designations are not typically shown on residential power floor plans. Also, outlets and other equipment are typically not connected by any lines.

Single-Line Diagrams

Single-line diagrams provide an overall view of the power in a building or facility. **See Figure 7-7.** Single-line diagrams depict the interconnection between the electrical service, feeders, distribution panels, branch-circuit panels, transformers, and motor control centers.

The general location of electrical equipment and the equipment designations are typically included on single-line diagrams.

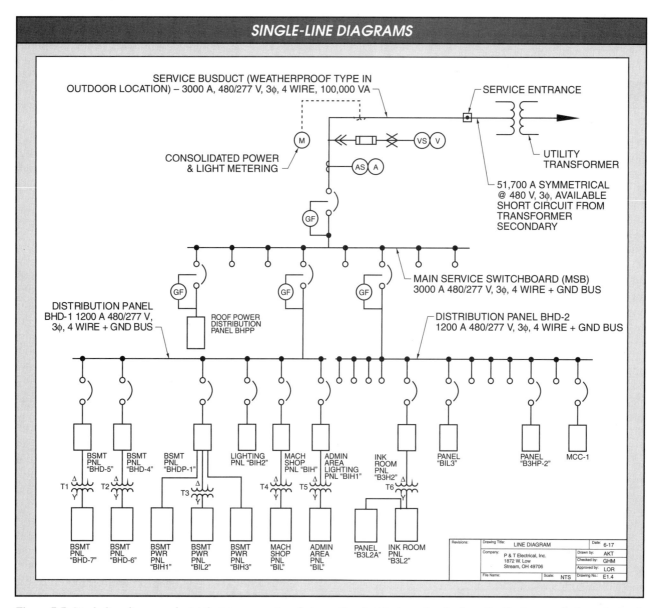

Figure 7-7. Single-line diagrams depict the interconnections between the electrical service, feeders, distribution panels, branch-circuit panels, transformers, and motor control centers by providing a simple view of the power in a building or facility.

A designation such as "PNL BH4 BSMT" represents Panel BH4, located in the basement. Sheet notes are also included to provide additional information. Some of the panels shown on a single-line diagram may be dedicated to a specific type of load, such as HVAC or lighting. Single-line diagrams do not provide specific location and/or mounting information.

Electrical Services. An *electrical service* is the point of electrical connection between the local power utility and the building. A building's electrical service can also be referred to as the main switchboard. Typically, the electrical service contains a main circuit breaker through which all power to a building passes, and one or more smaller circuit breakers. The smaller circuit breakers feed power distribution panels, branch-circuit panels, step-down transformers, or motor control centers. **See Figure 7-8.** In most commercial and industrial buildings, the electrical service is three-phase, 4 wire, 277/480 VAC. Where three-phase or single-phase 120/208 VAC is required, step-down transformers and branch-circuit panels are installed.

Single-line diagrams contain information on the ampacity, voltage, configuration, and metering of an electrical service. *Ampacity* is the current-carrying capacity (in amperes) of an electrical component. Information on service feeders and the utility source is also included on single-line diagrams. Information on the circuit breakers in the electrical service is contained in an electrical service schedule or a main switchboard schedule. Schedules are typically included on the same sheet as the single-line diagram.

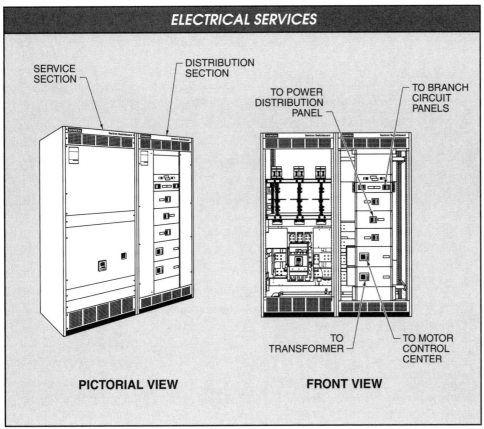

Figure 7-8. Electrical services contain a main circuit breaker through which all power to a building passes, and one or more smaller circuit breakers that feed power distribution panels, branch-circuit panels, step-down transformers, or motor control centers.

Feeders. A *feeder* is conduit and wire, cable tray and wire, metal-clad cable, or busduct that is used to supply power to electrical loads. For example, a feeder is used to connect a power distribution panel to a branch-circuit panel. **See Figure 7-9.** Feeders are shown as lines on a single-line diagram. The specific location, routing, and size of the feeders are not typically provided.

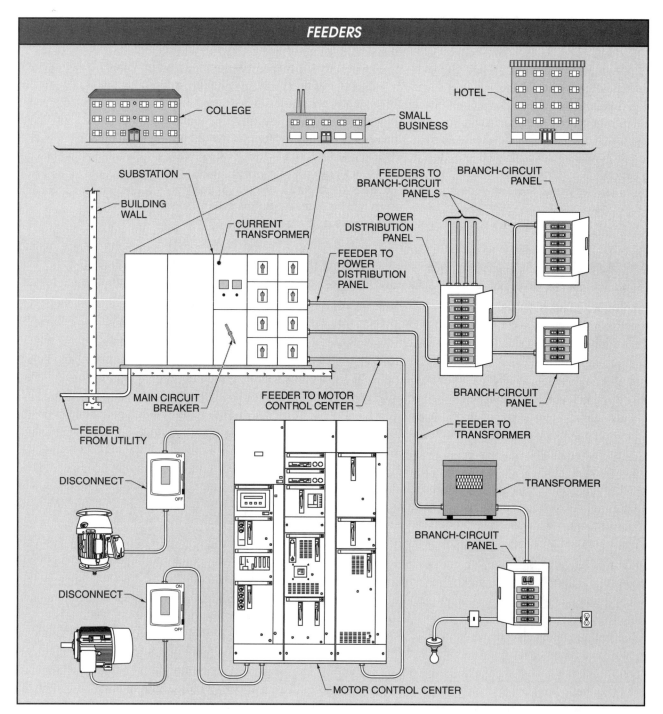

Figure 7-9. A feeder is used to connect power sources to electrical loads, such as using conduit and wire to connect a power distribution panel to a branch-circuit panel.

Feeders are either three-phase 277/480 VAC or three-phase 120/208 VAC. Feeders are found as three-phase, 3-wire circuits or three-phase, 4-wire circuits. The size of the feeder and the type of raceway used are found on the main switchboard schedule, power distribution panel schedule, transformer schedule, or feeder schedule.

Power Distribution Panels. A *power distribution panel* is a panel that can be located anywhere in a building to serve as a nearby power source for branch-circuit panels, transformers, or motor control centers. Power distribution panels receive power from circuit breakers located in the electrical service panel.

A power distribution panel is smaller in size and lower in ampacity than an electrical service panel, but larger in size and higher in ampacity than a branch-circuit panel. Single-line diagrams contain information on the ampacity, voltage, and configuration of power distribution panels. Information on the circuit breakers in the power distribution panels is contained in a power distribution panel schedule. **See Figure 7-10.**

Branch-Circuit Panels. Branch-circuit panels receive power from circuit breakers located in power distribution panels or electrical service panels, or from transformers. Branch-circuit panels are smaller in size and lower in ampacity than power distribution panels or electrical service panels. The circuit breakers in branch-circuit panels are the last overcurrent protection devices between the load and the power source.

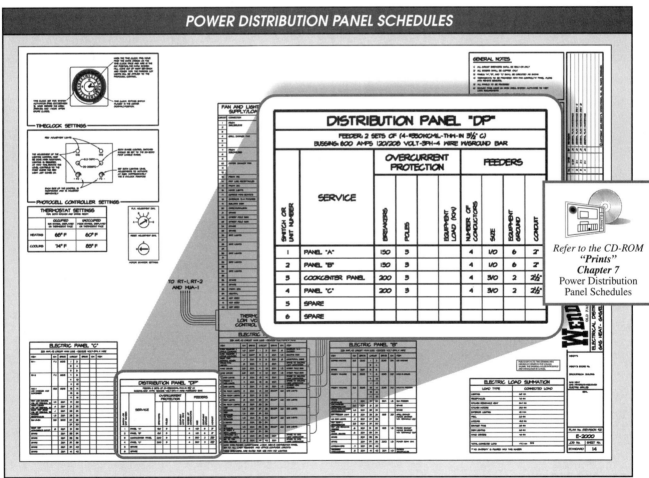

Wendy's International, Inc.

Figure 7-10. Schedules for power distribution panels contain information on the feeders, circuit breakers, and loads connected to the panel.

Branch-circuit panels are high-voltage or low-voltage. High-voltage panels provide power to single-phase and three-phase 277/480 VAC loads such as high-voltage motors, air conditioning equipment, step-down transformers, and lighting systems. Low-voltage panels provide power to single-phase and three-phase 120/208 VAC loads such as duplex outlets, lighting, and low-voltage motors. Information on the circuit breakers in branch-circuit panels is contained in a branch-circuit panel schedule. **See Figure 7-11.**

Transformers. The most common type of transformer in a residential or commercial building is a step-down transformer. In a residential building, step-down transformers are used to change 120 VAC to 48 VAC, 36 VAC, 24 VAC, or 12 VAC. The lower voltages are used for chimes, doorbells, low-voltage lighting, or security systems.

In a commercial building, step-down power transformers are typically used to change 480 VAC to 208 VAC or 120 VAC. The lower voltage is used for computers, fax machines, lighting, or other 120/208 VAC high-current loads. Typically, 480 VAC to 208 VAC or 120 VAC power transformers are mounted in an electrical room in close proximity to high-voltage and low-voltage electrical panels. Power transformers are fed power from power distribution panels, branch-circuit panels, or the electrical service panel. Power transformers are floor mounted, wall mounted, or have special enclosures when used for outdoor locations. **See Figure 7-12.**

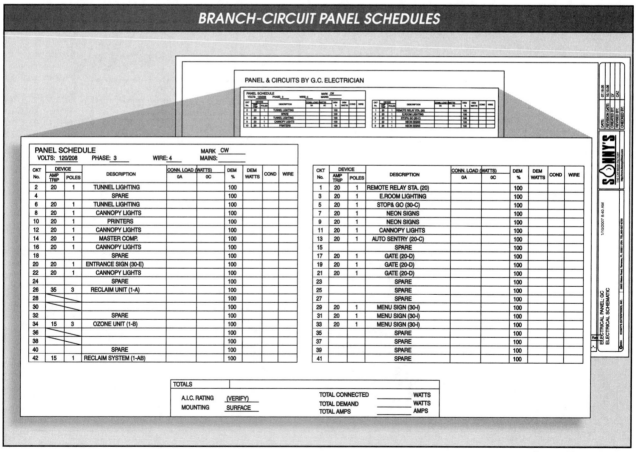

Sonny's Enterprises, Inc.

Figure 7-11. Branch-circuit panel schedules contain information on the circuit breakers in branch-circuit panels.

Chapter 7—Residential and Commercial Power and Lighting Systems   225

POWER TRANSFORMERS

FLOOR MOUNTED

WALL MOUNTED

Figure 7-12. Power transformers are typically step-down transformers that are floor mounted, wall mounted, ceiling mounted or in special enclosures. Power transformers are fed power from power distribution panels, branch-circuit panels, or service panels.

The output capability of a power transformer is rated in kilovolt-amperes (kVA). Single-line diagrams provide a transformer designation and information on the wiring configuration of the primary and secondary coils. The information on the wiring configuration is given as an abbreviation: Y for wye and Δ for delta. For example, T2 Δ/Y is an abbreviation for transformer 2, with a delta primary coil and wye secondary coil.

✓ SEPARATELY DERIVED SYSTEMS

A separately derived system (SDS) is a secondary power system that supplies electrical power that is derived, or taken from, transformers, storage batteries, solar photovoltaic systems, or generators. Most residential, commercial, and industrial separately derived systems use energy produced by the secondary coil of a distribution transformer. Separately derived systems must be grounded.

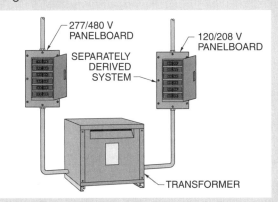

The NEC® requires non-current-carrying metal parts of a transformer to be grounded. Grounding provides an electrically conductive path that is designed and intended to carry current under ground fault conditions from the point of the fault to the electrical supply source. An SDS panel is grounded through either the main service panel or the transformer.

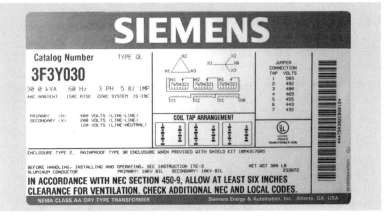

Siemens

A transformer nameplate describes the operating parameters of the transformer.

Additional information about the power transformer is found in a transformer schedule. **See Figure 7-13.** Transformer schedules contain information such as kVA rating, primary and secondary coil voltages, primary and secondary side feeders, and transformer mounting.

226 Printreading for Installing and Troubleshooting Electrical Systems

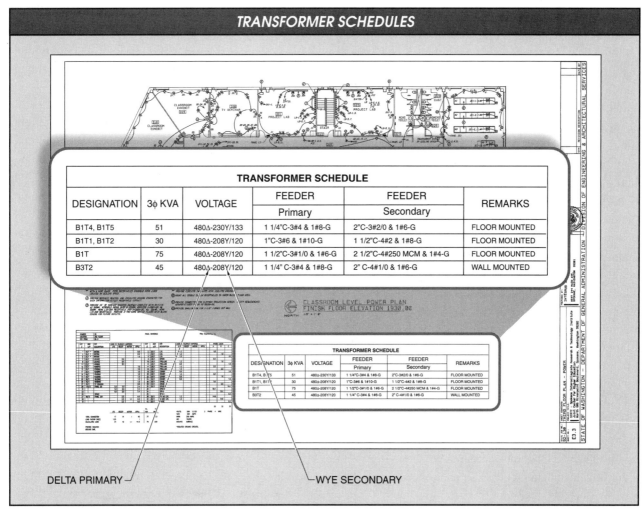

Figure 7-13. Transformer schedules contain information such as a transformer's kVA rating, primary coil and secondary coil voltages, primary side and secondary side feeder sizes, and type of mounting.

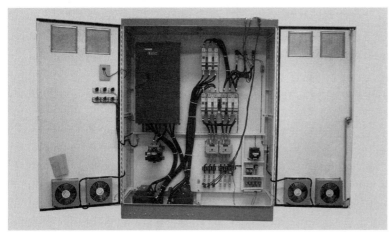

Many motor control centers and enclosures include electric motor drives to provide precise torque and speed control for the motor.

Motor Control Centers. A *motor control center (MCC)* is an electrical panel that is fed power from a power distribution panel or electrical service panel and contains several individual control units designed specifically for controlling motors. A set of vertical busbars runs from the top of the motor control center to the bottom. The individual control units are modular and plug into the vertical busbars. These individual units are sometimes referred to as buckets. Motor control centers are typically found in electrical rooms (E rooms) or mechanical rooms where motors for HVAC pumps and fans are located. **See Figure 7-14.**

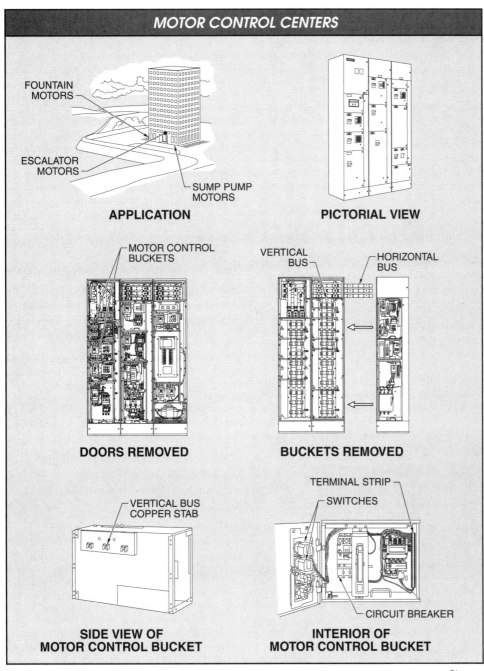

Figure 7-14. Motor control centers (MCC) are electrical panels that are fed power from power distribution panels or electrical service panels and contain several individual control units designed specifically for controlling motors.

A *bucket* is a motor control center module that consists of a circuit breaker, motor starter, and circuit control components. A terminal strip is located in the bucket to terminate wires from field devices such as limit switches and pushbuttons. Motor control schedules contain information on the sizes and types of motors, conductors, conduit, and motor starters. A motor control schedule may also refer to specific control diagrams for each motor.

LIGHTING PRINTS

The most common type of lighting print is a lighting floor plan. A lighting floor plan is similar to a power floor plan, but shows light fixtures, light switches, panels, and transformers for a specific floor or section of a building. Device and component locations, circuit numbers, and installation information are shown on lighting floor plans.

Other types of lighting prints include reflected ceiling plans and fixture schedules. Although not an electrical print, reflected ceiling prints are frequently used for fixture mounting locations. Fixture schedules provide detailed information on the various fixtures to be installed. The information typically includes the name of the fixture manufacturer, voltage rating, number of lamps, and type of fixture.

Lighting Floor Plans

A *lighting floor plan* is a print that shows light fixtures, exit lights, switches, panels, and transformers. A lighting plan also shows elevator shafts, equipment locations, doors, plumbing fixtures, rooms, stairways, walls, and windows. Both high-voltage (277/480 VAC) and low-voltage (120/208 VAC) lighting-load circuits are shown on lighting floor plans. A lighting floor plan shows the locations, circuit numbers, and installation information for lighting fixtures, exit lights, switches, panel, and transformers. **See Figure 7-15.**

A residential lighting floor plan is typically part of the residential project floor plan. A residential lighting floor plan shows outlets, directly connected loads such as HVAC equipment, and other construction information in addition to light fixtures and switches. A commercial lighting floor plan does not show outlets or directly connected loads such as HVAC equipment.

Depending on the type of installation, a lighting floor plan may not provide exact mounting location information for light fixtures or exit lights. A lighting floor plan for a commercial building may only provide a general fixture location. The exact mounting location information can be found on a reflected ceiling plan. A *reflected ceiling plan* is a plan view with the viewpoint of a mirror placed on the floor to reflect ceiling-mounted objects. **See Figure 7-16.**

SET OF ELECTRICAL PRINTS

Lighting floor plan (¼" = 1'-0" scale where possible)
Power floor plan showing receptacles, switches, and all outlets (¼" = 1'-0" scale where possible)
- Identify location of electric panels (new and existing)
- Identify location of each subpanel and size of feeders
- Provide service load calculations

Distribution riser diagram
- When existing electric service for proposed occupancy is of sufficient size for new loads, add note "Existing Electric Service is of Sufficient Ampacity & Capacity for New Load and is to Remain as Originally Installed"

One-line diagram of the complete electrical system
- Voltage, ampacity, phases, and overcurrent devices
- Maximum available fault current
- Conductor sizes, conduit size and number of conduits if parallel
- Sizes and type of all wire

Exterior lighting plan including fixture types, wattage, and conductor sizes

Nameplate rating of all motors, elevators, AC units, and equipment panel schedules
- Subpanel number
- Size of breakers
- Total load calculations

Identify any hazardous or classified areas per the NEC®

Ceiling-mounted objects such as air diffusers, light fixtures, and exit signs are indicated on reflected ceiling plans.

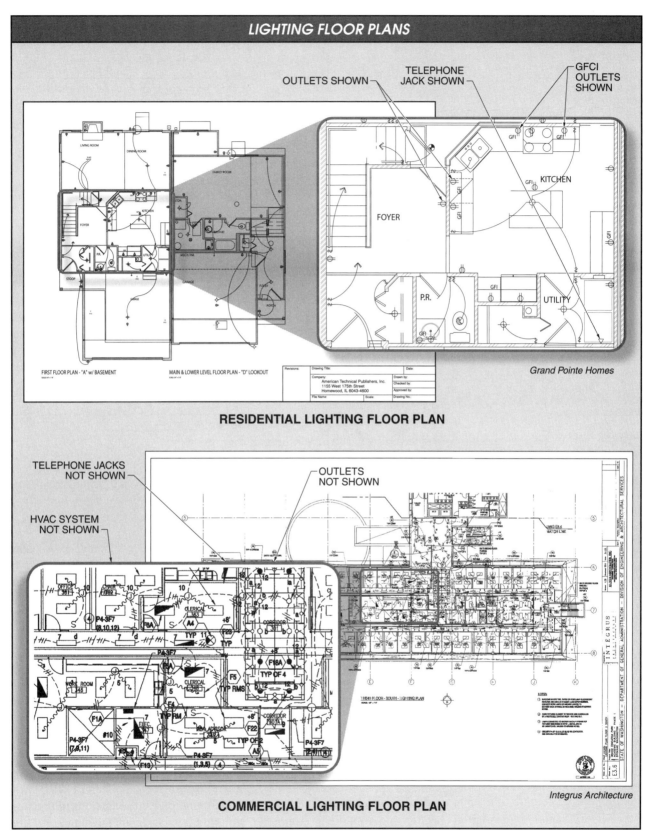

Figure 7-15. A lighting floor plan shows the locations, circuit numbers, and installation information for lighting fixtures, exit lights, switches, panels, and transformers.

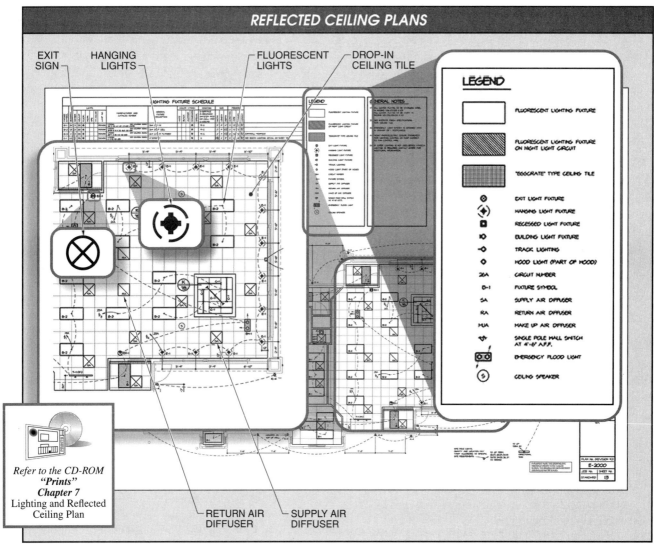

Figure 7-16. *A reflected ceiling plan shows the locations of acoustical tile (drop-in ceiling tile), return air diffusers, supply air diffusers, exit signs, fluorescent lights, recessed or hanging lights, and fire sprinkler heads.*

Symbols show locations for acoustical tile (drop-in ceiling tile), return air diffusers, supply air diffusers, exit signs, fluorescent lights, recessed or hanging lights, and fire sprinkler heads. The lighting floor plans for other types of installations, such as industrial buildings, show mounting heights, exact fixture locations, and spacing requirements between fixtures.

The locations of switches are typically shown on residential and commercial lighting floor plans. Typically, all switches are mounted at the same height, with the location of each indicated on the lighting floor plan. Prints or specifications provide the mounting height and whether the switches are to be installed exactly per a dimension or to the nearest stud. Some prints provide exact dimensions for every switch location. Typically, dimensions are only shown for switches that are not considered standard installations. On commercial lighting floor plans, circuit numbers and panel designations are shown in addition to the location of light fixtures and switches.

A lighting floor plan is drawn for each floor of a commercial building. More than one lighting floor plan may be necessary for a building floor that is extremely large. Lighting floor plans contain sheet notes and details that provide extra information about lighting fixtures, switches, lighting controls, and installation. Electricians must always consult the prints and specifications of a project to determine the proper location for commercial lighting fixtures and switches.

Light Fixture Schedules

Light fixture schedules provide detailed information on all of the lighting fixtures used on a particular construction project. Light fixture schedules include interior lighting fixtures, exterior lighting fixtures, emergency lighting fixtures, and illuminated exit signs. Light fixture schedules are found in all sets of residential and commercial electrical prints. **See Figure 7-17.** Schedules for commercial projects typically include more detailed information than schedules for residential projects.

> Electrical drawing software automatically updates prints, load calculations, and floor illumination calculations when a change is made to a light fixture schedule.

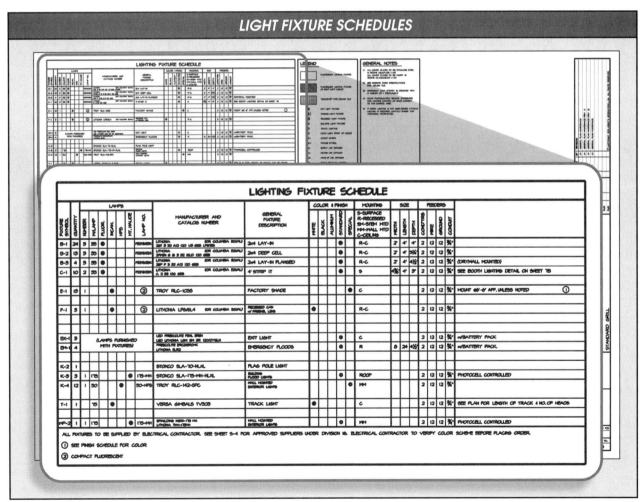

Wendy's International, Inc.

Figure 7-17. Light fixture schedules provide detailed information on all of the lighting fixtures used on a particular construction project.

Light fixture schedules may use the term "luminaire" instead of light fixture. A *luminaire* is a complete unit consisting of a lamp or lamps, the parts that connect the lamp to the power source, and the parts that distribute the light. Light fixture schedules typically include the description, letter or number, type, and voltage rating of the fixture. They may also include the manufacturer part number or series, the number of lamps and description of the lamp type, and the mounting method.

Typically, light fixture schedules are not included on lighting floor plans, but on a print that contains other electrical schedules or electrical details. When many different types of light fixtures are used on a construction project, the light fixture schedule may be a separate sheet entirely.

The various types of light fixtures are identified on a lighting floor plan by a letter or number next to the light fixture symbol. **See Figure 7-18.** On occasion, a circle or hexagon with a fixture letter or number inside is used to aid in identification. The fixture letter or number is also found on the fixture schedule. A leader line can also be used to point to a specific light fixture.

Since the difference between fixture types may be subtle, an electrician must use a light fixture schedule to ensure the correct fixture is installed in the proper location. For example, a 2 × 4 emergency fixture with a battery backup is almost identical in appearance to a 2 × 4 non-emergency fixture. Failure to install the correct fixture may result in project delays and cost overruns.

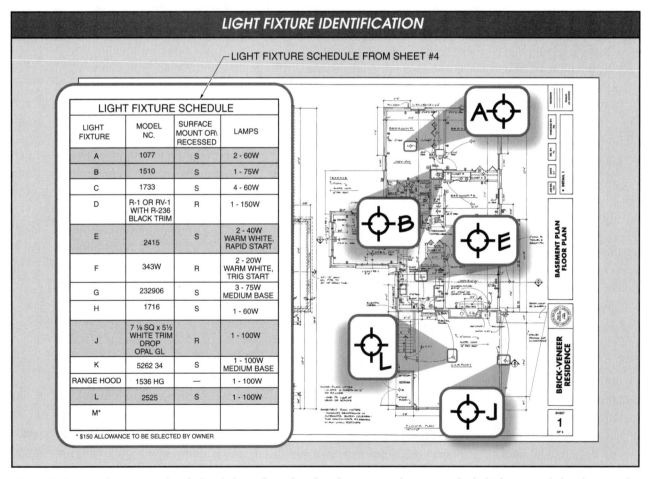

Figure 7-18. Light fixtures are identified on lighting floor plans by a letter or number next to the light fixture symbol and next to the fixture description on the schedule.

Chapter 7—Residential and Commercial Power and Lighting Systems **233**

ELECTRICAL DETAILS

An *electrical detail* is a precise drawing that provides in-depth information that cannot be shown on a site plan, power print, or lighting print. Electrical details are identified by a circle, with a horizontal line dividing the circle in half. The upper half of the circle contains the number of the detail. The bottom half contains the number of the sheet where the detail is found. **See Figure 7-19.**

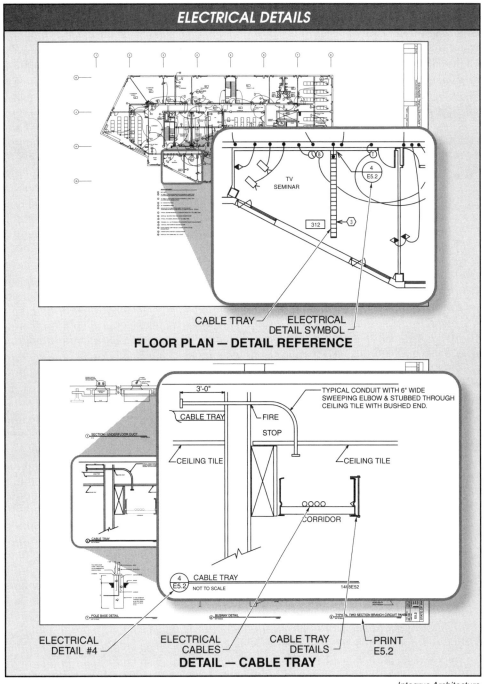

Refer to the CD-ROM
"Prints"
Chapter 7
School – 3rd Floor
Signal Plan

Integrus Architecture

Figure 7-19. *Electrical detail drawings provide in-depth information using elevation drawings, drawings with mounting and installation information, schedules, and diagrams for the connection of electrical control devices and components.*

Electrical details include a wide array of electrical prints. Electrical details consist of elevation drawings of electrical equipment, drawings that provide in-depth mounting and installation information, various schedules, and diagrams for the connection of electrical control components, such as switches, pushbuttons, motor starters, and photocells. Electrical detail drawings may or may not be drawn to scale. Electrical details can be on the same sheet with the site plan, power print, or lighting print if space permits. Multiple electrical detail drawings are typically grouped together as a separate print.

Electrical Elevation Drawings

Electrical prints typically include elevation drawings. An electrical elevation drawing is a scaled view of the vertical surface of an object. Electrical elevation drawings are typically provided for electrical service panels/main switchboards, power distribution panels, and motor control centers. Electrical elevation drawings show dimension information, mounting information, location of lugs for incoming power, and circuit breaker information for electrical equipment. **See Figure 7-20.**

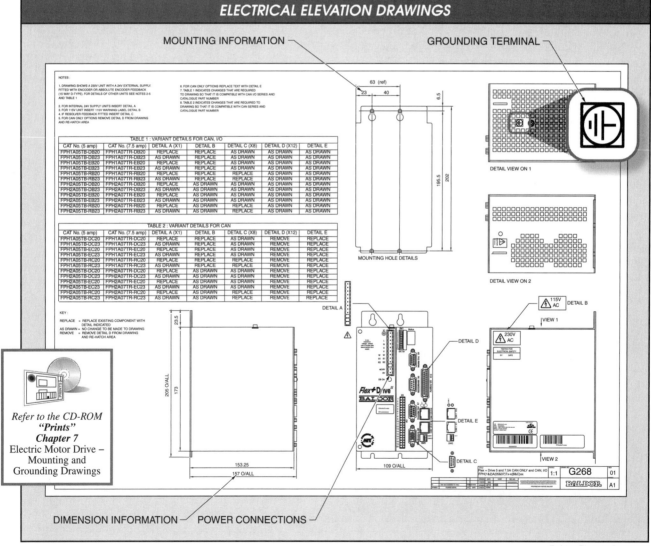

Figure 7-20. Electrical elevation drawings show the vertical surfaces, dimension information, mounting information, location of lugs for incoming power, and circuit breaker information of electrical equipment.

Dimension information is used to verify that the intended location has sufficient space and clearances for the electrical equipment. Electricians should consult Article 110 of the NEC® for required installation clearances. **See Figure 7-21.** Dimension information is also used to plan how the equipment will be moved from the delivery location to the installation location. For example, some door openings may not be large enough to accommodate some pieces of electrical equipment.

Electrical equipment can be fed power from the bottom or the top. The location of the lugs for incoming power to the electrical equipment is shown on elevation drawings. The location of the lugs helps determine the type of feeders (busduct, cable tray, or conduit) and the route the feeders will take. For example, a motor control center with incoming power lugs at the bottom will most likely be fed from conduit in the slab or from below the floor.

Mounting and Installation Details

A *mounting and installation detail drawing* is a product-specific drawing that provides additional information to an electrician about product mounting and installation. Common mounting and installation details include conduit mounting details, light-fixture installation details, and transformer mounting details. **See Figure 7-22.**

Mounting and installation detail drawings provide specific information about the size and type of hardware used and how it is installed. Hardware brand names and catalog numbers may be included in the detail. Electricians must not deviate from mounting and installation details unless a written approval is received from the project architect or engineer.

> When mounting electrical equipment that has the abbreviation AFF (above finished floor), the floor thickness must be added to the mounting height.

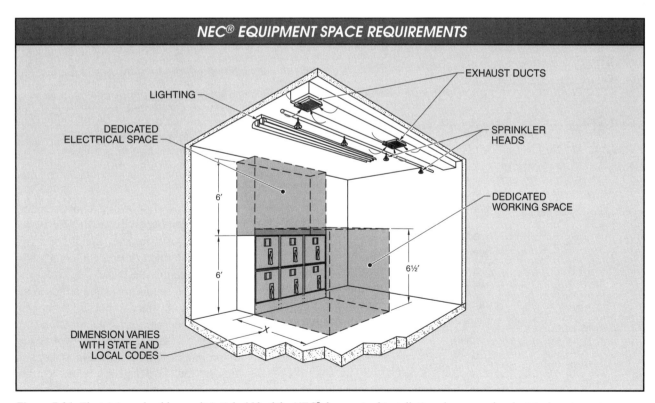

Figure 7-21. Electricians should consult Article 110 of the NEC® for required installation clearances for electrical equipment.

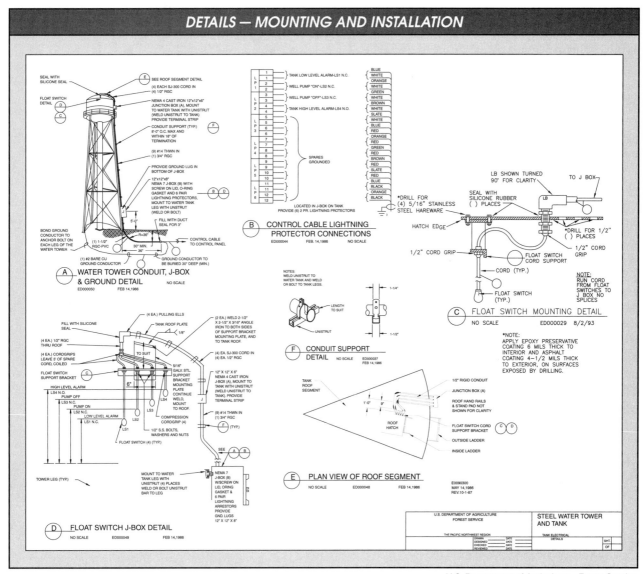

U.S. Department of Agriculture—Forest Service

Figure 7-22. *Mounting and installation detail drawings provide additional information to an electrician about the mounting and installation of specific electrical items.*

Schedules

Schedules for the main switchboard, feeders, power distribution panels, branch-circuit panels, transformers, and motor control centers are typically included as part of the electrical detail drawings. The schedules provide additional information about a specific subject area. **See Figure 7-23.** Electrical schedules contain information on circuit breaker location, circuit breaker size, feeder size, transformer size, and motor starter size.

Electricians use schedules to obtain information on conduit and wire size for power distribution panel feeders, primary and secondary transformer feeders, and motor loads. It is important that electricians verify that the information on a schedule corresponds to the information on other prints. For example, circuit numbers and loads on a branch-circuit panel schedule should correspond with the circuit numbers and loads on a power floor print. Some schedules will have a reference to a diagram, such as a motor control center schedule referencing a control circuit diagram.

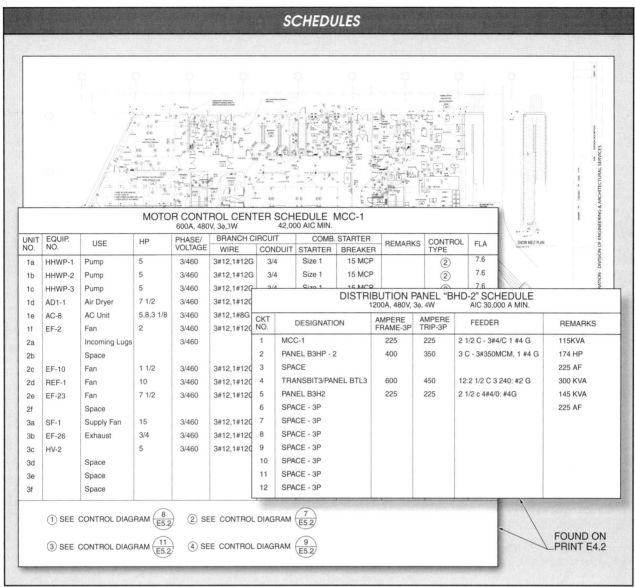

Figure 7-23. Electricians use electrical schedules to find information on circuit breaker location, circuit breaker size, feeder size, transformer size, and motor starter size.

Diagrams

Electrical detail drawings include diagrams. Typically, diagrams contain information on the connection and wiring of electrical control components. Common diagrams are motor starter diagrams, fan-control diagrams, and lighting-control diagrams. **See Figure 7-24.** Information found on a diagram includes the physical location of the control components, the make and model of the control components, a listing of all the units to be wired per diagram, and notes detailing any interconnection with other systems.

A separate diagram is drawn for each unique control scheme or situation. Although two diagrams may appear to be identical at first glance, minor differences will exist. To ensure proper operation, electricians must always consult the appropriate diagram before wiring a control circuit.

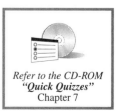

Refer to the CD-ROM "Quick Quizzes" Chapter 7

238 Printreading for Installing and Troubleshooting Electrical Systems

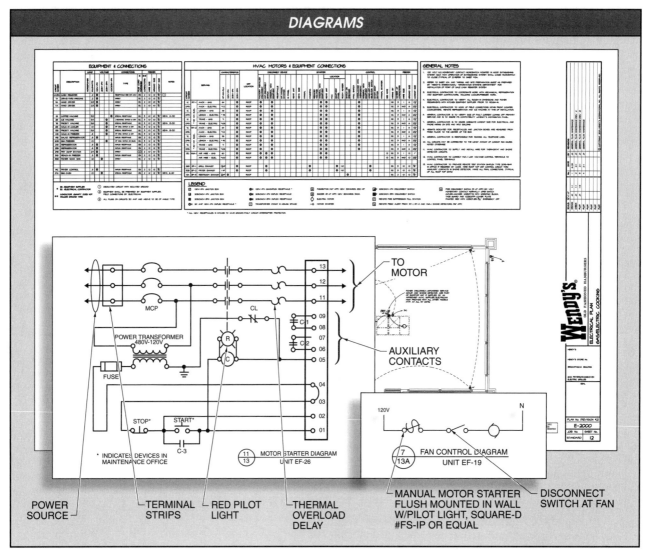

Figure 7-24. Diagrams contain information on the connection and wiring of electrical control components, such as the physical location of the components, make and model of the components, listing of all units to be wired per the diagram, and notes detailing any interconnection with other systems.

☑ Example—Installing Underground Conduit

7-1

Scenario:
Example 7-1 is an installation activity where the electrical service (underground electrical cables and conduit) for a new project must be installed in trenching.

Task:
Install underground electrical conduits and wiring for the electrical service of a new building and parking lot per the electrical site plan.

Required Information for Installing Electrical Underground Cables:
1. Highlight the route for the telephone company conduit(s).
2. What are the identifying number(s) of the general and sheet notes that refer to the telephone company conduit(s)? General _____ Sheet _____
3. How many telephone company conduit(s) will be installed underground? _____
   a. What is the size and type of the telephone company conduit? _____
   b. What is the minimum distance from the finished grade to the top of the plastic conduit spacers? _____
   c. What is the thickness of the concrete cover over the top of the conduit(s)? _____
4. Highlight the concrete communication handhole(s).
   a. What is the size of the concrete communication handhole(s)? _____
   b. What are the manufacturer and model numbers of the concrete communication handhole(s)? _____
5. Highlight the concrete lighting handhole(s) in the parking lot.
6. What is the minimum size of wire for the 120 V circuits shown on the electrical site plan? _____

ELECTRICAL UNDERGROUND SITE WORK

Reference Prints:

SHEET NOTES

1. ENTRANCE GATE OPERATION 460V. 3P.
2. WEATHERPROOF FUSED DISC SWITCH 30AS/15AF, 460V, ± 36"
3. AT UTILITY TRANSFORMER PADS CONNECT ALL 120V, DEVICES TO PANEL BILA-32, EACH LIGHTING FIXTURE TO BE CONTROLLED BY ITS LOCAL WEATHERPROOF SWITCH. FOR EXACT LOCATION OF DEVICES COORDINATE WITH OWNER IN THE FIELD.
4. TERMINATE EACH CONDUIT WITH 90 DEG. ELBOWS AT BOTH ENDS (5 FEET RADIUS). ALL BENDS TO BE ENCASED IN CONCRETE. PROVIDE 3/8" PULL ROPE IN EACH CONDUIT. ALL WORK TO BE FULLY COORDINATED WITH TELEPHONE UTILITY COMPANY STANDARDS.
5. EXCEPT AS NOTED ALL UNDERGROUND CONDUITS WILL BE PVC TYPE DB (DIRECT BURIAL).
6. EACH TAMPER SWITCH SHALL BE WIRED TO A SEPARATE FIRE ALARM ZONE.
7. WEATHERPROOF PEDESTAL MOUNTED JUNCTION BOX. VERIFY EXACT LOCATION.
8. NOT USED.
9. CONCRETE COMMUNICATION HANDHOLE - 3' X 3' X 3' H FORNI CORP. CAT. NO Z-80 WITH 33TC-TRAFFIC COVER WITH HOLD-DOWN BOLTS. LABEL COVER "COMMUNICATION".
10. NOT USED.
11. VIA PHOTOCELL CONTROLLED CONTRACTOR. SEE DIAGRAM.
12. CHRISTY, OR EQUAL #N9 CONCRETE BOX WITH #4121 BASE, #J16 TRAFFIC LID AND EXTENSION(S) AS REQUIRED. LABEL COVER "LIGHTING",U.O.N.
13. CONCRETE SPLICE BOX,3' X 5' X 3' – 6" HIGH-FORNI CORP. CAT. NO. P-185 WITH P-138 COVER, STEEL PEDESTRIAN TRAFFIC, GALVANIZED. DO NOT LOCATE BOX IN VEHICULAR TRAFFIC AREAS.
14. NOT USED.
15. MINIMUM SIZE WIRE FOR 120V. CIRCUITING THIS PLAN SHALL BE #10 AWG.
16. PROVIDE ACCESS OPENING ,3" X 5" IN THE LIGHT POLE AT + 18'- 0" (FOR MOUNTING AND WIRING TV CAMERA BY OTHERS).

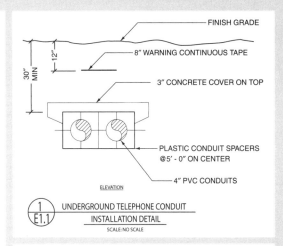

GENERAL NOTES

1. COORDINATE ALL UNDERGROUND INSTALLATION WITH THE LANDSCAPE, CIVIL AND PLUMBING WORK.
2. CONTACT CP&L INSPECTORS FOR OPEN TRENCH INSPECTION.
3. COMPLY WITH CP&L REQUIREMENTS AND SPECIFICATION FOR RELATED WORK MINIMUM BENDING FOR CONDUIT SHALL BE 36".
4. CONTACT TELEPHONE COMPANY INSPECTORS FOR OPEN TRENCH INSPECTION.

Example—Installing Underground Conduit

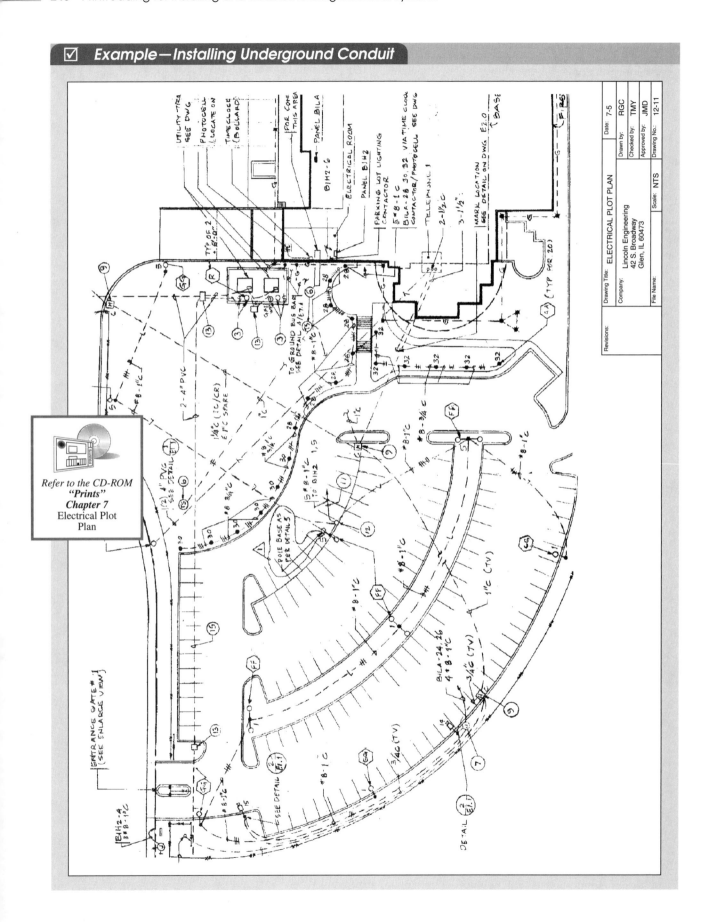

Refer to the CD-ROM
"Prints"
Chapter 7
Electrical Plot Plan

Example—Installing Underground Conduit

Step 1: Highlight the route of the telephone company conduit(s).

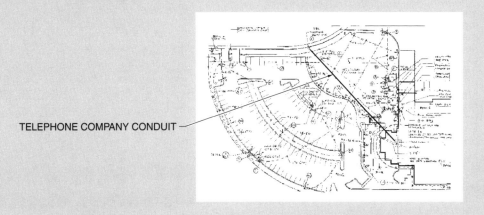

TELEPHONE COMPANY CONDUIT

Step 2: Determine which general notes and sheet notes refer to the telephone company underground conduit(s).

GENERAL NOTES

1. COORDINATE ALL UNDERGROUND INSTALLATION WITH THE LANDSCAPE, CIVIL AND PLUMBING WORK.
2. CONTACT CP&L INSPECTORS FOR OPEN TRENCH INSPECTION.
3. COMPLY WITH CP&L REQUIREMENTS AND SPECIFICATION FOR RELATED WORK MINIMUM BENDING FOR CONDUIT SHALL BE 36".
4. CONTACT TELEPHONE COMPANY INSPECTORS FOR OPEN TRENCH INSPECTION.

SHEET NOTES

1. ENTRANCE GATE OPERATION 460V. 3P.
2. WEATHERPROOF FUSED DISC SWITCH 30AS/15AF, 460V, ± 36"
3. AT UTILITY TRANSFORMER PADS CONNECT ALL 120V, DEVICES TO PANEL BILA-32, EACH LIGHTING FIXTURE TO BE CONTROLLED BY ITS LOCAL WEATHERPROOF SWITCH. FOR EXACT LOCATION OF DEVICES COORDINATE WITH OWNER IN THE FIELD.
4. TERMINATE EACH CONDUIT WITH 90 DEG. ELBOWS AT BOTH ENDS (5 FEET RADIUS). ALL BENDS TO BE ENCASED IN CONCRETE. PROVIDE 3/8" PULL ROPE IN EACH CONDUIT. ALL WORK TO BE FULLY COORDINATED WITH TELEPHONE UTILITY COMPANY STANDARDS.
5. EXCEPT AS NOTED ALL UNDERGROUND CONDUITS WILL BE PVC TYPE DB (DIRECT BURIAL).
6. EACH TAMPER SWITCH SHALL BE WIRED TO A SEPARATE FIRE ALARM ZONE.
7. WEATHERPROOF PEDESTAL MOUNTED JUNCTION BOX. VERIFY EXACT LOCATION.

Step 3: Determine the specifications for underground telephone conduit installation.

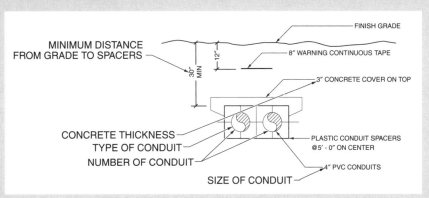

Example—Installing Underground Conduit

Step 4: Determine the concrete communication handhole(s) specifications.

SIZE OF COMMUNICATION HANDHOLES

5. EXCEPT AS NOTED ALL UNDERGROUND CONDUITS WILL BE PVC TYPE DB (DIRECT BURIAL).
6. EACH TAMPER SWITCH SHALL BE WIRED TO A SEPARATE FIRE ALARM ZONE.
7. WEATHERPROOF PEDESTAL MOUNTED JUNCTION BOX. VERIFY EXACT LOCATION.
8. NOT USED.
9. CONCRETE COMMUNICATION HANDHOLE - 3' X 3' X 3' H FORNI CORP. CAT. NO Z-80 WITH 33TC-TRAFFIC COVER WITH HOLD-DOWN BOLTS. LABEL COVER "COMMUNICATION".

COMMUNICATION HANDHOLE MODEL NUMBER

COMMUNICATION HANDHOLE MANUFACTURER

Step 5: Find the concrete lighting handhole(s) and the communication handhole(s).

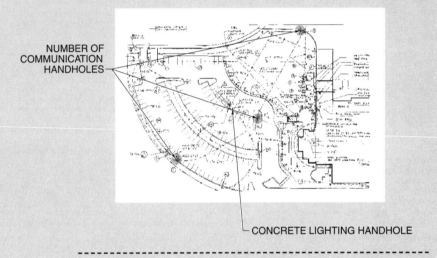

NUMBER OF COMMUNICATION HANDHOLES

CONCRETE LIGHTING HANDHOLE

Step 6: Determine the minimum wire size for 120 V circuits.

15. MINIMUM SIZE WIRE FOR 120V. CIRCUITING THIS PLAN SHALL BE #10 AWG.
16. PROVIDE ACCESS OPENING ,3" X 5" IN THE LIGHT POLE AT + 18'/0" (FOR MOUNTING AND WIRING TV CAMERA BY OTHERS).

120 V CIRCUIT MINIMUM WIRE SIZE

Chapter 7—Residential and Commercial Power and Lighting Systems

☑ Example—Troubleshooting a Power Distribution System

7-2

Scenario:
Example 7-2 is a troubleshooting activity where an infrared image indicates a possible problem in the power distribution system from the service entrance to Motor Control Center 1, circuit SF-1.

Task:
Track down the devices and wiring to Motor Control Center 1, circuit SF-1.

Required MCC-1 Circuit Information:
1. What type of raceway connects the service entrance to the Consolidated Power & Light transformer? _____
   WEATHERPROOF
2. What type of metering is provided at the service? _CONSOLIDATED POWER LIGHT_
3. What is the short circuit current available from the secondary coil of the utility transformer? _51,700 A SYMETRICAL 480V 3ø_
4. MCC-1 information:
   a. Source panel: _BHD-2_
   b. Number and size of feeder conduits: _2½"_
   c. Number and size of feeder conductors: _3 #4_
   d. What unit has the incoming power lugs: _2 A_
5. Feeder information for Panel BHD-2:
   a. Source panel: _MAIN SERVICE SWITCHBOARD_
   b. Number and size of feeder conduits: _3" 4G_
   c. Number and size of feeder conductors: _3 3/0_
6. SF-1 information:
   a. Motor size: _3A 15_
   b. Number and size of feeder conduit: _1"_
   c. Number and size of feeder conductors: _3 #10_
   d. Starter size: _2_
   e. Detail number of control diagram: _E.52_
   f. Bucket number: _3A_

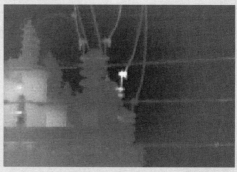

Snell Infrared
INFRARED IMAGE — 3ø TRANSFORMER

Reference Prints

DISTRIBUTION PANEL "BHD-2" SCHEDULE					
1200A, 480/277V, 3ø, 4W AIC 30,000 A MIN.					
CKT NO	DESIGNATION	AMPERE FRAME-3P	AMPERE TRIP-3P	FEEDER	REMARKS
1	MCC-1	225	225	2 ½" C - 3#4/C. 1 #4 G	115 KVA
2	PANEL B3HP-2	400	350	3" C - 3#350MCM. 1 #4 G	174 HP
3	SPACE				225 AF
4	TRANSF B1T3/PANEL B1L3	600	450	(2) 2 ½" C - 3#4/C. 1 #2 G	300 KVA
5	PANEL B3H2	225	225	(2) 2 ½" C - 4#4/C. 1 #4 G	145 KVA

MAIN SERVICE SWBD "MSB" SCHEDULE					
480/277V, 3ø, 4W AIC 100,000 A MIN 3000 A MAIN CB					
CKT NO	DESIGNATION	AMPERE FRAME-3P	AMPERE TRIP-3P	FEEDER	REMARKS
1	SPACE	225	225	2 ½" C - 3#4/C. 1 #4 G	115 KVA
2	PANEL BHP-2	400	350	3" C - 3#350MCM. 1 #4 G	174 HP
3	SPACE				800 AF
4	PANEL BHD-X 2	1200	1200	(3) 3 ½" C - 4# 600 MCM. 1 #3/0 G IN EACH	300 KVA

Example—Troubleshooting a Power Distribution System

MOTOR CONTROL CENTER SCHEDULE MCC-1
600A, 480V, 3φ, 3W   42,000 AIC MIN.

UNIT NO.	EQUIP. NO.	USE	HP	PHASE/ VOLTAGE	BRANCH CIRCUIT WIRE	BRANCH CIRCUIT CONDUIT	COMB. STARTER. STARTER	COMB. STARTER. BREAKER	REMARKS	CONTROL TYPE	FLA
1a	HHWP-1	PUMP	5	3/460	3#12, 1#12 G	¾"	SIZE-1	15 MCP			7.6
1b	HHWP-2	PUMP	5	3/460	3#12, 1#12 G	¾"	SIZE-1	15 MCP			7.6
1c	HHWP-3	PUMP	5	3/460	3#12, 1#12 G	¾"	SIZE-1	15 MCP			7.6
1d	AD1-1	AIR DRYER	7 ½	3/460	3#12, 1#12 G	¾"	SIZE-1	30 MCP	FEEDER CB		11
1e	AC-8	AC UNIT	5 8 3-½	3/460	3#4, 1#8 G	1 ¼"	–	100 A	FEEDER CB		47
1f	EF-2	FAN	2	3/460	3#12, 1#12 G	¾"	SIZE-1	7 MCP			3.4
2a		INCOMING LUGS									
2b		SPACE									
2c	EF-10	FAN	1 ½	3/460	3#12, 1#12 G	¾"	SIZE-1	7 MCP			2.6
2d	REF-1	FAN	10	3/460	3#12, 1#12 G	¾"	SIZE-1	35 MCP			14
2e	EF-23	FAN	7 ½	3/460	3#12, 1#12 G	¾"	SIZE-1	30 MCP			11
2f		SPACE									
3a	SF-1	SUPPLY FAN	15	3/460	3#5, 1#10 G	1"	2	60 MCP			21
3b	EF-26	EXHAUST	¾	3/460	3#12, 1#12 G	¾"	1	7 MCP			1.4
3c	HV-2		5	3/460	3#12, 1#12 G	¾"	1	15 MCP			7.5
3d		SPACE									
3e		SPACE									
3f		SPACE									

① SEE CONTROL DIAGRAM 8/E5.2   ② SEE CONTROL DIAGRAM 7/E5.2
③ SEE CONTROL DIAGRAM 11/E5.2   ④ SEE CONTROL DIAGRAM 9/E5.2

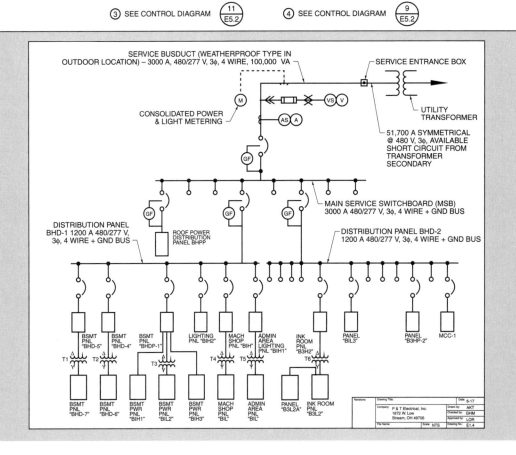

☑ Example—Troubleshooting a Power Distribution System

Step 1: Determine the type of service raceway used so that a visual inspection can be performed.

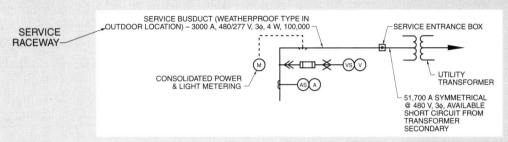

Step 2: Determine the metering used at the service entrance.

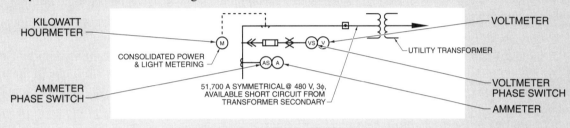

Step 3: Determine the short circuit current available from the secondary of the utility transformer.

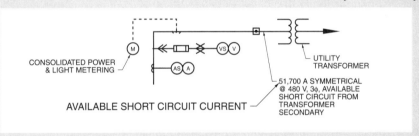

Step 4: Identify the feeder circuit devices and wiring to MCC-1.

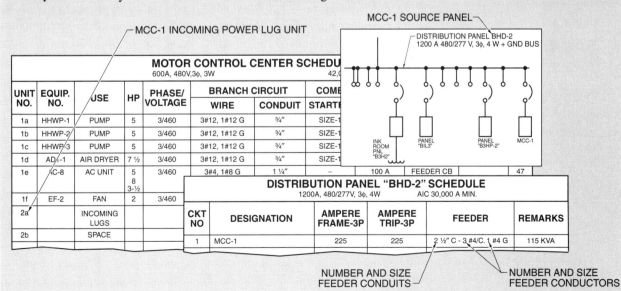

Example—Troubleshooting a Power Distribution System

Step 5: Identify the feeder circuit devices and wiring to BHD-2.

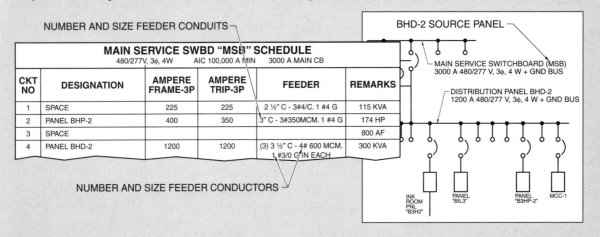

MAIN SERVICE SWBD "MSB" SCHEDULE 480/277V, 3φ, 4W AIC 100,000 A MIN 3000 A MAIN CB					
CKT NO	DESIGNATION	AMPERE FRAME-3P	AMPERE TRIP-3P	FEEDER	REMARKS
1	SPACE	225	225	2 ½" C - 3#4/C. 1 #4 G	115 KVA
2	PANEL BHP-2	400	350	3" C - 3#350MCM. 1 #4 G	174 HP
3	SPACE				800 AF
4	PANEL BHD-2	1200	1200	(3) 3 ½" C - 4# 600 MCM. 1 #3/0 G IN EACH	300 KVA

NUMBER AND SIZE FEEDER CONDUITS
NUMBER AND SIZE FEEDER CONDUCTORS

Step 6: Identify the devices and components of circuit SF-1.

MOTOR CONTROL CENTER SCHEDULE MCC-1 600A, 480V, 3φ, 3W 42,000 AIC MIN.										
UNIT NO.	EQUIP. NO.	USE	HP	PHASE/ VOLTAGE	BRANCH CIRCUIT		COMB. STARTER.		REMARKS	CONTROL TYPE	FLA
					WIRE	CONDUIT	STARTER	BREAKER			
1a	HHWP-1	PUMP	5	3/460	3#12, 1#12 G	¾"	SIZE-1	15 MCP		②	7.6
1b	HHWP-2	PUMP	5	3/460	3#12, 1#12 G	¾"	SIZE-1	15 MCP		②	7.6
1c	HHWP-3	PUMP	5	3/460	3#12, 1#12 G	¾"	SIZE-1	15 MCP		②	7.6
1d	AD1-1	AIR DRYER	7 ½	3/460	3#12, 1#12 G	¾"	SIZE-1	30 MCP	FEEDER CB		11
1e	AC-8	AC UNIT	5 8 3-½	3/460	3#4, 1#8 G	1 ¼"	—	100 A	FEEDER CB		47
1f	EF-2	FAN	2	3/460	3#12, 1#12 G	¾"	SIZE-1	7 MCP		①	3.4
2a		INCOMING LUGS									
2b		SPACE									
2c	EF-10	FAN	1 ½	3/460	3#12, 1#12 G	¾"	SIZE-1	7 MCP		②	2.6
2d	REF-1	FAN	10	3/460	3#12, 1#12 G	¾"	SIZE-1	35 MCP		①	14
2e	EF-23	FAN	7 ½	3/460	3#12, 1#12 G	¾"	SIZE-1	30 MCP		②	11
2f		SPACE									
3a	SF-1	SUPPLY FAN	15	3/460	3#10, 1#10 G	1"	2	60 MCP		②	21
3b	EF-26	EXHAUST	¾	3/460	3#12, 1#12 G	¾"	1	7 MCP		③	1.4
3c	HV-2		5	3/460	3#12, 1#12 G	¾"	1	15 MCP		①	7.5
3d		SPACE									
3e		SPACE									
3f		SPACE									

① SEE CONTROL DIAGRAM 8/E5.2   ② SEE CONTROL DIAGRAM 7/E5.2
③ SEE CONTROL DIAGRAM 11/E5.2   ④ SEE CONTROL DIAGRAM 9/E5.2

STARTER SIZE
CIRCUIT SF-1 MOTOR HP
SF-1 BUCKET NUMBER
NUMBER AND SIZE FEEDER CONDUCTORS
CONDUIT SIZE
DETAIL NUMBER CONTROL DIAGRAM

Activity—Pulling Conductors to Room Devices

7-1

Scenario:
Activity 7-1 is an installation activity requiring that conductors be pulled into the Bloom Education and Research Center (B.E.R.C.) room to connect switches with the light fixtures in the room.

Task:
Pull the appropriate amount of conductors through the B.E.R.C. room conduits to supply power to light fixtures and to connect the three-way switches at the doorways.

Required B.E.R.C. Room and Circuit Information:
1. What is the room number of the B.E.R.C. room? ____
2. List the number and type of each fixture found in the B.E.R.C. room. ____
3. What is the voltage used by the each type of fixture in the B.E.R.C. room? ____
4. What is the mounting method used for the light fixtures in the B.E.R.C. room? ____
5. What is the difference between the different types of fixtures used in the B.E.R.C. room? ____
6. How many light fixtures are controlled by the 3-way switches? ____
7. What circuit is used to feed power to the light fixtures in the B.E.R.C. room? ____
8. Add hash marks to the B.E.R.C. room to determine the number of conductors required between junction boxes and switches. Add hash marks for the hot and neutral feeding through to other rooms. Omit hash marks for flexible conduits feeding the light fixtures.

B.E.R.C. ROOM

Reference Prints:

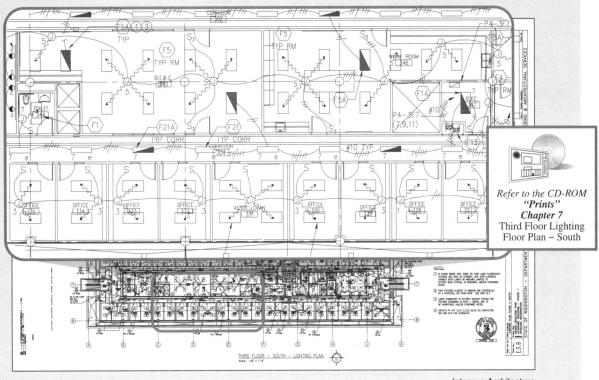

Refer to the CD-ROM "Prints" Chapter 7 Third Floor Lighting Floor Plan – South

Integrus Architecture

Activity—Pulling Conductors to Room Devices

LUM #	TYPE	VOLTAGE	LUMINAIRE DESCRIPTION	MANUFACTURER/ SERIES #	LAMPS & COLOR TEMP	FINISH	MOUNTING
F1	FLUOR	277V	2'x4' STATIC TROFFER FIXTURE W/AL DOOR FRAME AND SPEC GRADE ACRYLIC #12 PATTERN LENS STEEL HOUSING, POLYESTER ENAMEL FINISH	COLUMBIA #5PS	(2) FO32/31K (32 WATTS)	WHITE	RECESS
F1A	FLUOR	277V	SAME AS TYPE F1 EXCEPT W/EMERGENCY BATTERY BACKUP	SEE TYPE F1	SEE TYPE F1	WHITE	SEE TYPE F1
F2	FLUOR	277V	2'x4' STATIC TROFFER FIXTURE W/AL DOOR FRAME AND SPEC GRADE ACRYLIC #12 PATTERN LENS STEEL HOUSING, POLYESTER ENAMEL FINISH	COLUMBIA #5PS	(3) FO32/31K (32 WATTS)	WHITE	RECESS
F2A	FLUOR	277V	SAME AS TYPE F2 EXCEPT W/EMERGENCY BATTERY BACKUP	SEE TYPE F2	SEE TYPE F2	WHITE	SEE TYPE F2
F3	FLUOR	277V	2'x4' STATIC TROFFER FIXTURE W/AL DOOR FRAME AND SPEC GRADE ACRYLIC #12 PATTERN LENS STEEL HOUSING, POLYESTER ENAMEL FINISH	COLUMBIA #5PS	(4) FO32/31K (32 WATTS)	WHITE	RECESS
F4	FLUOR	277V	2'x4' STATIC TROFFER FIXTURE W/3" DEEP 16-CELL (2x8 CELLS) AL SEMI-SPECULAR LOW IRIDESCENT PARABOLIC LOUVER, STEEL HOUSING	COLUMBIA #P4-242	(2) FO32/31K (32 WATTS)	WHITE	RECESS
F4D	FLUOR	277V	SAME AS TYPE F4 EXCEPT W/DIMMABLE ELECTRONIC BALLAST	SEE TYPE F4	SEE TYPE F4	WHITE	SEE TYPE F4
F5	FLUOR	277V	2'x4' STATIC TROFFER FIXTURE W/3" DEEP 24-CELL (3x8 CELLS) AL SEMI-SPECULAR LOW IRIDESCENT PARABOLIC LOUVER, STEEL HOUSING	COLUMBIA #P4-243	(3) FO32/31K (32 WATTS)	WHITE	RECESS
F5A	FLUOR	277V	SAME AS TYPE F5 EXCEPT W/EMERGENCY BATTERY BACKUP	SEE TYPE F5	SEE TYPE F5	WHITE	SEE TYPE F5
F6	FLUOR	277V	2'x4' STATIC TROFFER FIXTURE W/3" DEEP 32-CELL (4x8 CELLS) AL SEMI-SPECULAR LOW IRIDESCENT PARABOLIC LOUVER, STEEL HOUSING	COLUMBIA #P4-244	(4) FO32/31K (32 WATTS)	WHITE	RECESS
F6A	FLUOR	277V	SAME AS TYPE F6 EXCEPT W/EMERGENCY BATTERY BACKUP	SEE TYPE F6	SEE TYPE F6	WHITE	SEE TYPE F6

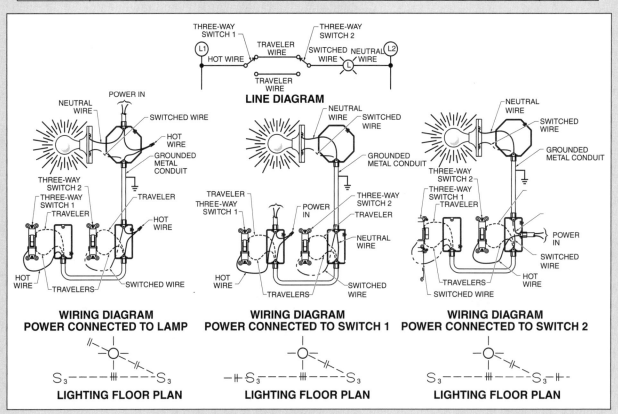

Name _____ Date _____

Activity—Installing Kitchen Outlets

7-2

Scenario:
Activity 7-2 is an installation activity requiring that kitchen outlets be installed.

Task:
Lay out the location of outlet boxes for mounting.

Required Kitchen Information:

1. How many outlets are located on the cooktop wall? _____

FINISHED KITCHEN

2. List the type and mounting height of each outlet:

TYPE	HEIGHT	TYPE	HEIGHT
_____	_____	_____	_____
_____	_____	_____	_____
_____	_____	_____	_____

3. The cooktop outlet must be located in the space beneath the cooktop. What is the center of the opening for the cooktop from the lazy Susan wall? _____

4. What is the height from the top of the countertop to the bottom of the cabinets? _____

5. The finished kitchen floor will be 1″ thick. What height above the subflooring must the center of countertop outlets be installed at to keep the faceplate of the outlets off of the backsplash? _____

6. The outlet that is located between the lazy Susan and the cooktop is to be centered between the lazy Susan and the cooktop. What is the distance from the lazy Susan wall to the center of the outlet? _____

Reference Prints:

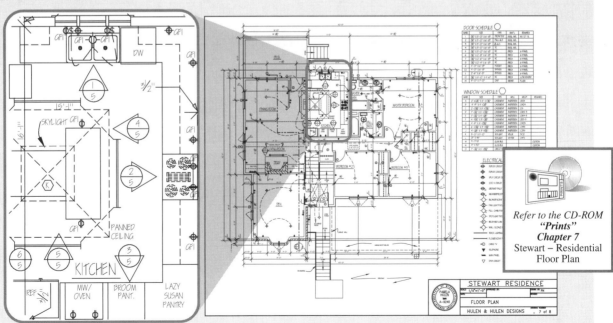

Refer to the CD-ROM
"Prints"
Chapter 7
Stewart – Residential
Floor Plan

Activity—Installing Kitchen Outlets

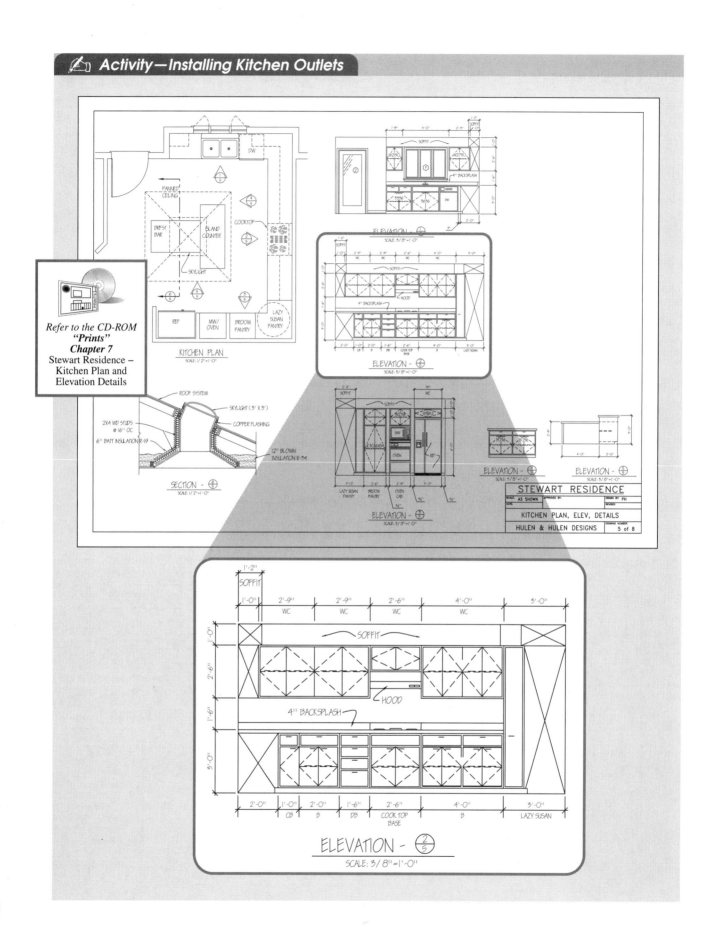

Refer to the CD-ROM "Prints" Chapter 7 Stewart Residence – Kitchen Plan and Elevation Details

Residential and Commercial VDV Systems

Printreading for Installing and Troubleshooting Electrical Systems

As cable and satellite TV, fax machines, the Internet, and personal computers have beome part of our daily lives, the need for low-voltage wiring for connecting these devices has grown rapidly. Voice/data/video (VDV) systems are found in many residential and commercial construction projects. VDV systems provide connections for phones (voice), computers (data), camera (video), and cable or satellite TV.

CSI MASTERFORMAT™— DIVISION 27

Division 27 (Communications) of the 2004 edition of the Construction Specifications Institute's Masterformat™ covers VDV systems. **See CD-ROM.** Prior to the 2004 edition, VDV systems were included in Division 16.

VDV SYSTEM PRINTS

The size, complexity, and type of construction project (residential or commercial) determine the number and type of prints used to depict a VDV system. A basic residential VDV system is shown on a residential floor plan along with power and lighting information. Large commercial VDV systems appear on a separate set of prints. VDV system prints are typically divided into VDV riser diagrams, VDV floor plans, and VDV detail drawings. **See Figure 8-1.** Other electrical prints may need to be consulted in addition to VDV-specific prints. For example, an electrical site plan may contain information on the conduit and cables connecting a building to the local telephone company.

VDV Symbols, Abbreviations, and Legends

A set of VDV prints includes a list of symbols, abbreviations, and a legend that are used throughout the set of prints. They provide a clear and concise method for conveying information without cluttering a print.

> The Telecommunications Industry Association (TIA) is a trade organization made up of companies that provide VDV products and services, and that develop standards for the installation of VDV cabling.

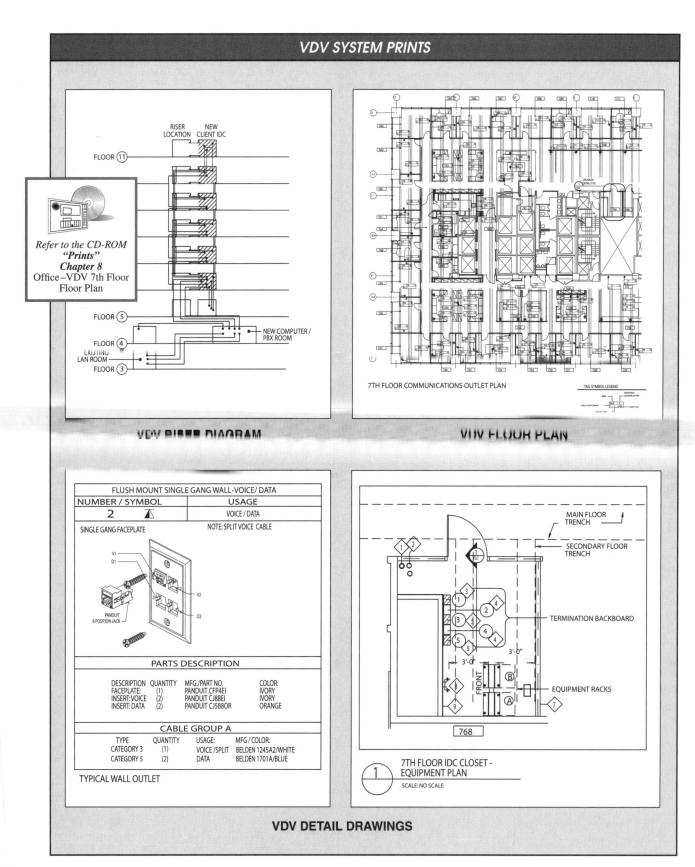

Figure 8-1. VDV system prints are typically divided into VDV riser diagrams, VDV floor plans, and VDV detail drawings.

A specific set of symbols is used to depict VDV components. **See Figure 8-2.** Typically, the symbols are part of a set of standards, such as ANSI (American National Standards Institute) standards. The symbols depict outlets (data, telephone, and television) and VDV-related equipment. The symbols are often modified with letters that provide additional information. Some VDV designers prefer to use a manufacturer-developed set of symbols. While a manufacturer-developed set of symbols may resemble standard VDV symbols, manufacturer-developed symbols often possess minor differences. **See Figure 8-3.**

Outlet addressing information may be included with VDV symbols. The outlet address is a unique system of letters and/or numbers used for identifying cables and their locations. The outlet addressing format used is generally specified by the VDV designer or the customer. Also, a unique set of abbreviations is used on all VDV prints. The VDV abbreviations are specific to cable TV, telecommunications, and wiring systems. **See Figure 8-4.**

DATA AND TELEPHONE OUTLET SYMBOLS

Device	Part	Symbol
DATA OUTLET	Wall Mounted Abbr = DOTLT	▽
	Floor Mounted Abbr = DOTLT	▽F Flush Mounted OR ▽S Surface Mounted
TELEPHONE OUTLET	Wall Mounted Abbr = TELOTLT	▼
	Floor Mounted Abbr = TELOTLT	▼F Flush Mounted OR ▼S Surface Mounted
TELEPHONE/ DATA OUTLET	Wall Mounted Abbr = TELDOTLT	▽
	Floor Mounted Abbr = TELDOTLT	▽F Flush Mounted OR ▽S Surface Mounted

Figure 8-2. Specific symbols are used to depict telephone outlets, data outlets, cable TV outlets, and related equipment in VDV systems.

VDV CABLE TEST METERS

When unshielded twisted pair (UTP) cable is installed, the cable must be tested. There are three primary types of test meters: certification, qualification, and verification.

Certification test meters perform a complex series of tests to determine if a cable installation meets TIA standards. A cable installation will only be guaranteed by the cable manufacturer if the installation meets these standards.

Qualification test meters allow a VDV technician to determine if existing cable can support certain network speeds. Qualification test meters can also be used for basic VDV troubleshooting, such as for continuity and wire mapping. A qualification test meter cannot be used to certify VDV cable for warranty purposes.

Verification test meters are used to check for continuity and that VDV cable is correctly connected. Verification test meters cannot be used to certify cable for warranty purposes or to determine if a VDV cable is capable of supporting specific network speeds.

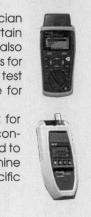

	Certif	Qualif	Verif
Meets warranty requirements from cabling manufacturers	•		
Provides Pass/Fail results compliant with TIA and ISO standards	•		
Provides documentation reports	•		
Advanced troubleshooting (NEXT, RL)	•		
Tests existing cable		•	
Network connectivity troubleshooting		•	
Troubleshooting: distance to problem	•	•	
Troubleshooting: distance to break	•	•	•
Continuity and wire map	•	•	•

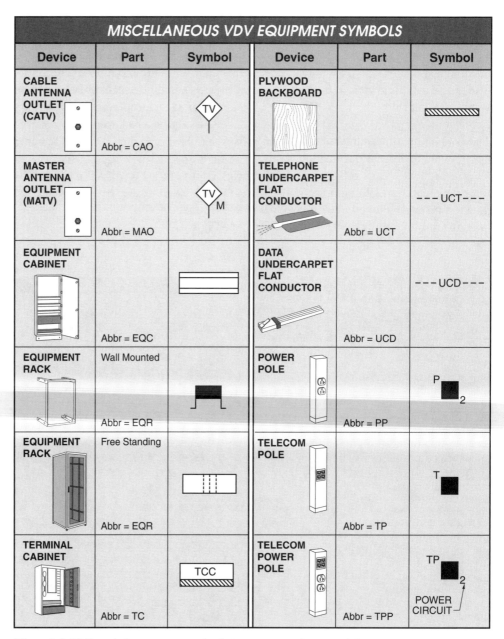

Figure 8-3. VDV symbols may vary per the designer or manufacturer of the equipment.

The main DVD standard for structured wiring (EIA/TIA 568) is typically known as "document 568." Document 568 is in its B revision, which includes sections on general specifications (B.1), copper (B.2), and fiber optics (B.3).

Legends are found on many VDV prints. Legends are generally included with VDV abbreviations and symbols, but can be on a separate print if necessary. Legends provide information about a VDV installation. Each item in a VDV system legend is associated with a number or letter. Often the number or letter is inside a geometric shape, such as a circle, diamond, or square. The number or letter is located throughout the various VDV prints where the specific piece of information applies. **See Figure 8-5.**

VDV SYSTEM ABBREVIATIONS

EQUIPMENT		EQUIPMENT	
110	Twisted pair termination block	TEC	Telecom entrance conduit
ADF	Area distribution facility	TEL	Telephone
BDF	Building distribution frame or facility	TELECOM	Telecommunications
BEF	Building entrance frame or facility	TERM	Terminal
CAB	Telecom cabinet or enclosure	TP	Twisted pair
CONN	Connector	TPB	Telecom pull box
CSC	Copper splice closure	TSL	Telecom wall or floor slot
CVE	Controlled environmental vault	TSV	Telecom conduit sleeve, vertical
FDF	Fiber distribution facility		
FF	Front facing		
FS	Fiber shelf/fiber termination panel	**WIRE AND CABLE**	
FSC	Fiber optic splice enclosure		
HH	Handhole		
IDC	Internet data center	AFMW	24 AWG bonded fill flooded twisted cable
IDF	Intermediate distribution frame or facility	ARMM	24 AWG riser armored bonded multipair cable
MDF	Main distribution frame	AWG	American wire gauge
MH	Manhole; Maintenance hole	CAT3	Category 3 twisted pair copper cable
MPOE	Minimum point of entry; Main point of entry	CAT6	Category 6 twisted pair copper cable
OCEF	Optical cable entrance facility	CM	Communications cable
PAV	Pavement	CMP	Communications plenum cable
PC	Plastic conduit	CMR	Communications riser cable
PG	Pair group	COAX	Coaxial cable
POP	Point of presence	FO	Fiber optic
PR	Pair	HDPE	High density polyethylene
PVC	Polyvinyl chloride	LTFF	Loose tube filled & flooded
R/W	Right-of-way	MDPE	Medium density polyethylene
SC	Splice closure	MM	Loose tube filled & flooded
SCS	Structured cabling system	MPP	Multipurpose plenum cable
SER	Serial	OFC	Optical fiber conductive cable
SM	Single mode fiber optic cable	OFCP	Optical fiber conductive plenum cable
SMR	Surface mounted raceway	OFCR	Optical fiber conductive riser cable
SS	Fiber splice shelf	OFNR	Optical fiber non-conductive riser cable
TC	Telecom closet	OFN	Optical fiber non-conductive cable
TCC	Telecom conduit	OFNP	Optical fiber non-conductive plenum cable
TCH	Telecom conduit sleeve, horizontal	STP	Shielded twisted pair
TCR	Telecom horizontal & vertical riser conduit	TB	Tight buffered
TCT	Telecom cable tray	UTP	Unshielded twisted pair

Figure 8-4. The VDV abbreviations used on VDV prints are specific to cable TV, telecommunications, and wiring systems.

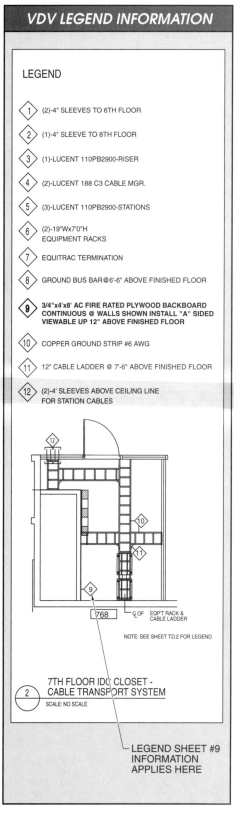

Figure 8-5. Each item on a legend is associated with a number or letter. The number or letter is used on the prints to save space.

VDV Riser Diagrams

Riser diagrams are provided for VDV installations in multistory buildings. Multistory buildings can be commercial office space, high-rise residential condominiums, retail stores, or mixed use (a combination of office space, residential, and retail). VDV riser diagrams are not provided when constructing multistory single-family residences.

A VDV installation in a multistory building has vertical (backbone) cabling and horizontal cabling. Backbone cabling runs from the main telecommunications/data equipment room to the telecommunications/data equipment room on each floor. The backbone cabling is terminated at connector blocks in the main equipment room and in each floor's equipment room. Horizontal-cabling runs from the equipment room on each floor to individual workstations or desks and is terminated at the connector blocks in the equipment room. The backbone and horizontal cables are cross-connected in the equipment room on each floor.

VDV riser diagrams provide an overview of the backbone cabling in a building. **See Figure 8-6.** For a new construction project, all of the floors in a building are shown on a VDV riser diagram. For a retrofit project on an existing building, a VDV riser diagram may omit floors that are not involved in the retrofit. In addition to VDV cables, an installation may require a dedicated telecommunications grounding backbone. Telecommunication equipment and cables that require grounding are connected to the grounding backbone, which is shown on a VDV riser diagram.

A VDV riser diagram shows the room designations of telecommunications/data equipment rooms, the type of cables run (copper or fiber-optic), and the size of the cables run (number of pairs for copper, or number of fibers for fiber-optic). VDV riser diagrams are not drawn to scale. VDV riser diagrams do not typically provide the size of the conduit or raceway for the backbone cables nor do they show the exact route of the cables. Electrical prints for the building must be consulted for information on conduit and raceway size.

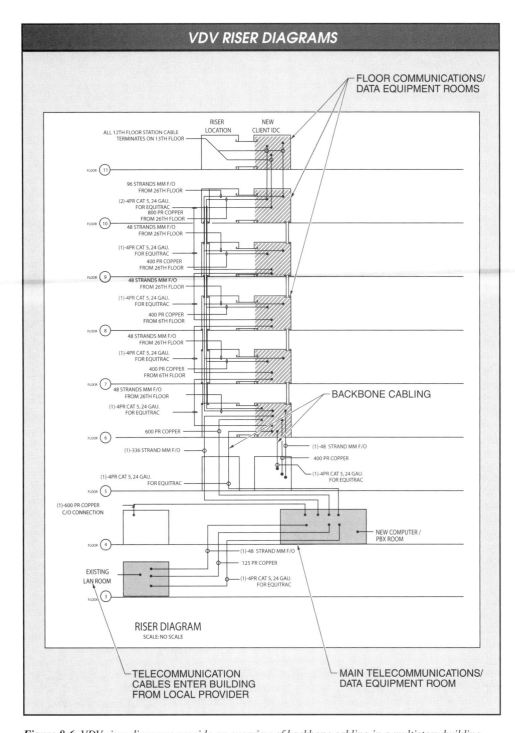

Refer to the CD-ROM
"Prints"
Chapter 8
VDV Riser Diagram

Figure 8-6. *VDV riser diagrams provide an overview of backbone cabling in a multistory building.*

VDV Floor Plans

A floor plan is a drawing that provides an overhead view of a floor of a building. A typical commercial floor plan shows cellular metal floor raceways, elevator shafts, equipment locations, doors, modular furniture partitions and power poles, plumbing fixtures, rooms, stairways, walls, and windows. A description and/or room number is shown for each room, including closets and corridors.

A VDV floor plan shows these items and also shows the location of VDV outlets and the telecommunications/data equipment room (when present). **See Figure 8-7.** VDV floor plans are drawn for both commercial and residential projects, and contain much of the same information. However, there are also significant differences between the two.

Appliances are available today that can be connected to a network. In order to make these appliances compatible with a VDV network, a residence should have VDV outlets and UTP cables installed near the refrigerator, laundry area, oven, and cooktop.

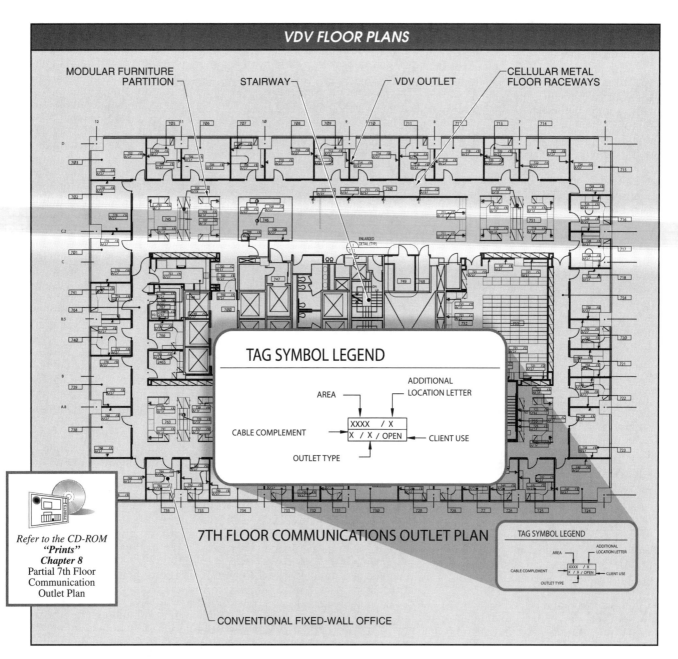

Figure 8-7. VDV floor plans for large commercial and industrial projects show the location of VDV outlets, but not power outlets or other devices typically shown on floor plans.

Commercial VDV Floor Plans. Commercial VDV floor plans are drawn for office buildings, banks, schools, retail stores, and other commercial buildings. Large commercial projects have a separate set of floor plans for VDV systems, with a floor plan drawn for each floor. More than one VDV floor plan may be necessary when a floor is extremely large. Small commercial projects include VDV system elements on the power floor plan. **See Figure 8-8.**

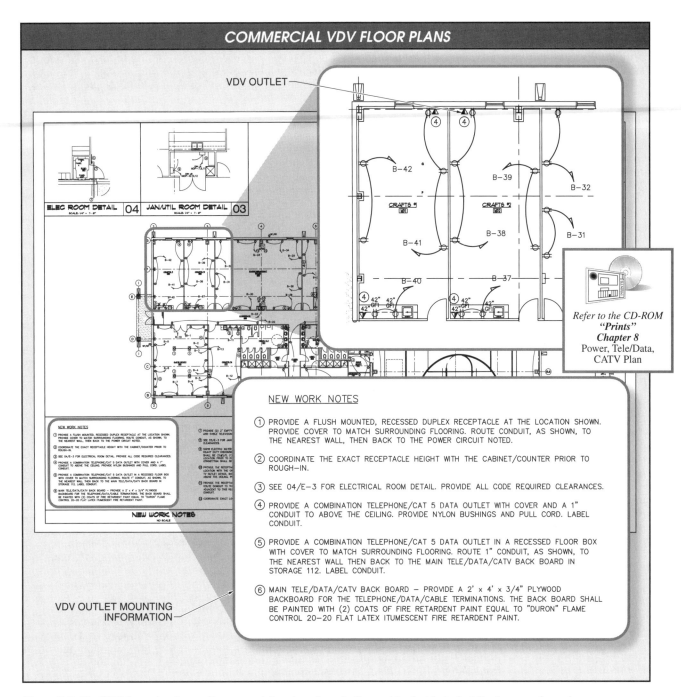

Figure 8-8. The VDV floor plan for small commercial projects is typically combined with the building's power floor plan.

It is important when connecting cabling to patch panels that cables (Cat 5e) be pulled from a box without kinking, be pulled with less than 25 lb of force, be supported to prevent stress, and not be tied so tight with cable ties as to distort the jacket of the cable.

Typically, all VDV outlets in a commercial building are mounted at the same height, with the location of each indicated on the print. The prints or specifications provide the mounting height and whether the outlets are to be installed exactly per a dimension or to the nearest stud. Some prints provide exact dimensions for every VDV outlet. In general however, dimensions are only shown for VDV outlets that are not considered typical installations.

In addition to the location of VDV outlets, commercial VDV floor plans contain sheet notes. Sheet notes provide information about VDV outlet addressing and installation of VDV components. Unlike power devices and outlets on a power floor plan, VDV outlets do not have lines representing communication cables drawn from each outlet back to the telecommunications/data equipment room. The number and type of cables from each outlet can be found in the specifications, as part of the outlet symbol, or as a sheet note.

A typical commercial VDV installation consists of four unshielded twisted pair (UTP) cables run from the telecommunications/data equipment room to each outlet. Each outlet has four separate jacks. The UTP cables are used for voice and/or data. **See Figure 8-9.**

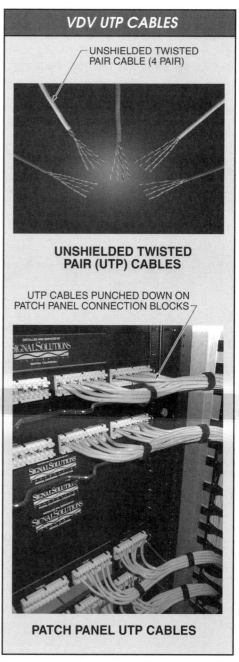

Figure 8-9. Unshielded twisted pair (UTP) cables used for voice and data transmission terminate at connector blocks on the rear of patch panels.

Residential VDV Floor Plans. In general, there is only one floor plan for a residential construction project. A residential VDV floor plan is shown on the project floor plan. Like a commercial VDV floor plan, the residential project floor plan provides the location of VDV outlets. **See Figure 8-10.**

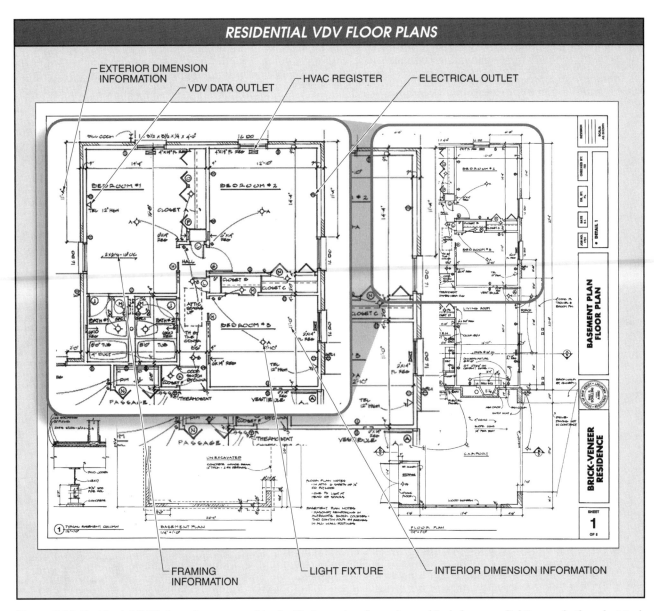

Figure 8-10. Residential VDV floor plans are combined with the project floor plan and include power, lighting, and other electrical, mechanical, and architectural elements.

Because it is a project floor plan, the residential VDV floor plan may show the locations of electrical outlets and devices, light fixtures and switches, and HVAC registers. It may also show interior and exterior building and room dimensions and give information on plumbing fixtures, doors and windows, floor and/or roof framing, and floor and wall coverings.

As on a commercial project, most residential VDV outlets are mounted at the same height, with the location of each being indicated on the print. Information on VDV outlet mounting height and mounting location is per a dimension or to the nearest stud. Unlike with commercial VDV prints, VDV outlet addressing information is not provided.

A common residential VDV installation consists of two UTP cables and two coaxial cables that run from a network center to a single outlet in each room. **See Figure 8-11.**

A network center serves as the connection point for cables from the local telephone company, local cable provider, and the UTP and coaxial cables routed to various rooms in a residence. The VDV outlets in the rooms are designed to accommodate multiple jacks of various configurations (voice, data, and/or video).

UTP cables are used for voice or data, and the coaxial cable is used for video. Certain cable makers manufacture a cable with two UTP cables and two coaxial cables under a common insulation jacket for residential applications. **See Figure 8-12.** This combination cable speeds up installation time and helps with cable identification. Some residential projects also include cable for speakers and home-theater systems.

Figure 8-11. A common residential VDV installation consists of unshielded twisted pair (UTP) and coaxial cables that run from a network center to each room with an outlet. Each outlet has multiple jacks of various configurations (voice, data, and/or video).

Figure 8-12. A combination cable has two individual UTP cables and two individual coaxial cables under a common insulation jacket and is used for residential applications only.

VDV Detail Drawings

VDV detail drawings provide additional information that cannot be shown on VDV riser diagrams or VDV floor plans.

The number and type of VDV detail drawings depends on the size of the construction project. A small commercial office building may have a few detail drawings. A large campus with multiple buildings would have many detail drawings. VDV detail drawings are rarely found on prints for single-family residences.

The most common types of VDV detail drawings include equipment room details, VDV outlet details, and fire-stopping details. The precise information helps ensure that a VDV installation meets all applicable codes and all customer requirements.

> As of 2005, the IEEE Ethernet committee added the "af" provisions to the 802.3 Ethernet standards for powering VDV devices off the spare pairs in a 4-pair UTP cable. Since Ethernet up to 100Base-TX uses pairs 2 and 3, pairs 1 and 4 are available to provide power. Pair 1 (pins 4/5) are the + conductors, pair 4 (pins 7/8) are the – conductors.

Equipment Room Details. A telecommunications/data equipment room detail shows backbone and vertical cabling, connector blocks, cable management and support devices, telephone equipment, and network equipment. Several equipment room detail drawings are required for complex installations. VDV technicians must consult equipment room detail drawings before installing any equipment.

Equipment room detail drawings are needed to show the location and mounting method for VDV equipment. Detail drawings show where and how raceways are mounted in ceiling spaces and between floors. Elevations and plan views are used to show the location of cable ladders, equipment racks, termination backboards, connector blocks, and raceways. **See Figure 8-13.** A legend is supplied to avoid cluttering the drawings.

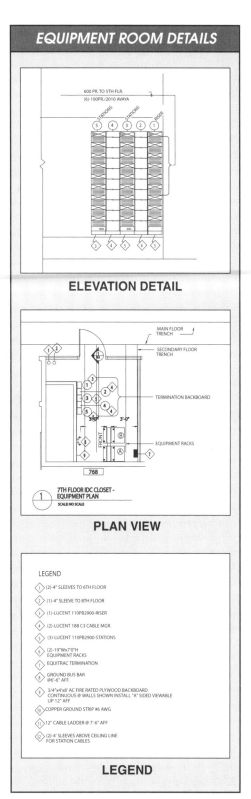

Figure 8-13. Equipment room details include elevations, plan views, and legends that show the locations of cable ladders, equipment racks, termination backboards, connector blocks, and raceways.

VDV Outlet Details. VDV outlets can be single or double gang outlets and can be wall mounted, floor mounted, or mounted on modular furniture. Also, the number and types of jacks can vary. A VDV outlet with two telephone jacks and two data jacks is typically found in commercial installations.

VDV outlet details provide specific information for each type of VDV outlet, such as the exact termination scheme for the UTP cable to the jacks, the outlet-addressing format being used, the VDV outlet symbol, and the number and type of cables and jacks. Some VDV outlet details show cable part numbers, faceplates, and jacks.

See Figure 8-14. Additional information found in VDV outlet details includes information on the type of label used for the faceplate of a VDV outlet, such as size of the label, label font, and label font size.

Fire-Stopping Details. Holes (penetrations) in fire-rated walls and floors must be sealed to maintain the required fire rating. This procedure is called fire-stopping and is an extremely important element of any VDV project. There are many variables to consider when choosing a fire-stopping method, such as the initial rating of the wall or floor, the material the wall or floor is constructed of, and the type of raceway penetrating the wall or floor.

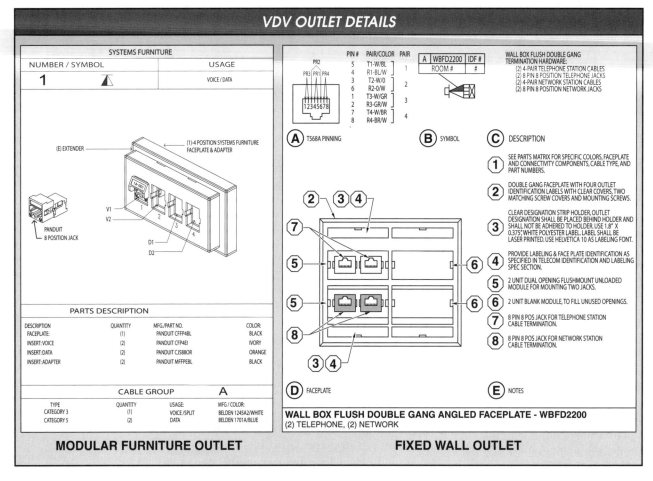

Figure 8-14. VDV outlet details provide information on the number and type of jacks, termination, and labeling for each type of VDV outlet.

Fire-stopping details provide a section view of the fire-rated wall or floor and the raceway penetration. **See Figure 8-15.** Fire-stopping details also provide exact information for each particular fire-stopping situation. Information includes the type of floor or wall, dimensions for locating the opening, the maximum permissible diameter of the opening, the type of raceway penetrating the wall of floor, and the type of fire-stopping material used (pliable putties, caulks, foams, fireblocks, pillows, and cement type compounds) and how to apply it.

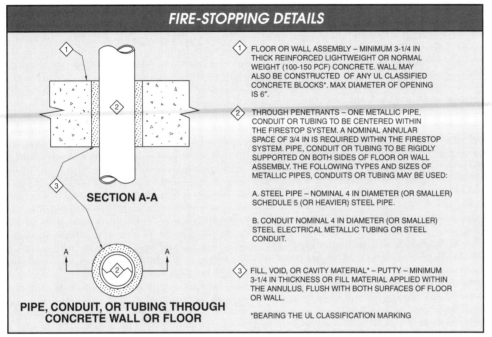
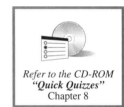

Figure 8-15. Fire-stopping details provide information on how to correctly fire-stop penetrations in walls and floors.

Review Questions and Activities

Residential and Commercial VDV Systems

Name _____ Date _____

True-False

T F 1. Each item in a VDV system legend is associated with a number or letter that is located on the print where the information applies.

T F 2. VDV riser diagrams always provide the size of the conduit or raceway for the backbone cables and show the exact route of the cables.

T F 3. VDV floor plans are drawn for both commercial and residential projects and contain much of the same information.

T F 4. Most residential VDV outlets are mounted at the same height, with the location of each being indicated on the print.

T F 5. VDV riser details provide general information for each type of VDV outlet.

T F 6. A basic residential VDV system is typically shown on a residential floor plan along with power and lighting information.

T F 7. VDV abbreviations are specific to cable TV, telecommunications, and wiring systems.

T F 8. Sheet notes are always included with VDV abbreviations and symbols, but can be on a separate print if necessary.

T F 9. Residential VDV floor plans are drawn for office buildings, banks, schools, retail stores, and other such buildings.

T F 10. Additional information found in VDV outlet details includes precise information on the type of label used for the faceplate of a VDV outlet, such as size of the label, label font, and label font size.

Completion

_____ 1. A set of ___ includes a list of symbols, abbreviations, and a legend that are used throughout the set of prints.

_____ 2. A common ___ VDV installation consists of two UTP cables and two coaxial cables that run from a distribution cabinet (network center) to a single outlet in each room.

_____ 3. ___ detail drawings are needed to show the location and mounting method for VDV equipment.

_____ 4. ___ and elevations are used to show the location of cable ladders, equipment racks, termination backboards, connector blocks, and raceways.

_____ 5. VDV ___ can be single or double gang, and can be wall mounted, floor mounted, or mounted on modular furniture.

Multiple Choice

_____ 1. A(n) ___ is a unique system of letters and/or numbers used for identifying VDV cables and their termination locations.
   A. VDV tag
   B. outlet address
   C. plot plan
   D. abbreviation

_____ 2. VDV ___ provide an overview of the backbone cabling in a building.
   A. power floor plans
   B. electrical diagrams
   C. elevation plans
   D. riser diagrams

_____ 3. ___ provide information about VDV outlet addressing and installation of VDV components.
   A. Path panels
   B. Floor plans
   C. Sheet notes
   D. all of the above

_____ 4. VDV ___ provide additional information that cannot be shown on VDV riser diagrams or VDV floor plans and provide a high degree of precise information.
   A. detail drawings
   B. plot plans
   C. section view drawings
   D. single-line diagrams

_____ 5. The number and type of VDV detail drawings depends on the ___ of the construction project.
   A. cost
   B. size
   C. location
   D. none of the above

Activity—Planning Residential VDV Installation 8-1

Scenario:
Activity 8-1 is an installation activity where an electrician must review the floor plan of a residence for VDV outlets required in each room of the house. VDV outlets can be combined in any combination of voice, data, and video using composite cables, which saves on installation costs.

Task:
Determine where in the residence combination VDV outlets can be used and where to use composite cables to feed VDV boxes.

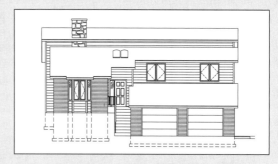

STEWART RESIDENCE — SOUTH ELEVATION

Required VDV Information:
1. What is the total number of telephone outlets required? _____
2. What is the total number of data outlets required? _____
3. What is the total number of cable TV outlets required? _____
4. Are there any conditions where VDV outlets in adjacent locations can use composite cables to feed the VDV boxes? _____
5. List any locations where VDV outlets can be combined and a composite cable used to feed single VDV outlets. _____

Reference Prints:

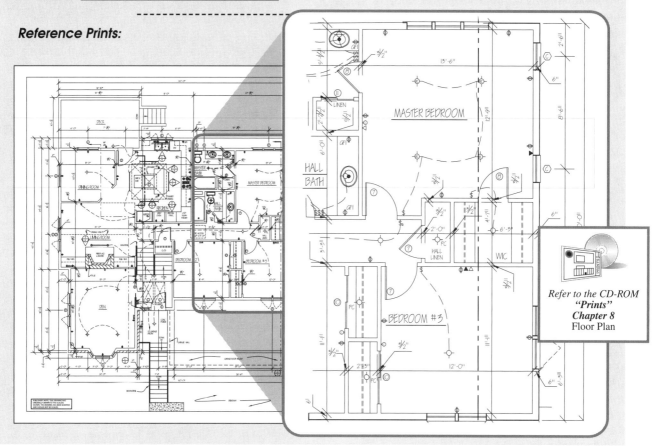

Refer to the CD-ROM "Prints" Chapter 8 Floor Plan

Activity—Planning Residential VDV Installation

STEWART RESIDENCE — FLOOR PLAN ENLARGEMENT

COMBINATON VDV OUTLET
Cooper Wiring Devices

DUAL COMBINATON VDV OUTLET

COMPOSITE VDV CABLE

Activity—Planning Commercial VDV Installation

8-2

Scenario:
Activity 8-2 is an installation activity where an electrician must install VDV support equipment in an IDC closet using prints and multiple detail drawings.

Task:
Determine which VDV items need to be installed and where the items will be located in the 7th floor IDC closet.

Required IDC Closet Information:
1. What is the manufacturer name and part number for the connector block assembly where the voice riser cable terminates? _____
2. How wide is the cable ladder that will be installed in the 7th floor IDC closet? _____
3. What is the mounting height of the cable ladder? _____
4. How are the horizontal cables that are routed through the 7th floor ceiling brought into and out of the 7th floor IDC closet? _____
5. What is the mounting height of the ground bus in the 7th floor IDC closet? _____
6. What is the distance between the top of the equipment rack and the bottom of the cable ladder? _____

7th FLOOR - IDC CLOSET

Reference Print

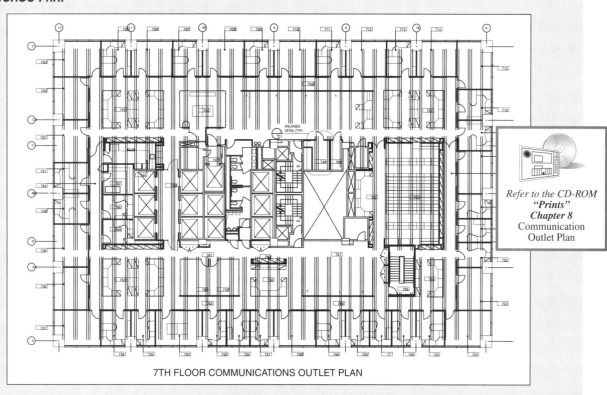

7TH FLOOR COMMUNICATIONS OUTLET PLAN

Refer to the CD-ROM "Prints" Chapter 8 Communication Outlet Plan

Activity—Planning Commercial VDV Installation

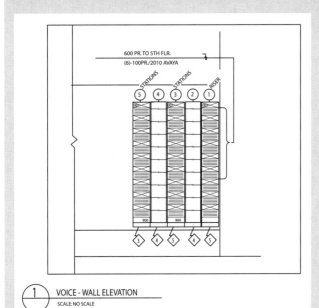

1 VOICE - WALL ELEVATION
SCALE: NO SCALE

LEGEND
1. (2)-4" SLEEVES TO 6TH FLOOR
2. (1)-4" SLEEVE TO 8TH FLOOR
3. (1)-LUCENT 110PB2900-RISER
4. (2)-LUCENT 188 C3 CABLE MGR.
5. (3)-LUCENT 110PB2900-STATIONS
6. (2)-19"Wx7'0"H EQUIPMENT RACKS
7. EQUITRAC TERMINATION
8. GROUND BUS BAR @6'-6" AFF.
9. 3/4"x4'x8' AC FIRE RATED PLYWOOD BACKBOARD CONTINUOUS @ WALLS SHOWN INSTALL "A" SIDED VIEWABLE UP 12" AFF
10. COPPER GROUND STRIP #6 AWG
11. 12" CABLE LADDER @ 7'-6" AFF
12. (2)-4' SLEEVES ABOVE CEILING LINE FOR STATION CABLES

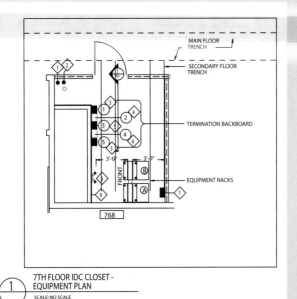

1 7TH FLOOR IDC CLOSET - EQUIPMENT PLAN
SCALE: NO SCALE

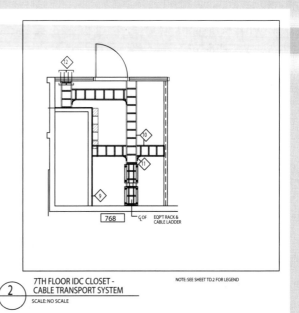

2 7TH FLOOR IDC CLOSET - CABLE TRANSPORT SYSTEM
SCALE: NO SCALE

NOTE: SEE SHEET TD.2 FOR LEGEND

Trade Competency Test
Residential and Commercial VDV Systems

Name _____ Date _____

REFERENCE PRINT #1 (Riser Diagram)

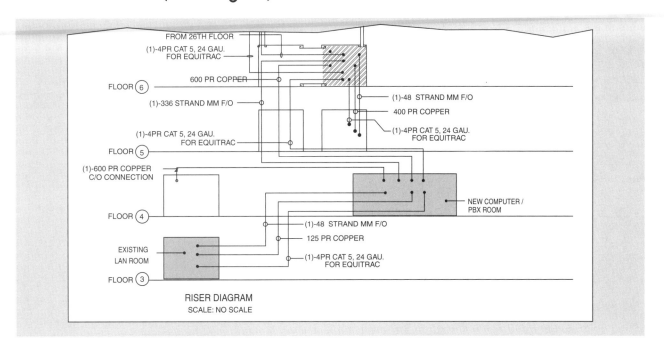

Questions – Reference Print #1

T F 1. The dimensions for the new Computer/PBX Room are shown on the print.

T F 2. The riser diagram is drawn to scale.

T F 3. The 125 pair copper cable that runs between the existing LAN Room and the new Computer/PBX Room is 24 gauge.

_____ 4. The existing LAN Room is located on the ___ floor.

_____ 5. A(n) ___ pair copper cable is the C/O connection to the new Computer/PBX room.

_____ 6. A 4 pair CAT 5 cable for ___ runs between the existing LAN Room and the new Computer/PBX Room.

_____ _____ 7. A(n) ___ strand and ___ strand multi-mode fiber optic cable exits the new Computer/PBX Room to the floors above.

REFERENCE PRINT #2 (Power, Tele/Data, and CATV)

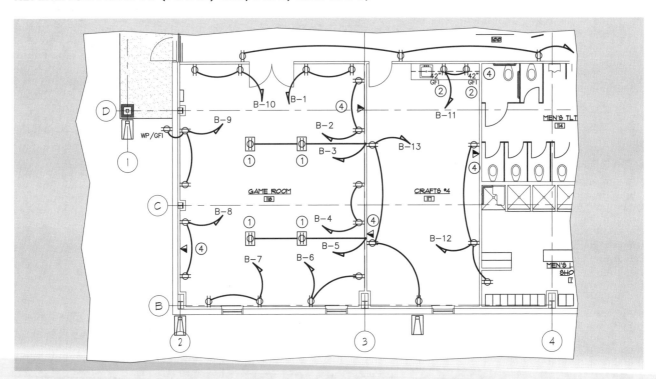

Questions – Reference Print #2

_____ 1. The Game Room is room number ___.

_____ 2. The WP/GFI receptacle on the exterior wall of the Game Room is fed power from circuit ___.

T   F 3. Note 4 refers to the telephone/data outlets.

T   F 4. There are multiple entrances to the Game Room.

_____ 5. The game room has ___ telephone/data outlets.

_____ _____ 6. The Game Room is located between column number ___ and column number ___.

_____ _____ 7. Note 1 refers to receptacles fed power from circuit ___ and ___.

_____ 8. The circuits feeding the Game Room have a maximum of ___ receptacles each (not including the WP/GFI receptacle).

Advanced CD-ROM Print Questions (Power, Tele/Data, and CATV)

_____ 1. ___" conduit is used for the telephone service entrance and the cable television service entrance.

_____ 2. Panel MDP is located in room number ___.

____ ____ ____ 3. Electric water heater #1 is fed power from Panel MDP, circuits ___, ___, and ___.

_____ 4. The Utility/Janitor Room is located in room number ___.

Fire Alarm, Life Safety, and Security Systems

Printreading for Installing and Troubleshooting Electrical Systems

Fire alarm, life safety, and security systems are found in residential, commercial, and industrial projects. Fire alarm and life safety systems are installed to protect personnel and property from fire. Security systems are installed to protect building occupants, building contents, and the building itself from unauthorized entry, harm, or theft. There are a wide variety of fire alarm, life safety, and security systems available, and these systems are shown on many different types of prints.

CSI MASTERFORMAT™— DIVISION 28

Fire alarm and life safety prints include riser diagrams, floor plans, and detail drawings. Security system prints include floor plans and detail drawings. The 2004 edition of the Construction Specifications Institute's MasterFormat™ has a division for fire alarm, life safety and security systems: Division 28—Electronic Safety and Security. **See Figure 9-1.** Prior to the 2004 edition of the CSI MasterFormat™, fire alarm and life safety systems were included in the Division 16—Electrical.

FIRE ALARM AND LIFE SAFETY SYSTEMS

A *fire alarm system* is a fire protection system designed to emit a sound and/or signal to building occupants and fire fighting personnel on detection of a fire by combustion monitoring devices. A *life safety system* is a building or facility system that aids occupants and fire fighting personnel in the event of a fire, terrorist attack, or other catastrophe that could endanger the lives of building personnel. Fire alarm and life safety systems both use initiating devices and notification components. **See Figure 9-2.** Initiating devices include pull stations, smoke detectors, and sprinkler flow switches. Notification components include horns, strobes, combination horn/strobes, and bells.

Typically life safety systems are found in buildings that are above a certain height. Height is used as a system criterion because taller buildings present special challenges to fire department personnel and equipment.

Refer to the CD-ROM "CSI MasterFormat" Numbers & Titles

280 Printreading for Installing and Troubleshooting Electrical Systems

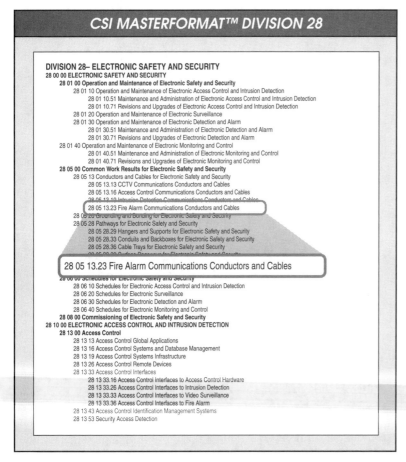

Figure 9-1. Division 28 of the 2004 CSI MasterFormat™ covers fire alarm, life safety, and security systems.

Alarms that are part of a fire detection system typically notify the local fire department automatically when a fire suppression system is activated.

Many fire alarm and life safety systems have addressable devices. The addresses are set by dip switches or hexadecimal switches, or by using programming devices. Each device has a unique address and is connected to the fire alarm control panel via a data communication line. The status of each individual unit is monitored by the control panel.

Fire Alarm and Life Safety Abbreviations and Symbols

Fire alarm and life safety prints include a list of abbreviations and symbols on the legend sheet. The abbreviations and symbols are used throughout the set of prints. The abbreviations and symbols provide detailed information without cluttering the print.

Fire alarm and life safety prints have abbreviations that are unique to these safety systems. **See Figure 9-3.** Some of the abbreviations used in fire alarm and life safety prints are also used in power and lighting prints. Other fire alarm and life safety abbreviations are only found on prints for a retrofit or replacement project.

A specific set of symbols is used to depict fire alarm and life safety system devices. There is a significant variation in these symbols. In fire alarm or life safety systems, a device can be represented by two different symbols. **See Figure 9-4.** The difference between the two symbols may be slight. For example, a smoke detector can be represented by the letters "SD" inside a square or the letter "S" inside a hexagon. On the other hand, some conditions make the difference between the symbols extreme, such as the choice of graphic used, and the symbols bear no resemblance to each other.

There are a number of reasons for the variation in fire alarm, life safety, and security system symbols. The fire alarm and life safety equipment provider may have developed specific symbols; the symbols

Chapter 9—Fire Alarm, Life Safety, and Security Systems   **281**

for addressable system devices may need to convey more information; or the symbols may be a result of the personal preference of the fire protection engineer.

Symbols may be modified with letters or numbers to provide additional information. **See Figure 9-5.** The additional information can include, for example, the light output of a strobe in candelas (cd), the sound output of a horn in decibels (db), and special enclosures for outdoor locations, such as weatherproof (WP).

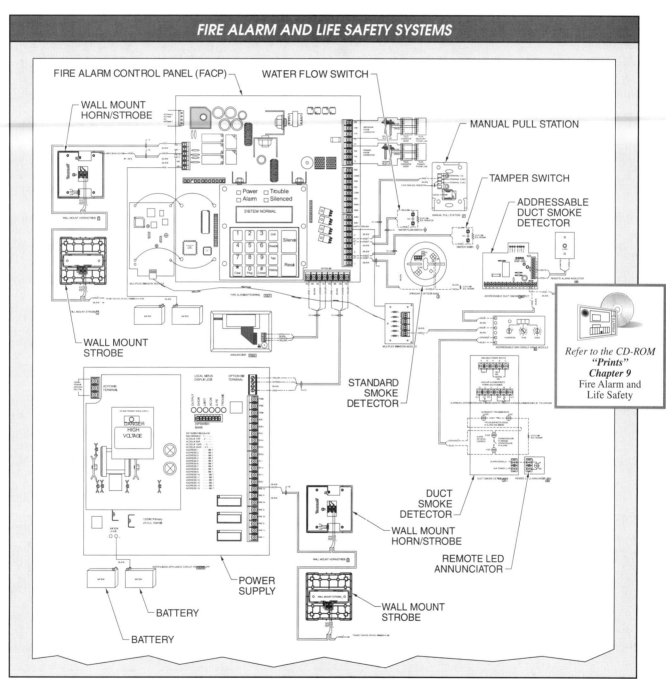

Benson Systems, Inc.

Figure 9-2. *Fire alarm and life safety systems include a fire alarm control panel, pull stations, notification components (horns), fan shutdown circuits for smoke control, elevator recall, and fireman's phone systems.*

FIRE ALARM AND LIFE SAFETY ABBREVIATIONS

RETROFIT

(E)	Existing
(ED)	Existing to be demolished
(ER)	Existing to be replaced
(N)	New
(P)	Plenum mounted device
(R)	Relocated
(RD)	Device to be relocated
(RR)	Remove and replace

CONSTRUCTION

AFF	Above finished floor
ALRM	Alarm
BPM	Beats per minute
BOM	Bill of materials
CONC	Concrete
DET	Detector
ECP	Extinguishing control panel
EOL	End of line resistor
EOLR	End of line relay
EMO	Emergency machine off switch
EPO	Emergency power off
EP	Explosionproof
F	Flush mounted
FACP	Fire alarm control panel
HT	Height
HVAC	Heating ventilation & air conditioning
LA	Low air
MAX	Maximum
MIN	Minimum
N/A	Not applicable
N/C	Normally closed
N/O	Normally open
NIC	Not in contract
NOC	Network operation center
NTS	Not to scale
S	Surface mounted
SMK	Smoke
SPRV	Supervisory
TRBL	Trouble
TYP	Typical
UON	Unless otherwise noted
VT	Valve tamper
WF	Waterflow
WP	Weatherproof

Figure 9-3. Common electrical abbreviations, fire alarm abbreviations, and life safety abbreviations are found on fire alarm and life safety prints.

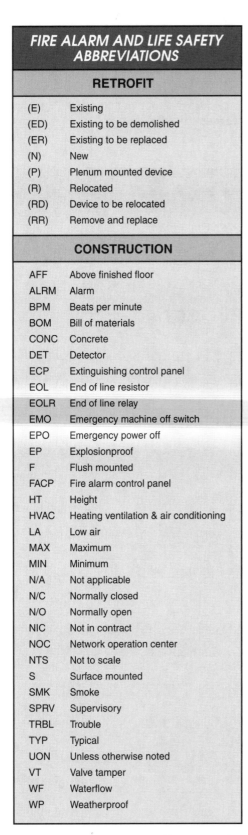

Figure 9-4. Due to the wide variety of symbols used, electricians must familiarize themselves with fire alarm and life safety system symbols and prints.

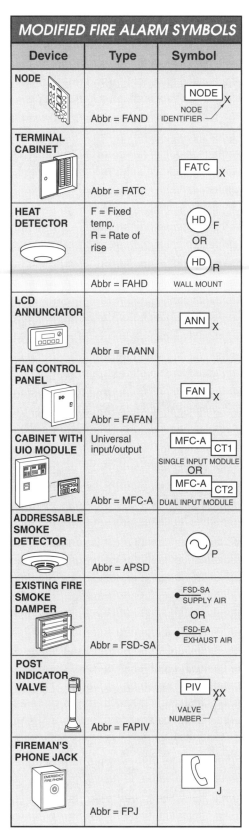

Figure 9-5. Fire alarm symbols are modified with letters or numbers to provide additional information.

FIRE ALARM AND LIFE SAFETY WIRING

There are several methods used to install conductors and cables for fire alarm and life safety systems. The methods include using steel conduit (EMT), fire alarm and life safety metal-clad cable (MC cable), and fire alarm and life safety cable. The specific method used depends on local building codes and project specifications.

Fire alarm and life safety EMT has a bright red topcoat for instant identification (other colors are available for VDV systems) and has an E-Z Pull® interior finish that provides a smooth raceway for wire pulling.

Fire alarm and life safety MC cable is color-coded bright red with solid copper conductors. MC cable is fully plenum-rated, available with twisted shielded pairs, and ready for hazardous locations (Class I, Division 2; Class II, Division 2; and Class III, Divisions 1 and 2).

Fire alarm and life safety cable is available in a wide variety of gauge sizes, shielding configurations, and jacketing materials.

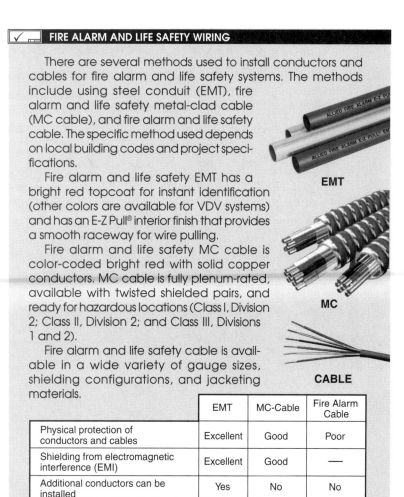

	EMT	MC-Cable	Fire Alarm Cable
Physical protection of conductors and cables	Excellent	Good	Poor
Shielding from electromagnetic interference (EMI)	Excellent	Good	—
Additional conductors can be installed	Yes	No	No
Installation cost	Most	Moderate	Least

Fire Alarm and Life Safety Riser Diagrams

Riser diagrams are typically provided for fire alarm and life safety installations in multistory buildings. A riser diagram for a multistory building shows all of the floors in the building. Multistory buildings may be commercial office buildings, high-rise residential condominiums, retail stores, or manufacturing facilities. On occasion, riser diagrams are also provided for large single-story buildings.

A riser diagram for a fire alarm and life safety system shows all of the fire alarm and life safety devices and equipment located on each floor. **See Figure 9-6.** When there are several devices of the same type and limited print space, a riser diagram typically shows one device with the abbreviation "TYP" adjacent to the device.

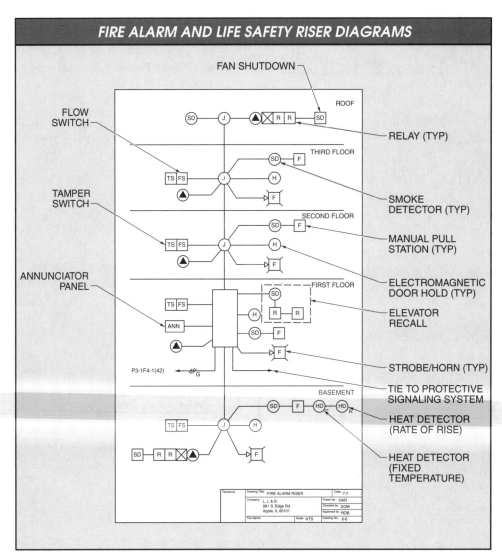

Figure 9-6. A fire alarm and life safety riser diagram provides a view of all the floors in a building and the fire alarm and life safety devices and components on each floor.

Lines are used to represent the conduit/conductors that connect the devices together on each floor and connect one floor to another. The exact location of the devices and the conduit route are not shown on riser diagrams. The type of conductors or cables may or may not be shown on a riser diagram. Fire alarm and life safety riser diagrams for multistory buildings are not drawn to scale.

Occasionally, riser diagrams are provided for large single-story buildings. **See Figure 9-7.** Single-story riser diagrams are similar to elevation drawings and depict all of the rooms and locations that contain fire alarm and life safety devices and components. Instead of a vertical riser, lines are used to represent the conduit/conductors that connect the devices and components together on a single floor. The exact location of the devices and the route of the conduit are not shown on a riser diagram. The type of conductors or cables may or may not be shown. Fire alarm and life safety riser diagrams for single-story buildings are not drawn to scale.

Chapter 9—Fire Alarm, Life Safety, and Security Systems **285**

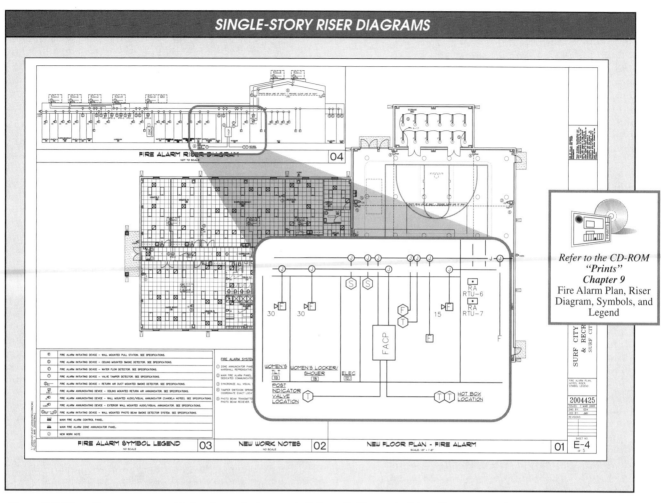

Figure 9-7. Riser diagrams are provided on large single-story buildings. Unlike vertical riser diagrams, single-story riser diagrams show the individual rooms in which devices and components are located.

Additional information found on a fire alarm and life safety riser diagram includes sheet notes covering specific installation requirements; the power source for the fire alarm control panel (FACP), panel designations and circuit numbers; and the connection of the FACP to the monitoring company or agency.

Fire Alarm and Life Safety Floor Plans

A typical floor plan for a commercial building indicates elevator shafts, equipment locations, doors, modular furniture partitions, plumbing fixtures, rooms, stairways, walls, and windows. A description and/or room number is shown for each room, including closets and corridors. Large commercial projects generally have a separate set of floor plans for fire alarm and life safety systems, with each floor having a separate plan. Small projects often include fire alarm and life safety system information on the power floor plan.

A fire alarm and life safety system floor plan shows the location of initiating devices, notification components, control panels, and equipment that the system interfaces with, such as fire smoke dampers and electric door locks. Lines may or may not be drawn on a fire alarm and life safety floor plan to depict the conduit/conductors that connect the devices and components together. **See Figure 9-8.**

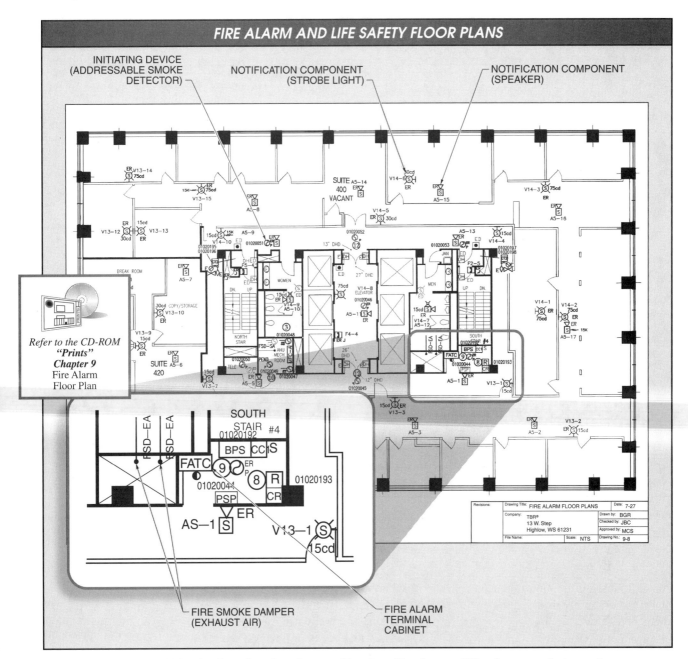

Figure 9-8. Fire alarm and life safety floor plans show the general location of fire alarm and life safety system devices and components.

The mounting location, mounting height, and number of initiating devices and notification components are governed by very specific codes and regulations. Electrical codes include the National Electrical Code® (NEC®), the National Fire Alarm Code®, (NFPA 72), and local building codes. Initiating devices such as smoke detectors must be located where the devices can detect smoke at the earliest onset of a fire. Notification components, such as horns and strobes, must be located where the components can be seen and heard by all of the occupants of a floor. The location of notification components must also take into account the needs of the disabled, hearing-impaired, sight-impaired, and those in wheelchairs.

Most fire alarm and life safety floor plans provide a general location for initiating devices and notification components. The exact location is obtained by consulting the sheet notes, specifications, and detail drawings. Electricians typically have to consult mechanical prints for the proper placement of devices that mount to mechanical systems, such as duct smoke detectors and water flow switches. In addition to sheet notes, fire alarm and life safety floor plans have legends for symbols, cable types, and device and component addressing. **See Figure 9-9.**

Some fire alarm and life safety prints use a reflected ceiling plan to show the location of devices and components on a specific floor. A reflected ceiling plan is a plan view with the viewpoint located beneath the object. In addition to showing acoustical tile (drop-in ceiling tile), return and supply air diffusers, exit signs, fluorescent lights, recessed-down lights, and fire sprinkler heads, a reflected ceiling plan also shows the location of initiating devices, notification components, control panels, and interface equipment. Sheet notes and a symbol legend are also included on a reflected ceiling plan. **See Figure 9-10.** Lines depicting the conduit/conductors connecting the fire alarm and life safety equipment together are not shown on a reflected ceiling plan.

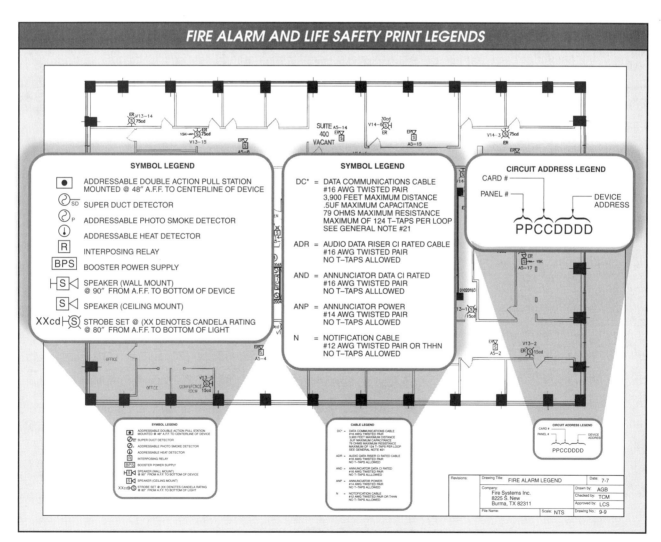

Figure 9-9. Along with the sheet notes, symbol, cable, and address legends are all found on fire alarm and life safety floor plans.

288 Printreading for Installing and Troubleshooting Electrical Systems

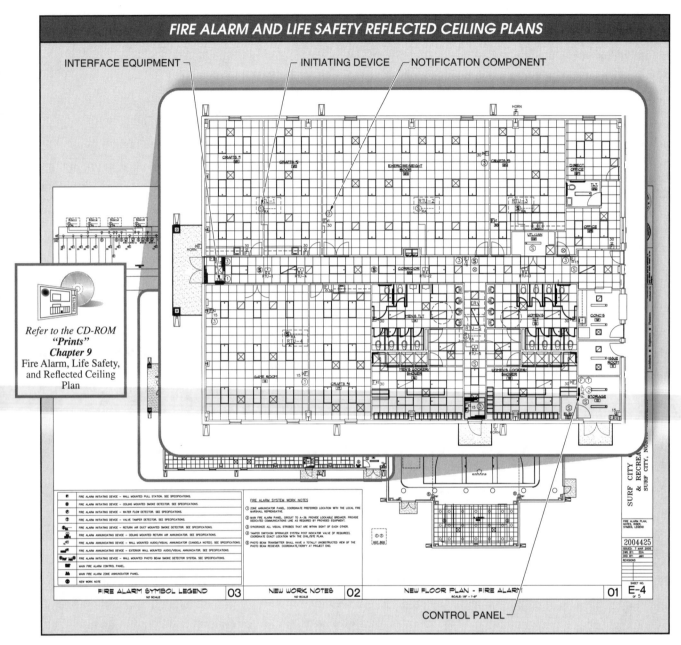

Figure 9-10. Reflected ceiling plans are used in place of floor plans when drop ceilings are used to show the location of fire alarm and life safety devices and components.

Fire Alarm and Life Safety Detail Drawings

Fire alarm and life safety detail drawings provide additional information that does not fit on fire alarm and life safety riser diagrams or floor plans. The amount and types of fire alarm and life safety detail drawings depend on the size of the construction project. If the project is large, many detail drawings are needed. Common types of fire alarm and life safety detail drawings include splicing details, mounting details for initiating devices and notification components, fire-stopping details, fire smoke damper installation notes, device and component wiring details, and fan status notes.

Fire alarm and life safety prints typically have a large number of detail drawings.

For example, the placement of duct smoke detectors varies depending on the location of fire smoke dampers, air inlets, and bends in the duct. Fire alarm and life safety detail drawings provide a high degree of specificity. Fire alarm and life safety detail drawings ensure that an electrician knows exactly how a system is installed and that the installation meets applicable codes.

Mounting Detail Drawings. Mounting detail drawings are provided for most fire alarm and life safety devices and components. Detail drawings provide information on device and component dimensions, type of box a device or component is mounted to, part numbers, and explanatory notes. Fire alarm and life safety mounting details may or may not be drawn to scale. **See Figure 9-11.**

> To avoid confusion on multifunctional prints (power floor plan, fire alarm, and life safety), highlight the power floor plan elements, fire alarm, and life safety elements in different colors.

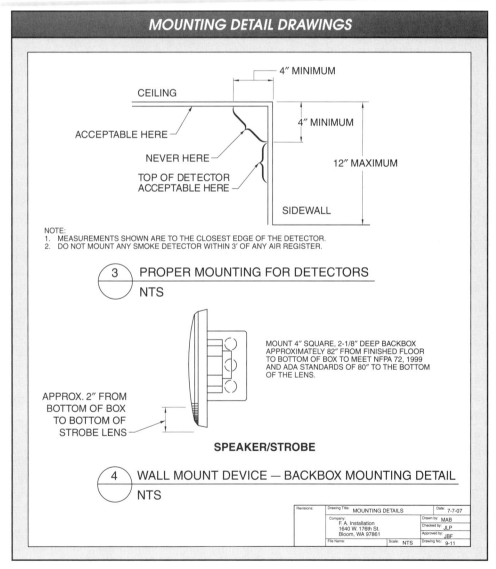

Figure 9-11. Mounting details are typically provided for fire alarm and life safety initiating devices and notification components.

Smoke detectors have the greatest number of detail drawings because of their numerous locations and applications. General mounting requirements for smoke detectors include the location of the smoke detector, such as a wall or ceiling. There are also specific requirements for the number and location of smoke detectors based on the design of a building. Building design requirements come into play when components such as smoke detectors are used for fire-door control.

Wiring Detail Drawings. Different types and sizes of conductors and cables are used for fire alarm and life safety systems. Wiring details are found on most fire alarm and life safety prints. Wiring details provide information on how conductors and cables are spliced and how the conductors and cables are terminated to specific devices and components. Separate wiring details are provided for each type of conductor termination. Terminations vary by conductor or cable used and by the device or component the conductor terminates at. **See Figure 9-12.**

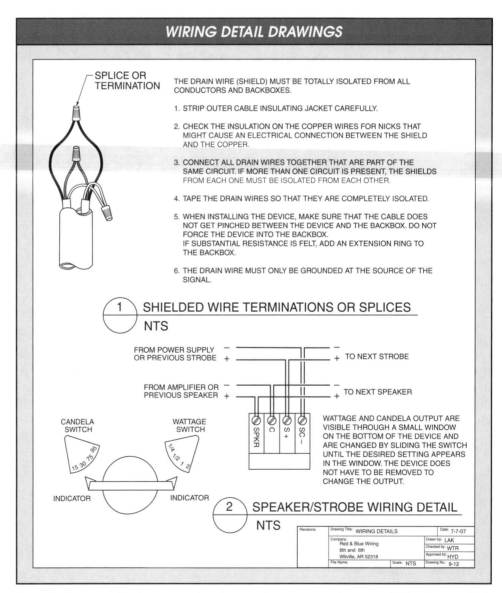

Figure 9-12. Wiring details give specific splicing and termination information for fire alarm and life safety conductors and cables.

Wiring details also include switch settings for devices and components. Switch settings are used to set the sound output of a speaker or the light output of a strobe. Close attention must be paid to wiring details because wiring mistakes are the most common errors made in the installation of fire alarm and life safety systems.

> Security closed-circuit television (CCTV) applications with multiple cameras should be powered from a central location. When cameras are powered from multiple locations, they may be powered from different phases of the power distribution system. The phase difference when powering different cameras can result in transmission problems.

SECURITY SYSTEMS

There are a wide variety of security systems available. Security systems range from simple home alarm systems to large systems with video cameras, biometric access control, and remote monitoring. The need for enhanced security in private and public buildings has grown in recent years, and security systems have extended to airports, highways, and seaports.

Security System Abbreviations and Symbols

A separate set of abbreviations and symbols is used to depict security system devices and components. Typically, the security system abbreviations and symbols used on a print are part of the American National Standards Institute (ANSI) symbol standards. Security system abbreviations and symbols depict cameras, card readers, motion detectors, security control panels, and other security-related equipment. **See Figure 9-13.**

SECURITY SYSTEM ABBREVIATIONS	
CPS	Camera power supply
CR	Card reader
MTV	CCTV monitor outlet
DB	Doorbell
DBZ	Door buzzer
DRCN	Door contact
DVR	Digital video recorder
ELDO	Electric door opener
ELDS	Electric door strike
ELMG	Electromagnetic lock
FPMPV	Flat panel monitor public view
GB	Glassbreak
HU	Hold-up button
IC	Intercom unit
ICM	Intercom master station
ICS	Intercom substation
KS	Keyswitch
MI	Master intercom and directory unit
ML	Magnetic lock-door alarm
MON	Monitor
MD	Motion detector
RB	Release button
DC	Security door contacts
SP	Surge protector
TA	Trailer alarm
TS	Touch sense bar

Figure 9-13. Security system abbreviations depict card readers, motion detectors, security door contacts, and other security-related equipment.

Although standards exist for security system symbols, some security system designers and manufacturers prefer to use their own sets of symbols. While some manufacturers' symbols resemble standard symbols, most possess minor differences. **See Figure 9-14.** Security system abbreviations and symbols with definitions are found on the legend sheet of a security system project.

Security System Floor Plans

The size and complexity of a security system determines whether or not a separate floor plan needs to be provided. A large, complex security system requires a separate security system floor plan. A small simple security system may be depicted on a power floor plan or on a fire alarm and life safety floor plan. A security system is more likely to be shown on a fire alarm and life safety floor plan than a power floor plan because the two systems are similar in nature.

A security system floor plan uses the plan view of the building floor. In addition to showing elevator shafts, doors, plumbing fixtures, rooms, walls, and windows, the security system floor plan also shows the location of door contacts, motion detectors, cameras, security control panels, and items that interface with the security system. **See Figure 9-15.** In order to conserve space, other drawings related to the security system, such as detail drawings, elevation drawings, and/or schedules, may be shown on the security system floor plan.

A security system floor plan provides information about the location and mounting of devices and components that are connected to the security system. Often the security system floor plans do not have lines drawn from the devices back to the security system control panel.

Special care must be taken to ensure that all devices and components are mounted according to the print. The effectiveness of the security system depends upon the proper placement of the devices. For example, a motion detector must be properly positioned in order to detect the presence of an intruder. Additional information that may be found on a security system floor plan includes security camera viewing angles, security system zones, manufacturer's name and part numbers for security-related items, security hardware for doors, and outdoor floodlight locations.

SECURITY SYSTEM SYMBOLS

Device	Type	Symbol
CCTV CAMERA 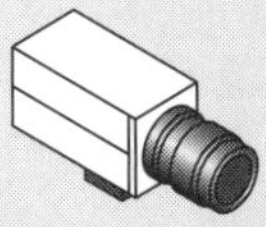	Closed circuit fixed camera Abbr = FCTV	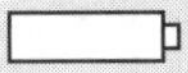
	Fixed camera in dome housing Abbr = FCTV	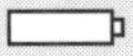
	Weatherproof exterior camera Abbr = TVWP	C WP
CCTV COAXIAL CABLE OUTLET 	Abbr = CCCO	CCTV
DOOR CHIME 	Abbr = DC	B
SIREN 	Abbr = SRN	
FLAT PANEL MONITOR	Abbr = FPM	OR PV PUBLIC VIEW
KEYPAD	Abbr = SKPD	KPD OR K
CONTROL PANEL 	Abbr = SCP	SCP
CARD READER	Abbr = SCR	CR OR CR WP

Figure 9-14. Symbols for access control and closed circuit TV (CCTV) that come from the manufacturer can possess minor differences.

Chapter 9—Fire Alarm, Life Safety, and Security Systems   293

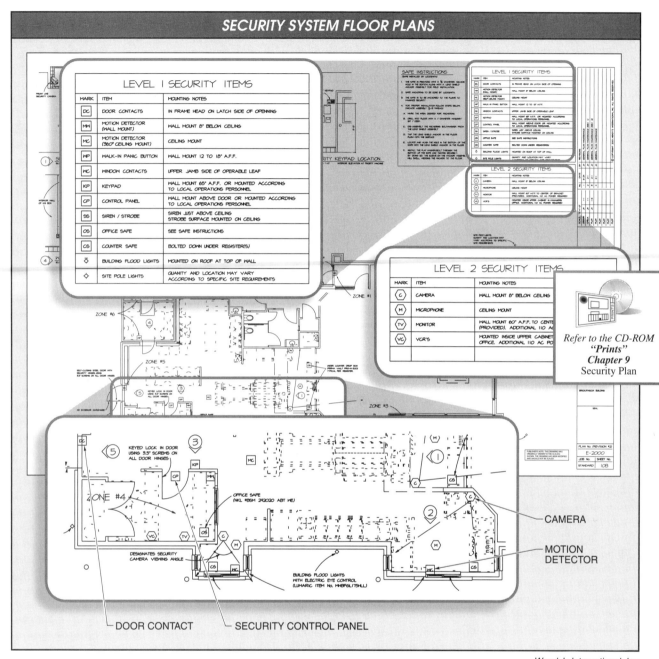

Figure 9-15. Security system devices and components are shown on security system floor plans.

Wendy's International, Inc.

NFPA 731, *Standard for the Installation of Electronic Premises Security Systems*, is the standard that covers the application, location, installation, performance, testing, and maintenance of physical security systems and their components.

Security System Detail Drawings

Security system detail drawings provide additional information that cannot be shown on a security system floor plan. Common types of security system detail drawings include elevations, installation instructions, and schedules. **See Figure 9-16.**

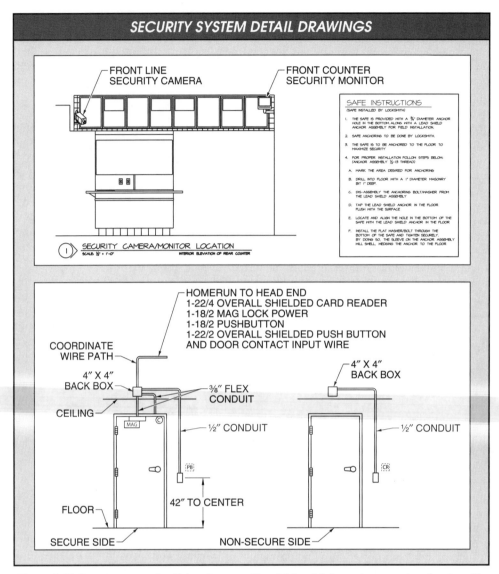

Figure 9-16. Security system detail drawings provide additional information to assist when installing or troubleshooting security systems.

Elevation drawings provide a view of how security devices, such as door contacts and electric door strikes, interface with building components. Elevation drawings may or may not be drawn to scale.

Installation instructions for security systems provide a step-by-step guide for installing a specific security-related item. The tradesperson must not deviate from the instructions unless authorized to do so by the security system designer or architect.

Schedules provide a summary of the installation information for security devices and components. A schedule shows the symbols for devices and components, symbol definitions, and relevant installation and mounting information.

Review Questions and Activities

Fire Alarm, Life Safety, and Security Systems

Name _____ Date _____

True-False

T F 1. Fire alarm and life safety prints include riser diagrams, floor plans, and detail drawings.

T F 2. A specific set of symbols is used to depict fire alarm and life safety system devices.

T F 3. Fire alarm and life safety riser diagrams for multistory buildings are always drawn to scale.

T F 4. Small commercial fire alarm and life safety systems typically have separate floor plans for each floor of a building.

T F 5. Most fire alarm and life safety floor plans provide a general location for initiating devices and notification components.

T F 6. Lines depicting the conduit/conductors connecting the fire alarm and life safety equipment together are not shown on a reflected ceiling plan.

T F 7. Common types of fire alarm and life safety detail drawings include splicing details.

T F 8. Mounting detail drawings are provided for most fire alarm and life safety devices and components.

T F 9. Although standards exist for security system symbols, some security system designers and manufacturers prefer to use their own sets of symbols.

T F 10. The location and cost of a security system determine whether or not a separate security floor plan needs to be provided.

Completion

_____ 1. ___ diagrams are typically provided for fire alarm and life safety installations in multistory buildings.

_____ 2. ___ components must be located where they can be seen and heard by all occupants of a building floor.

_____ 3. ___ details provide information on how conductors and cables are spliced and how the conductors and cables are terminated at specific devices and components.

_____ 4. A(n) ___ system is more likely to be shown on a fire alarm and life safety floor plan than on a power floor plan because the two systems are similar in nature.

_____ 5. ___ drawings provide a view of how security devices, such as door contacts and electric door strikes, interface with building components.

Multiple Choice

_____ 1. When there are several devices of the same type and limited print space, a riser diagram typically shows one device with the abbreviation "___" adjacent to the device.
    A. TYP
    B. TS or FS
    C. CSI
    D. AFF

_____ 2. Additional information found on a fire alarm and life safety riser diagram includes sheet notes covering specific installation requirements of the ___.
    A. power source for the fire alarm control panel (FACP)
    B. panel designations and circuit numbers
    C. connection of the FACP to the monitoring company or agency
    D. all of the above

_____ 3. Some fire alarm and life safety prints use a ___ plan to show the location of devices and components on a specific floor.
    A. riser
    B. reflected ceiling
    C. lighting fixture
    D. detail

_____ 4. ___ have the greatest number of detail drawings because of their numerous locations and applications.
    A. Sprinkling systems
    B. Fire extinguishers
    C. Smoke detectors
    D. none of the above

_____ 5. Security system ___ provide additional information that cannot be shown on a security system floor plan.
    A. abbreviations
    B. detail drawings
    C. electrical plans
    D. riser plans

Name _____ Date _____

Activity—Installing a Security System

9-1

Scenario:
Activity 9-1 is an installation activity requiring that security system devices and components be installed in a new Wendy's restaurant. The general contractor overseeing the security system work is Miller & Associates.

Task:
Use the Security Floor Plan for a Wendy's restaurant to determine the security system devices and components that must be installed in the new Wendy's building. Fill out any required paperwork. All information is needed in five days to keep the project on schedule.

CLOSED CIRCUIT TELEVISION CAMERA (CCTV)

Required Security System Information:
1. What type of security system devices will be installed in Zone 6? _____
2. What is the mounting height of the security system devices found in Zone 6? _____
3. How many CCTV cameras are to be installed in the Wendy's restaurant? _____
4. Where will the security system keypad be installed? _____
5. What height will the keypad be mounted at? _____
6. Where is the keypad mounting height information found? _____
7. Where will the security system control panel be installed? _____
8. Which circuit feeds power to the security system control panel? _____
9. If any information cannot be found, send an RFI to the architects.

Reference Print:

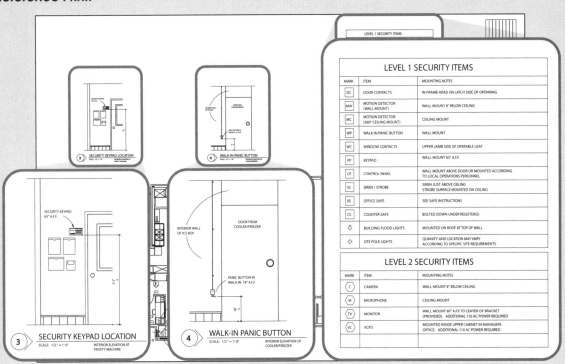

Activity—Installing a Security System

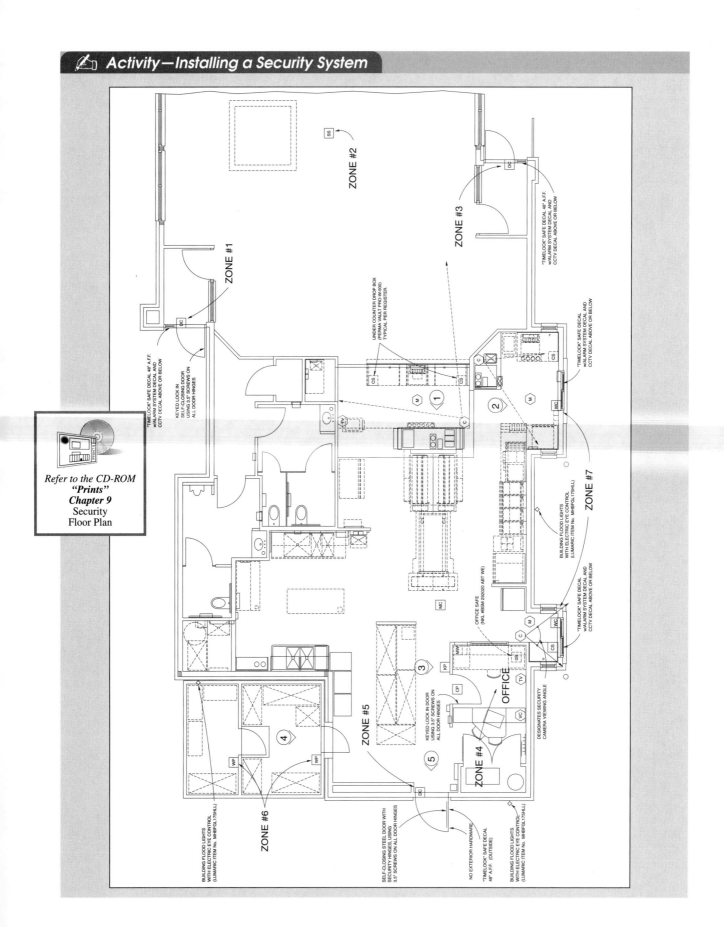

Refer to the CD-ROM
"Prints"
Chapter 9
Security
Floor Plan

Name _____ Date _____

Activity—Troubleshooting Fire Alarm and Life Safety Systems 9-2

Scenario:
Activity 9-2 is a troubleshooting activity requiring that fire alarm and life safety system devices and components be installed in a recreation center. Rooftop units (RTU-6 and RTU-7) have duct-mounted smoke detectors that send a signal to the FACP when smoke is detected. The FACP activates the notification devices (horns, speakers, and strobes). Upon finishing the fire alarm and life safety system installation, testing shows that the system is not functioning correctly.

Task:
Use the fire alarm and life safety system prints to the recreation center to determine if the fire alarm and life safety system devices and components were installed correctly.

FIRE ALARM CONTROL PANEL AND FIRE SPRINKLER RISER

Required Fire Alarm and Life Safety System Troubleshooting Information:

1. How many audio/visual annunciators need to be checked in the recreation center? _____
2. What is the mounting height of the audio/visual annunciators in the gym? _____
3. What should the candela rating be for the audio/visual annunciators in the gym? _____
4. How many wall-mounted photo-beam smoke detectors should there be in the gym? _____
5. Which room is the FACP located in? _____
6. Which power circuit feeds the FACP? _____
7. To which terminals of the smoke detector are the DATA OUT (–) and DATA OUT (+) wires connected? _____
8. To which terminal of the smoke detector is the remote alarm indicator connected? _____
9. To which terminals should the control-panel power for the RTUs be connected in order to shut down the RTUs when the duct smoke detector senses smoke? _____

Reference Print:

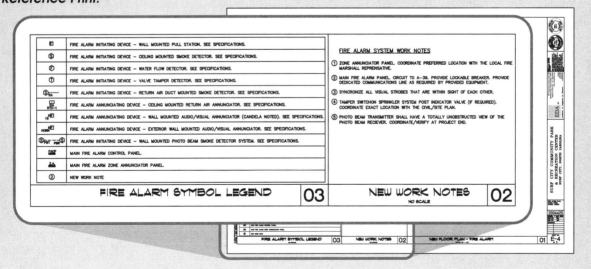

Activity—Troubleshooting Fire Alarm and Life Safety Systems

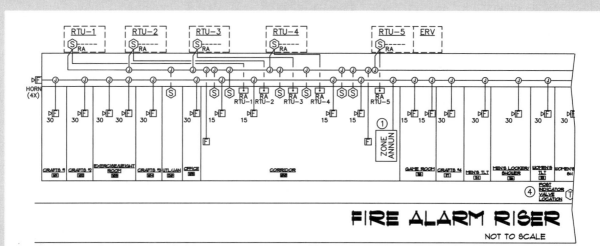

FIRE ALARM RISER
NOT TO SCALE

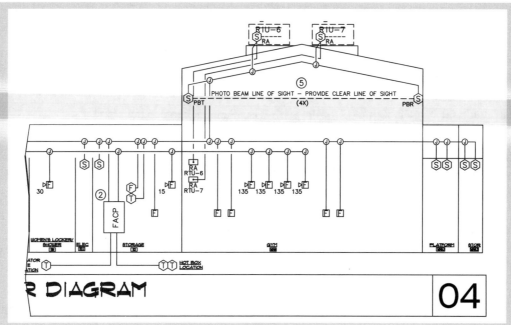

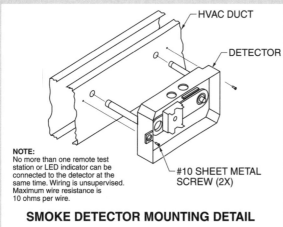

SMOKE DETECTOR MOUNTING DETAIL

NOTE:
No more than one remote test station or LED indicator can be connected to the detector at the same time. Wiring is unsupervised. Maximum wire resistance is 10 ohms per wire.

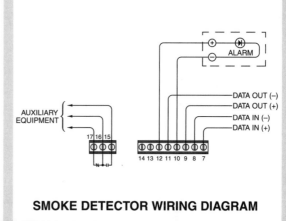

SMOKE DETECTOR WIRING DIAGRAM

Trade Competency Test
Fire Alarm, Life Safety, and Security Systems

Name _____ Date _____

REFERENCE PRINT #1 (Security Plan)

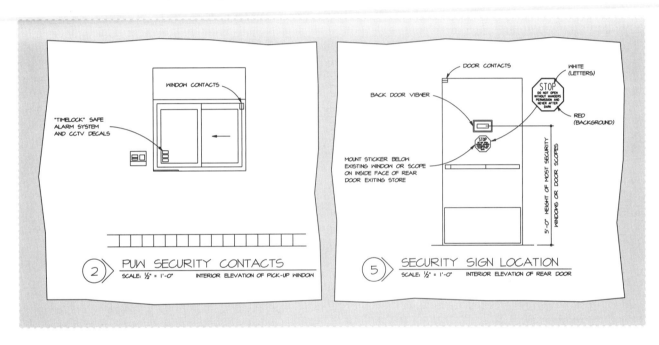

Questions – Reference Print #1

_____ _____ 1. Most Rear Door Security Windows or Door Scopes are located at ___' ___" from the bottom of the door to the center of the window or scope.

T   F   2. Detail 5 is the exterior elevation of the building rear door.

T   F   3. The sign on the rear door is located below the security window or door scope.

_____ 4. The scale for Detail 5 is ___.

_____ _____ 5. The sign on the rear door has a(n) ___ background with ___ letters.

_____ 6. Detail 2 is of the ___ elevation of the Pick-Up Window.

T   F   7. The scale for Detail 2 is ½" = 1'-0".

_____ 8. There are ___ decals on the Pick-Up Window.

301

REFERENCE PRINT #2 (Fire Alarm Plan, Notes, Riser, Symbol Legend)

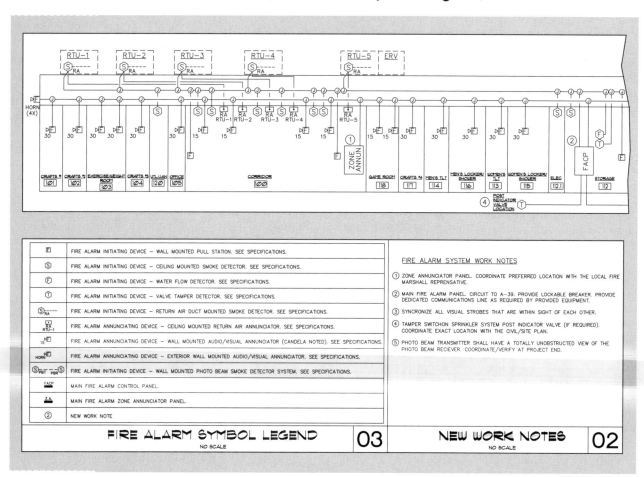

Questions – Reference Print #2

_____ 1. The Corridor is room number ___.

_____ 2. There are ___ fire alarm pull stations in the corridor.

     T   F   3. The Zone Annunciator is located in the Corridor.

_____ 4. The location of the Zone Annunciator is to be coordinated with the ___.

_____ 5. The wall mounted audio/visual annunciators in the corridor are set at ___ candela.

_____ 6. There are ___ ceiling mounted return air annunciators located in the corridor.

ADVANCED CD-ROM PRINT QUESTIONS (Fire Alarm Plan, Notes, Riser, Symbol Legend)

_____ 1. Fire sprinkler system valve tamper detectors are located in the ___ box.

_____ 2. The gym storage area (Room 108.2) has twelve ___.

_____ 3. There are four ___ smoke detectors located in the gym.

_____ 4. The UTL/JAN closet (Room 120) has one ceiling mounted ___.

HVAC Systems

Printreading for Installing and Troubleshooting Electrical Systems

Heating, ventilating, and air conditioning (HVAC) systems are found in residential, commercial, and industrial buildings. HVAC systems provide building occupants with a comfortable environment in which to live or work.

HVAC systems require controls to monitor a building's temperature and adjust the heating and cooling system. In the past, most HVAC control systems were pneumatic, consisting of pneumatic thermostats, pneumatic controllers, pneumatic valves, and air compressors. HVAC control system devices and components are now electrical or electronic and are connected via a data network. Modern HVAC control systems provide precise control, monitor and record environmental data, and interface with other building systems.

CSI MASTERFORMAT™— DIVISION 25

Prior to the 2004 edition of the CSI MasterFormat™, HVAC control prints were not typically included with electrical drawings. Instead, a separate set of prints was dedicated to the HVAC control system. Now, CSI has a separate division, Division 25, Integrated Automation, which covers HVAC control systems. **See Figure 10-1.** HVAC system installers may need to consult the electrical and mechanical prints of a building in order to install an HVAC control system.

HVAC CONTROL SYSTEMS

Building automation is a broad term that includes many different types of automated systems in a building. Large equipment manufacturers design and engineer automated systems for integration into a building that incorporate several subsystems with different functionalities. Building subsystems include the HVAC control system, fire alarm and life safety system, access control, closed-circuit TV, and energy management.

Refer to the CD-ROM "CSI MasterFormat™" Numbers & Titles

Most HVAC software programs can provide high-quality reports that include a title page, project data, detailed building and system load summaries, room and zone summary data, various load reports, sales proposals, and different types of pie charts and bar graphs covering system operation.

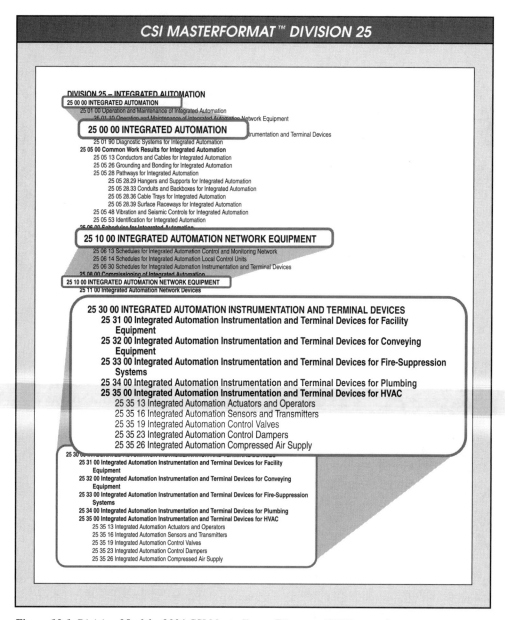

Figure 10-1. Division 25 of the 2004 CSI MasterFormat™ covers HVAC control systems.

A wide variety of prints are used for HVAC control systems. HVAC control system prints consist of abbreviations and symbols, wiring diagrams, detail drawings, and sequence-of-operation narratives. **See Figure 10-2.** On small projects, a single sheet may contain more than one type of HVAC control system print. On large projects, such as projects that include chillers, there are separate sheets for each type of HVAC control system print.

HVAC Control System Abbreviations and Symbols

HVAC control system prints include a list of abbreviations and symbols that are used throughout a set of HVAC prints. Abbreviations and symbols provide detailed information while using a minimum of space on a print.

HVAC Abbreviations. A unique set of abbreviations is used for HVAC control system prints. The abbreviations cover both electrical and mechanical terms. Some of the abbreviations used in HVAC control system prints can be represented in two different ways. **See Figure 10-3.** For example, the abbreviation for valve is either V or VLV. The abbreviations are also used on other electrical prints. HVAC control system abbreviations and definitions for a project are found on the legend and abbreviation sheets of a set of prints.

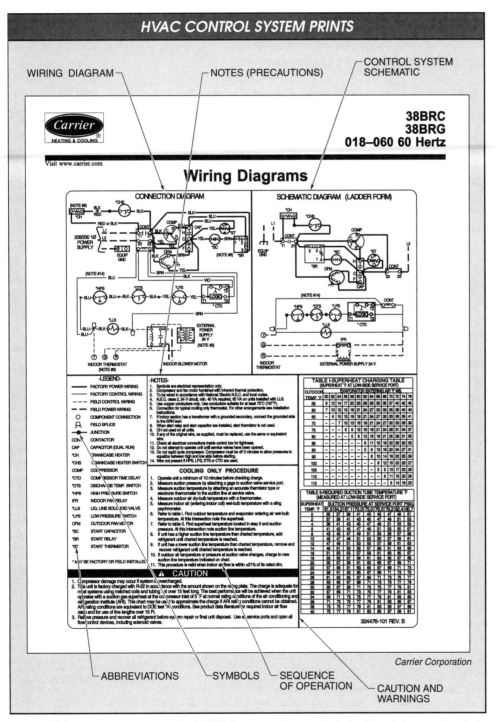

Refer to the CD-ROM "Prints" Chapter 10 Central Air Conditioning – Wiring Diagram

Figure 10-2. On some projects, a single HVAC system print may include abbreviations and symbols, wiring diagrams, detail drawings, and sequence-of-operation narratives.

HVAC SYSTEM ABBREVIATIONS			
AHU	air handling unit	MA-LL	mixed air low limit
ASD	adjustable speed drive	NO	normally open
BLR	boiler	NC	normally closed
C	common	OA	outside air
CC	cooling coil	P	pump
CHWR	chilled water return	PE	pneumatic electric switch
CHWS	chilled water supply	PI	pressure indicator
CLG	cooling	R	relay
CWR	condenser water return	RA	return air
CWS	condenser water supply	RF	return fan
DA-T	duct air temperature sensor	RTU	roof top unit
DMPR	damper	SA	supply air
DP	differential pressure	SF	supply fan
DPS	differential pressure sensor	SP	static pressure
EA	exhaust air	S-SP	sensor static pressure
EP	electric pneumatic switch	T	temperature
HC	heating coil	TSP	twisted shielded pair cable
HWR	hot water return	V	valve
HWS	hot water supply	VFD	variable frequency drive
LL	low limit	VLV	valve
MA	mixed air	VSD	variable speed drive

Figure 10-3. HVAC control system prints include abbreviations for both mechanical and electrical terms.

Some HVAC control system abbreviations can have special significance depending upon the specific application. NO (normally open) and NC (normally closed) typically refer to electrical contacts but are also used for mechanical dampers and valves. The significance of NO and NC depends on whether the abbreviations refer to an electrical or mechanical component. **See Figure 10-4.**

HVAC Symbols. A set of symbols is used to depict HVAC control system devices and components. The symbols depict mechanical and electrical devices and components that are part of the HVAC control system, such as a pneumatic damper actuator or a low-voltage thermostat. **See Figure 10-5.** HVAC symbols also depict mechanical elements of a system such as sheet metal ducts and filters.

Although some of the symbols used on HVAC control system prints come from symbol standards, many do not. The use of nonstandard symbols is the result of HVAC control system manufacturers developing their own sets of unique symbols. Many manufacturer-developed symbols resemble the items they represent. **See Figure 10-6.** HVAC control system symbols and definitions are found on the legend and abbreviation sheets of a set of prints.

ABBREVIATION SIGNIFICANCE — NORMALLY OPEN AND NORMALLY CLOSED		
	NORMALLY OPEN (NO)	**NORMALLY CLOSED (NC)**
Electrical (relay, motor starter)	Contacts are open: no flow of electricity	Contacts are closed: flow of electricity
Mechanical (damper, valve)	Mechanical device is open: flow of fluid or gas	Mechanical device is closed: no flow of fluid or gas

Figure 10-4. The abbreviations normally open (NO) and normally closed (NC) have specific significance when the abbreviations refer to electrical or mechanical equipment.

Chapter 10—HVAC Systems 307

HVAC SYMBOLS

Device	Type	Symbol
COMBINATION	Magnetic Motor Starter and Disconnect	
ADJUSTABLE SPEED DRIVE	Abbr = ASD	ASD
AUTOMATIC TEMP. CONTROL PANEL	Abbr = ATC	ATC
SENSORS	Aquastat Abbr = AST	A
	Firestat Abbr = FST	F
	Humidistat Abbr = HST	H
THERMOSTAT	Abbr = TMST	T
	Line Voltage Abbr = LVT	T L
	Low Voltage Abbr = LWVT	T LV
TIME SWITCH	Abbr = TSW	TS
PIPE TRACE HEATER	Abbr = PTHR	—/\/\/\/—

Figure 10-5. Symbols for mechanical and electrical devices and components are used on HVAC control system prints.

NON STANDARD HVAC SYMBOLS

Device	Type	Symbol
SHEET-METAL DUCT	Abbr = SMD	OR
AUTOMATIC VALVES	2-Way	
	3-Way	
	Solenoid Actuated Abbr = SOL	S
DAMPERS	Pneumatic Abbr = PNDM	
	Electric Abbr = ELDM	
FAN	Abbr = FAN	
PUMP	Abbr = PMP	
ELECTRIC PNEUMATIC SWITCH	Electric Controlling Pneumatic Abbr = EPSW	
PNEUMATIC ELECTRIC SWITCH	Pneumatic Controlling Electric Abbr = PESW	
TWISTED SHIELDED PAIR CABLE	Abbr = TSPC	
HEATING COIL	Abbr = HC	H/C
COOLING COIL	Abbr = CC	C/C
FILTER	Abbr = FLT	

Figure 10-6. The use of nonstandard symbols is the result of HVAC control system manufacturers developing their own sets of symbols, which often look like the items the symbols represent.

HVAC SYSTEM PRINTS

An HVAC system print represents the mechanical and electrical devices and components that make up an HVAC system. Separate HVAC system prints are provided for each subsystem of an HVAC system in a building. For example, separate prints are used for boiler systems, chiller and cooling tower systems, and each type of air-handling system. **See Figure 10-7.** An HVAC system print provides the following items:

- a condensed view of the sheet metal ductwork, including the direction of airflow in each duct
- a condensed view of the mechanical piping system, including the direction of liquid flow within the piping of the system
- the general location of HVAC control system devices and components in relation to the mechanical equipment, sheet metal duct, and mechanical piping systems—for example, the location of a temperature sensor in the hot water supply piping from a boiler to the heating coils
- the addresses of sensors and actuators that are part of a networked HVAC control system
- the general location of control system component terminations
- notes covering contractor installation responsibilities

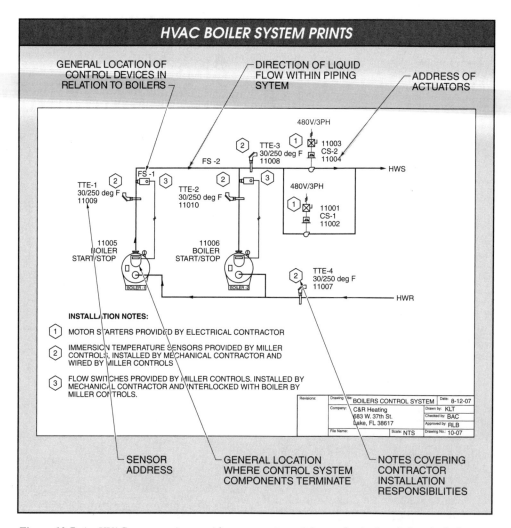

Figure 10-7. An HVAC system print provides an overview of the mechanical and electrical elements included in an HVAC subsystem such as a boiler.

HVAC system drawings do not provide exact routing, mounting, or location information for electrical equipment such as conduit, control panels, disconnects, motor starters, or variable frequency drives (VFDs). Nor do HVAC system drawings provide specifics on conduit and wire size, motor starter or VFD sizing, control wiring, or power wiring. The electrical prints of a project must be consulted for this information.

Also, HVAC system drawings do not provide the exact routing, mounting, or location information for mechanical equipment such as boilers, air handling units, mechanical piping, or sheet metal ducts. Nor do HVAC system drawings provide specifics on duct size, pipe size, air handler output, or boiler output. The mechanical prints of a project must be consulted for this information.

There are several different contractors involved in the installation of an HVAC control system, including the electrical contractor, the mechanical contractor, the HVAC equipment contractor, and the HVAC controls contractor. Often the HVAC controls contractor employs an electrician that specializes in HVAC control system wiring. The responsibilities of each contractor are typically spelled out in the specifications and the installation notes.

INSTALLATION NOTES:
1. MOTOR STARTERS BY DIVISION 16.
2. FLOW SWITCHES FURNISHED BY MILLER CONTROLS, INSTALLED BY MECHANICAL CONTRACTOR. FLOW SWITCH INTERLOCKED WITH CHILLER PER CHILLER MFR. RECOMMENDATION. INTERLOCK WIRING BY MILLER CONTROLS TERMINAL 1TB1-10&12 FOR CHILLED WATER SWITCH AND 11&13 FOR CONDENSER WATER SWITCH.
3. FLOW SWITCHES PROVIDED BY CHILLER MFR.
4. PUMP MOTOR STARTERS INTERLOCKED WITH CHILLER CONTROL PANEL PER CHILLER MFR. RECOMMENDATIONS. INTERLOCK WIRING BY MILLER CONTROLS.
5. IMMERSION TEMPERATURE SENSORS FURNISHED BY MILLER CONTROLS INSTALLED BY MECHANICAL CONTRACTOR.
6. LAND ON 1U5 TERMINALS 4&5

HVAC Wiring Diagrams

An *HVAC wiring diagram* is a diagram that shows the connections between sensors, actuators, HVAC control panels, motor starters, VFDs, and mechanical equipment control panels. **See Figure 10-8.** Some HVAC wiring diagrams show the entire control panel, while other wiring diagrams only show the terminal strip of the control panel. The abbreviations, symbols, and addresses that appear on HVAC system prints also appear on HVAC wiring diagrams.

HVAC control systems interface with mechanical equipment and other systems that operate at different voltage levels. Consequently, various voltage levels are found on HVAC wiring diagrams such as 24 VDC, 24 VAC, 120 VAC, and 460 VAC. Most HVAC control systems typically operate at 24 VAC. The actuators and sensors that connect directly to the HVAC control panel also operate at 24 VAC. When an HVAC control system must control a component operating at a different voltage, a relay is used. Typical information found on HVAC wiring diagrams includes the following:

- the location of the terminal strip for connecting conductors to the HVAC control panels
- the location of the terminal strip for connecting the conductors to related equipment such as a boiler control panel, interposing relay, or duct smoke detector
- the type and size of conductors used to connect control devices and components
- the addresses of the sensors and actuators that are part of the networked HVAC control system
- the voltage information for the HVAC control system and related equipment
- any notes covering contractor installation responsibilities and the locations of control components
- any section of an HVAC system print needed for clarification

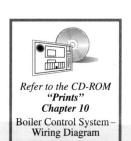

*Refer to the CD-ROM
"Prints"
Chapter 10
Boiler Control System –
Wiring Diagram*

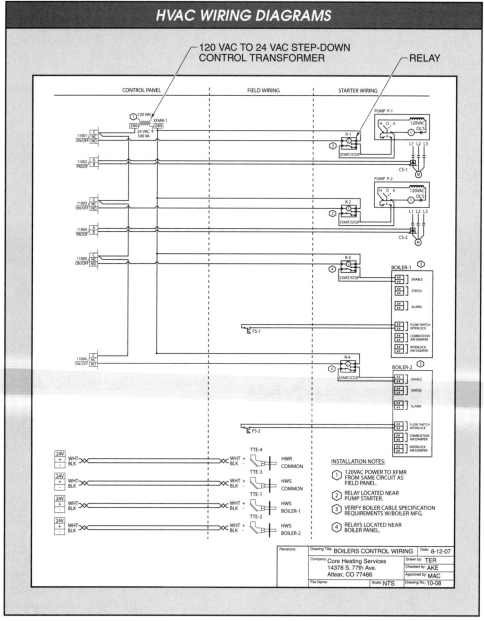

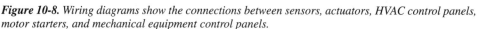

Figure 10-8. Wiring diagrams show the connections between sensors, actuators, HVAC control panels, motor starters, and mechanical equipment control panels.

HVAC panel wiring diagrams are drawn to clearly show the terminal strips for components that are internal to HVAC control panels and mechanical equipment panels and how the devices and components are wired together. HVAC panel wiring diagrams also show the terminal strips for actuators and sensors located in the field. **See Figure 10-9.**

The installation of an HVAC control system requires the use of both HVAC system prints and HVAC wiring diagrams to determine which devices and components must be wired together. HVAC wiring diagrams like HVAC system prints do not provide specifics on mounting, location, or cable routing. The electrical and mechanical prints must be consulted for this information.

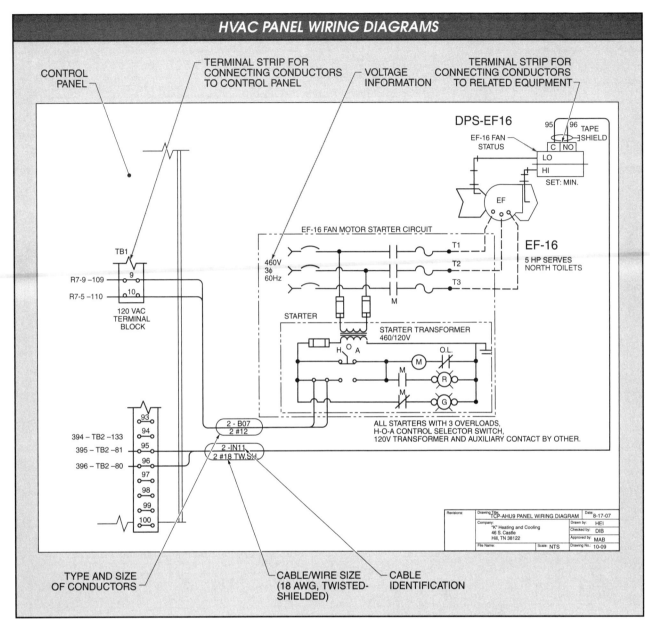

Figure 10-9. An HVAC panel wiring diagram shows the connections for devices and components connected in the panel.

HVAC Detail Drawings

HVAC detail drawings provide additional information that, due to print space, cannot be shown on HVAC system prints or HVAC wiring diagrams. Although HVAC detail drawings are not the most frequently used prints, they contain essential information for installing an HVAC control system. Common types of HVAC detail drawings include panel details and wiring details.

Panel Detail Drawings. There are a variety of panel detail drawings used for HVAC control systems. Panel detail drawings include panel layout details, panel enclosure details, and panel mounting details. **See Figure 10-10.**

A *panel layout detail* is a detail drawing that provides an overall view of the components mounted inside a control panel. Controllers, relays, terminal strips, transformers, and wireways are shown in their proper locations attached to the back panel. The dimensions of the back panel are also provided.

Refer to the CD-ROM "Prints" Chapter 10 HVAC Fan Panel Wiring Detail

Refer to the CD-ROM "Prints" Chapter 10 HVAC Fan Panel Mounting Detail

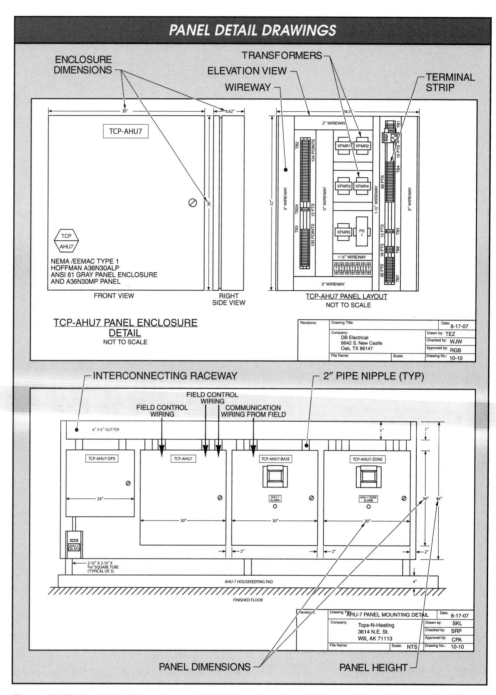

Figure 10-10. *Panel detail drawings provide information about the internal panel layout, panel dimensions, and panel mounting.*

A *panel enclosure detail* is a detail drawing that provides an elevation view of the front and side of a control system panel. Dimensions of the enclosure are typically provided. A panel enclosure detail can also include the enclosure's NEMA rating, power rating, and type of disconnect.

A *panel mounting detail* is a detail drawing that provides an enlarged view of the mounting supports of a control panel. Information on the mounting height, mounting support structure, panel dimensions, and interconnecting raceway or gutter is also provided.

Many different voltage levels are used in HVAC control systems. The various voltages levels may or may not be permitted in the same conduit or raceway. Consult the HVAC control system specifications and the NEC® before installing conductors with different voltage levels in the same conduit or raceway.

Wiring Detail Drawings. Wiring detail drawings are provided for most HVAC control systems. A *wiring detail drawing* is a detail drawing that provides information on terminating conductors at specific sensors or actuators. Wiring detail drawings are not provided for every sensor or actuator but are typically provided for sensors and actuators that have several termination possibilities or that connect to multiple voltage sources. The sensor or actuator, the HVAC control panel terminal strip, and the conductors connecting the terminal strips are shown on wiring detail drawings. **See Figure 10-11.**

Many devices and components, such as terminal units in an HVAC system, require wiring detail drawings.

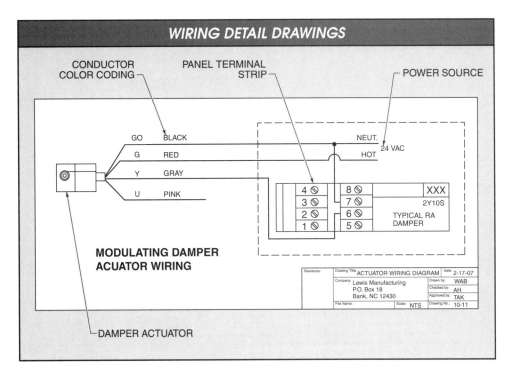

Figure 10-11. Wiring detail drawings provide termination information for sensors, actuators, and HVAC control panels.

Bill of Materials. Some HVAC control system prints include a bill of materials. Typically, each HVAC control system panel has a specific bill of materials that is organized like a spreadsheet. **See Figure 10-12.** A bill of materials lists the major items that make up a control panel along with the sensors and actuators that are mounted in the field. The items that typically make up a control panel include the enclosure, the control transformers, and the relays. Sensors mounted in the field include temperature, pressure, and smoke sensors, and dampers include damper actuators.

A bill of materials provides an abbreviation and/or symbol, a quantity, manufacturer model information, and a description for each item. When an item is missing or damaged, ordering information can be obtained from the bill of materials. Miscellaneous hardware such as screws, washers, and nuts are not included on a bill of materials.

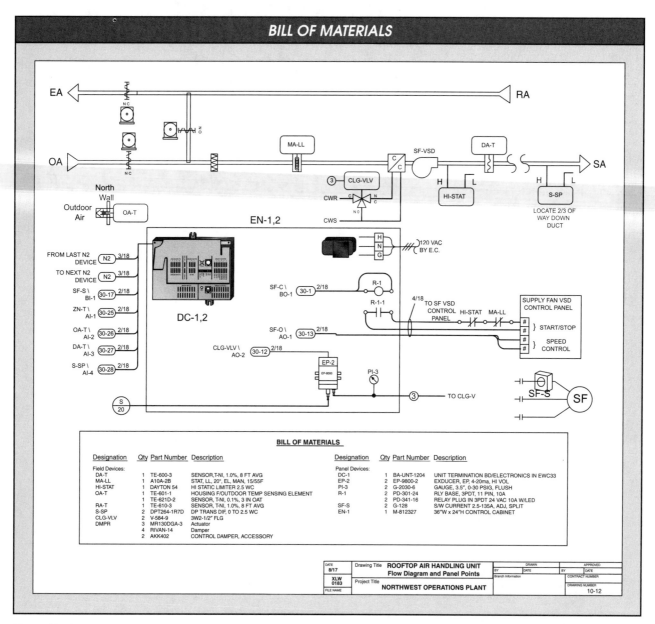

Figure 10-12. *A bill of materials lists the major items in a HVAC control panel as well as the sensors and actuators mounted in the field.*

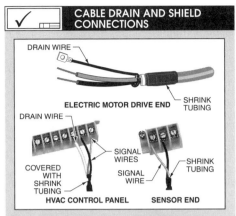

Analog inputs (sensors) are connected to HVAC control panels using twisted shielded pair (TSP) cable. TSP cable is used to connect sensors and actuators to HVAC control panels. The shield (drain) of the twisted pair cable is terminated at one end only, and taped off (insulated) at the other end. Typically the shield is only terminated at the HVAC control panel end and taped off at the sensor or actuator. The pairs also should be twisted as close as possible to the terminals.

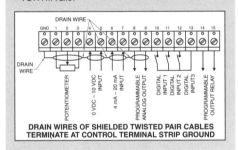

SEQUENCE OF OPERATION

Most HVAC control system prints include a sequence of operation. A sequence-of-operation narrative is also provided for each subsystem of an HVAC system. For example, sequence-of-operation narratives are typically provided for air handlers, boilers, chillers, and cooling towers. When a building contains multiple air handlers, boilers, chillers, or cooling towers with different control strategies, separate sequence-of-operation narratives are provided for each unit. A sequence of operation may appear on the same sheet as an HVAC system print or on a separate sheet. **See Figure 10-13.**

The material in an HVAC sequence of operation is written in an easy-to-understand format. The step-by-step language and graphics used in a sequence of operation make it easy for a control service technician to understand the operation of an HVAC system.

A sequence of operation provides specific details on how a subsystem of an HVAC system operates. A typical HVAC sequence of operation includes the following:
- the permissives that must be satisfied before a system can be enabled—for example, a hot water pump must be ON and water flowing before a boiler is enabled
- how a system is controlled in manual mode or automatic mode
- the time schedule for the system, which may correspond to building use and includes daily, weekly, monthly, and annual schedules
- how control valves, dampers, and the speed of motors respond to the cooling or heating requirements of a building
- connections and interlocks to other building systems—for example, most air handlers have duct smoke detectors that are connected to the fire alarm and life safety system, and the sequence of operation describes how the duct smoke detectors are interlocked with the air handler
- features designed to protect the system from damage—for example, many sheet metal ducts have a high duct static pressure switch and the sequence of operation describes how this switch will shut down a fan motor

During the startup of a HVAC control system, the sequence of operation is used to verify that the newly installed system functions per system prints. When it is necessary to troubleshoot an existing HVAC control system, the sequence of operation provides the troubleshooter with a guide to how the HVAC system should work.

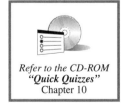

Refer to the CD-ROM "*Quick Quizzes*" Chapter 10

316 Printreading for Installing and Troubleshooting Electrical Systems

Refer to the CD-ROM "Prints" Chapter 10
Boiler Control Diagram and Sequence of Operations

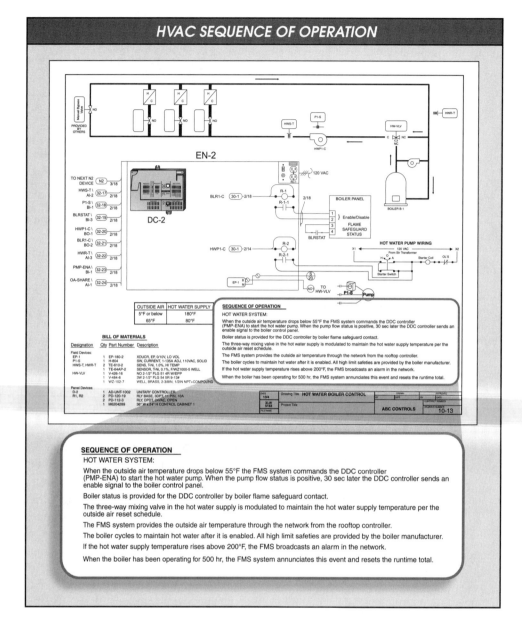

Figure 10-13. *A sequence-of-operation narrative provides an overview of how an HVAC subsystem, such as an air handler, functions.*

Review Questions and Activities

HVAC Systems

Name _____ Date _____

True-False

T F 1. On small projects, a single sheet may contain more than one type of HVAC control system print.

T F 2. Some HVAC control system abbreviations can have dual meanings depending upon the application.

T F 3. HVAC symbols do not depict mechanical elements of a system such as sheet metal ducts and filters.

T F 4. Separate HVAC prints are provided for each subsystem of an HVAC system in a building.

T F 5. A condensed view of a mechanical piping system is provided on an HVAC system logic print.

T F 6. The addresses of sensors and actuators that are part of a networked HVAC control system are provided on HVAC system prints.

T F 7. HVAC system drawings always provide exact routing, mounting, or location information for electrical equipment.

T F 8. HVAC system drawings always provide the exact routing, mounting, or location information for mechanical equipment.

T F 9. The voltage information for the HVAC control system and related equipment is provided on HVAC wiring diagrams.

T F 10. Wiring detail drawings are provided for every sensor and actuator.

Completion

_____ 1. HVAC control system symbols and definitions are found on the legend and ___ of a set of prints.

_____ 2. HVAC ___ diagrams are drawn to clearly show the terminal strips for components that are internal to HVAC control panels and mechanical equipment panels and how the devices and components are wired together.

_____ 3. ___ drawings include panel layout details, enclosure details, and mounting details.

_____ 4. A(n) ___ detail drawing is a detail drawing that provides information on terminating conductors at specific sensors or actuators.

_____ 5. When it is necessary to troubleshoot an HVAC control system, the ___ provides the troubleshooter with a guide.

Multiple Choice

_____ 1. Some HVAC ___ show the entire control panel, while others only show the terminal strip of the control panel.
   A. electrical prints
   B. wiring diagrams
   C. system sheet notes
   D. panel detail drawings

_____ 2. The installation of an HVAC control system requires the use of both HVAC ___ and HVAC wiring diagrams to determine which devices and components must be wired together.
   A. system prints
   B. plot plans
   C. sheet notes
   D. detail drawings

_____ 3. HVAC ___ provide additional information that, due to print space, cannot be shown on HVAC system prints or HVAC wiring diagrams.
   A. system prints
   B. layout diagrams
   C. detail drawings
   D. none of the above

_____ 4. A panel ___ detail is a detail drawing that provides an elevation view of the front and side of an HVAC control system panel.
   A. enclosure
   B. wiring
   C. mounting
   D. terminal

_____ 5. A(n) ___ lists the major items that make up a control panel along with the sensors and actuators that are mounted in the field.
   A. blueprint
   B. bill of materials
   C. HVAC system diagram
   D. panel mounting detail

Activity—Installing HVAC System Components 10-1

Scenario:
Activity 10-1 is an installation activity requiring that a chiller, cooling tower, pump, and controls be wired together in a bank building. Each sensor and actuator of the HVAC system that is connected to the system control microprocessor requires a unique address, which is used in the programming of the microprocessor. Work must begin in four days.

Task:
Determine which devices and components must be installed and wired and by which trade. Determine the addresses of the devices and components in the HVAC system. Unanswered questions or contradictory answers require that an RFI be submitted.

CHILLER CH-1

Required HVAC System Installation Information:
1. Who provides the motor starters for the cooling tower fans? _____
2. What is the address and designation of the temperature sensor for the chilled water return? _____
3. Who is responsible for installing the immersion temperature sensor? _____
4. What piece of equipment is the condenser water flow switch electronically connected to? _____
5. What is the designation for the condenser water flow switch? _____
6. What is the function and coil voltage of RE-5? _____
7. What is the address and designation of the current sensor for the chilled water pump? _____
8. Who is responsible for wiring the vibration switch? _____
9. What is the address for the chiller enable? _____
10. How many cooling tower motors are shown on the Chiller Control System print? _____
11. How many cooling tower motors are shown on the Chiller and Cooling Tower Wiring Diagram? _____

Reference Prints:

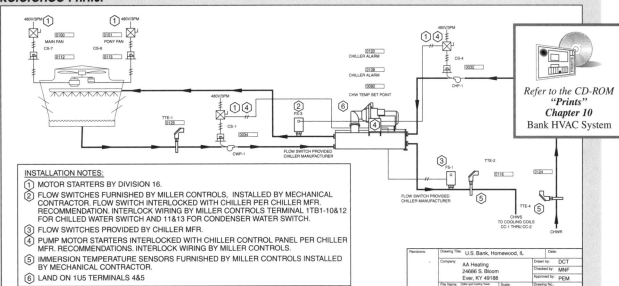

Refer to the CD-ROM "Prints" Chapter 10 Bank HVAC System

Activity—Installing HVAC System Components

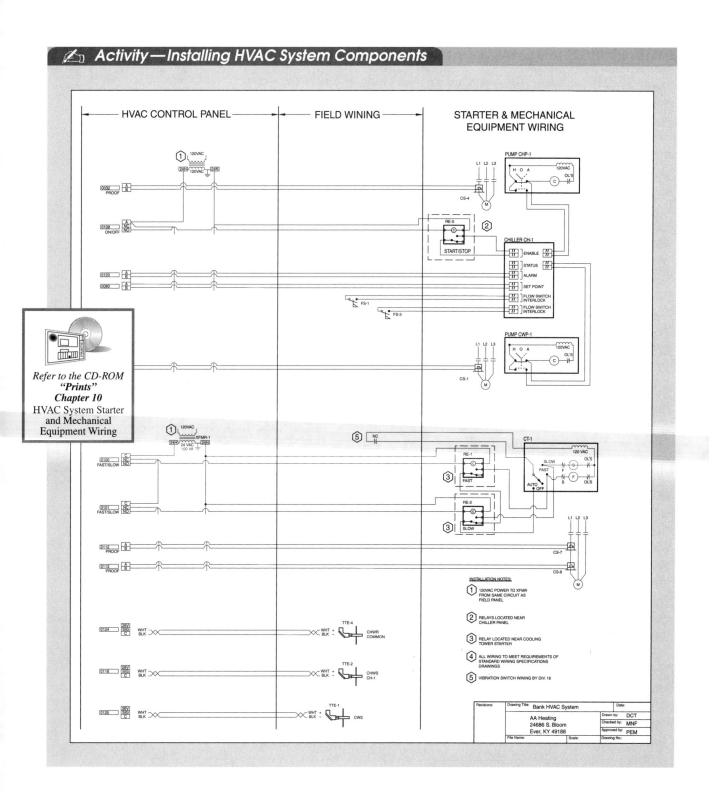

Refer to the CD-ROM "Prints" Chapter 10 HVAC System Starter and Mechanical Equipment Wiring

Activity—Troubleshooting a Central Air Conditioning System 10-2

Scenario:
Using the connection diagram, schematic diagram, notes, and procedures, troubleshoot the central air conditioning system.

Task:
Using the connection diagram, schematic diagram, notes, and procedures, troubleshoot the central air conditioning system.

Required Troubleshooting Information:
1. What is the minimum VA rating of the 24 V external power supply (transformer)? _____
2. What is the minimum VA rating of the 24 V external power supply (transformer) when the liquid solenoid valve is installed in the air conditioner?
3. Highlight the wire that will not be used because the compressor has a time-delay relay.
4. How many conductors run from the air conditioner unit to the indoor thermostat? _____
5. What terminal does L1 terminate at? _____
6. What terminal does L2 terminate at? _____
7. During normal operation, is it possible for the crankcase heater to be ON when the compressor is not running? _____
8. What color thermostat wire provides 24 V to the thermostat? _____
9. What color thermostat wire provides power to the coil of the indoor fan relay? _____
10. What color thermostat wire provides power to the compressor time-delay relay or to the compressor contactor? _____

Reference Print

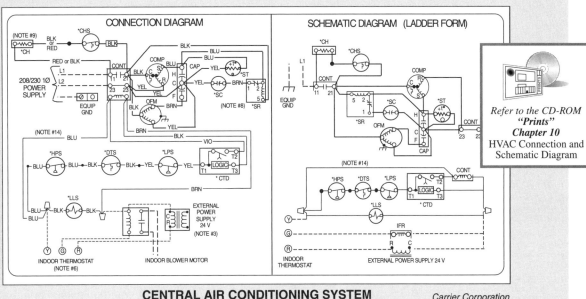

CENTRAL AIR CONDITIONING SYSTEM Carrier Corporation

Activity—Troubleshooting a Central Air Conditioning System

-LEGEND-

Symbol	Description
———	FACTORY POWER WIRING
———	FACTORY CONTROL WIRING
- - - -	FIELD CONTROL WIRING
- · - ·	FIELD POWER WIRING
○	COMPONENT CONNECTION
⊥	FIELD SPLICE
●	JUNCTION
CONT	CONTACTOR
CAP	CAPACITOR (DUAL RUN)
*CH	CRANKCASE HEATER
*CHS	CRANKCASE HEATER SWITCH
COMP	COMPRESSOR
*CTD	COMPRESSOR TIME DELAY
*DTS	DISCHARGE TEMP. SWITCH
*HPS	HIGH PRESSURE SWITCH
IFR	INDOOR FAN RELAY
*LLS	LIQ. LINE SOLENOID VALVE
*LPS	LOW PRESSURE SWITCH
OFM	OUTDOOR FAN MOTOR
*SC	START CAPACITOR
	RELAY
	THERMISTOR
	RY OR FIELD INSTALLED.

-NOTES-

1. Symbols are electrical representation only.
2. Compressor and fan motor furnished with inherent thermal protection.
3. To be wired in accordance with National Electric N.E.C. and local codes.
4. N.E.C. class 2, 24 V circuit, min. 40 VA required, 60 VA on units installed with LLS.
5. Use copper conductors only. Use conductors suitable for at least 75°C (167°F).
6. Connection for typical cooling only thermostat. For other arrangements see installation instructions.
7. If indoor section has a transformer with a grounded secondary, connect the grounded side to the BRN lead.
8. When start relay and start capacitor are installed, start thermistor is not used.
9. CH not used on all units.
10. If any of the original wire, as supplied, must be replaced, use the same or equivalent wire.
11. Check all electrical connections inside control box for tightness.
12. Do not attempt to operate unit until service valves have been opened.
13. Do not rapid cycle compressor. Compressor must be off 3 minutes to allow pressures to equalize between high and low side before starting.
14. Wire not present if HPS, LPS, DTS or CTD are used.

COOLING ONLY PROCEDURE

1. Operate unit a minimum of 10 minutes before checking charge.
2. Measure suction pressure by attaching a gage to suction valve service port.
3. Measure suction temperature by attaching an accurate thermistor type or electronic thermometer to the suction line at service valve.
4. Measure outdoor air dry-bulb temperature with a thermometer.
5. Measure indoor air (entering indoor coil) wet-bulb temperature with a sling psychrometer.
6. Refer to table I. Find outdoor temperature and evaporator entering air wet-bulb temperature. At this intersection note the superheat.
7. Refer to table II. Find superheat temperature located in step 6 and suction pressure. At this intersection note suction line temperature.
8. If unit has a higher suction line temperature than charted temperature, add refrigerant until charted temperature is reached.
9. If unit has a lower suction line temperature than charted temperature, remove and recover refrigerant until charted temperature is reached.
10. If outdoor air temperature or pressure at suction valve changes, charge to new suction line temperature indicated on chart.
11. This procedure is valid when indoor air flow is within ±21% of its rated cfm.

⚠ CAUTION

damage may occur if system is overcharged.

ctory charged with R-22 in accordance with the amount shown on the rating plate. The charge is adequate for using matched coils and tubing not over 15 feet long. The best performance will be achieved when the unit a suction gas superheat at the compressor inlet of 5 °F at normal rating conditions of the air conditioning nstitute (ARI). This chart may be used to approximate the charge if ARI rating conditions cannot be obtained. nditions are equivalent to DOE test "A" conditions. See product data literature for required indoor air flow rates and for use of line lengths over 15 Ft.
3. Relieve pressure and recover all refrigerant before system repair or final unit disposal. Use all service ports and open all flow-control devices, including solenoid valves.

TABLE I-SUPERHEAT CHARGING TABLE
(SUPERHEAT °F AT LOW-SIDE SERVICE PORT)

OUTDOOR TEMP °F	EVAPORATOR ENTERING AIR °F WB.														
	50	52	54	56	58	60	62	64	66	68	70	72	74	76
55	9	12	14	17	20	23	26	29	32	35	37	40	42	45
60	7	10	12	15	18	21	24	27	30	33	35	38	40	43
65	--	6	10	13	16	19	21	24	27	30	33	36	38	41
70	--	--	7	10	13	16	19	21	24	27	30	33	36	39
75	--	--	--	6	9	12	15	18	21	24	28	31	34	37
80	--	--	--	--	5	8	12	15	18	21	25	28	31	35
85	--	--	--	--	--	--	8	11	15	19	22	26	30	33
90	--	--	--	--	--	--	5	9	13	16	20	24	27	31
95	--	--	--	--	--	--	--	6	10	14	18	22	25	29
100	--	--	--	--	--	--	--	--	8	12	15	20	23	27
105	--	--	--	--	--	--	--	--	5	9	13	17	22	26
110	--	--	--	--	--	--	--	--	--	6	11	15	20	25
115	--	--	--	--	--	--	--	--	--	--	8	14	18	23

TABLE II-REQUIRED SUCTION TUBE TEMPERATURE °F
(MEASURED AT LOW-SIDE SERVICE PORT)

SUPERHEAT TEMP. °F	SUCTION PRESSURE AT SERVICE PORT PSIG.							
	61.5	64.2	67.1	70.0	73.0	76.0	79.2	82.4	85.7
0	35	37	39	41	43	45	47	49	51
2	37	39	41	43	45	47	49	51	53
4	39	41	43	45	47	49	51	53	55
6	41	43	45	47	49	51	53	55	57
8	43	45	47	49	51	53	55	57	59
10	45	47	49	51	53	55	57	59	61
12	47	49	51	53	55	57	59	61	63
14	49	51	53	55	57	59	61	63	65
16	51	53	55	57	59	61	63	65	67
18	53	55	57	59	61	63	65	67	69
20	55	57	59	61	63	65	67	69	71
22	57	59	61	63	65	67	69	71	73
24	59	61	63	65	67	69	71	73	75
26	61	63	65	67	69	71	73	75	77
28	63	65	67	69	71	73	75	77	79
30	65	67	69	71	73	75	77	79	81
32	67	69	71	73	75	77	79	81	83
34	69	71	73	75	77	79	81	83	85
36	71	73	75	77	79	81	83	85	87
38	73	75	77	79	81	83	85	87	89
40	75	77	79	81	83	85	87	89	91

324476-101 REV. B

Refer to the CD-ROM "Resource Library" Prints – Chapter 10 HVAC Connection and Schematic Diagram

Name _____ Date _____

REFERENCE PRINT #1 (Furnace Connection Diagram & Schematic)

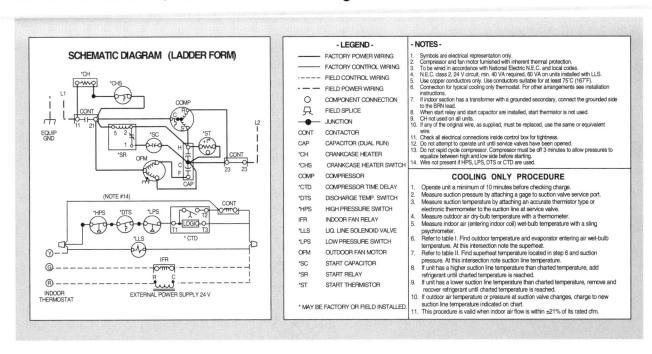

Questions – Reference Print #1

_____ 1. The Compressor Time Delay is a(n) ___ delay timing relay.

T   F   2. The compressor has an internal temperature switch/sensor.

T   F   3. The Discharge Temperature Switch is connected in parallel with the High Pressure Switch.

T   F   4. The external power supply is 24 V.

_____ _____   5. The coil of the start relay is connected to terminal ___ and terminal ___.

_____ _____ _____   6. The indoor thermostat is connected to terminals ___, ___, and ___.

_____ 7. The Liquid Line Solenoid is ___ V.

_____ 8. There are ___ temperature switches/sensors in the schematic diagram.

REFERENCE PRINT #2 (Hot Water Boiler Control)

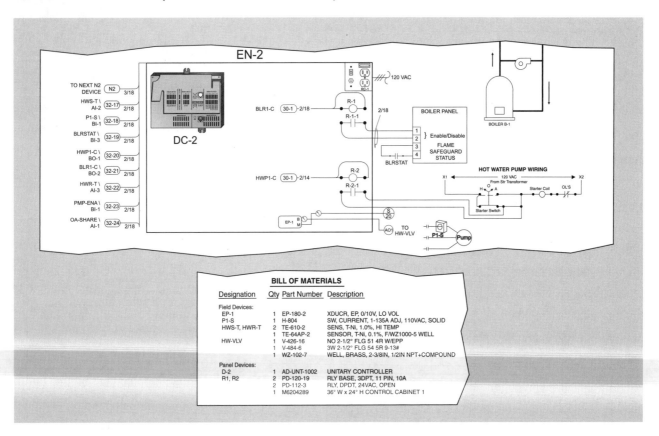

Questions – Reference Print #2

_____ 1. The coil voltage of R-1 is ___ V.

_____ 2. Two ___ gauge conductors are connected to the coil of R-1.

T   F   3. The NO contacts of R-2 are connected to the Boiler Panel.

T   F   4. R-1 and R-2 are open-type relays.

_____ 5. The air supply pressure for EP-1 is ___ psi.

_____ 6. Two ___ gauge conductors are connected to the coil of R-2.

_____ _____   7. The control cabinet is ___″ wide × ___″ high.

_____ 8. The part number of the Unitary Controller is ___.

ADVANCED CD-ROM PRINT QUESTIONS (Hot Water Boiler Control)

_____ 1. All boiler high limit safeties are provided by the ___.

_____ 2. There are ___ heating coils connected to the boiler.

_____ 3. Terminal 3 and 4 of the boiler's panel provide ___.

_____ 4. The status of the Hot Water Pump is monitored by P1-S, which is a(n) ___.

Industrial Control Systems

Printreading for Installing and Troubleshooting Electrical Systems

Most industrial electrical prints have two major parts: the power circuit and the control circuit. A power circuit includes the high-powered electrical components that produce most of the required work for an application. Power circuit components include motors, heating elements, and large power lamps. A control circuit includes the control devices and components that control the power circuit components. Control circuit components include motor starters, heating contactors, and lighting contactors.

POWER AND CONTROL CIRCUITS

Power circuits include electrical devices and components that draw as little as a few amps to thousands of amps. Power circuit devices and components typically have a voltage rating greater then 208 V such as 230 V, 460 V, or 600 V.

Control circuits include devices and components that control power circuit components such as motor starters and contactors, but can also include some low-power-consumption components, such as solenoids, low-wattage lamps (colored indicating lamps), and low-powered heating elements. **See Figure 11-1.**

The link between a power circuit and a control circuit is the control transformer. A *control transformer* is a transformer used to step-down the voltage of a power circuit to 120 VAC or 24 VAC and provide power for the control circuit loads (motor starters, lamps, and solenoids). Control transformers are located in the control circuit panel and provide power for all loads in the control circuit. In contrast, a power transformer supplies power for an entire system (both power and control circuits), a section of a building, or an entire building.

Water Tower Application

In order to understand the difference between a power circuit and a control circuit, the different numbering systems, rules, and circuit logic that are used when designing, reading, and using prints must be understood. A dual pump control circuit can be used as both a power circuit and control circuit example. For example, a dual pump power and control circuit can be used to represent a basic pump control circuit. The pumps maintain the water level in a water tower. **See Figure 11-2.**

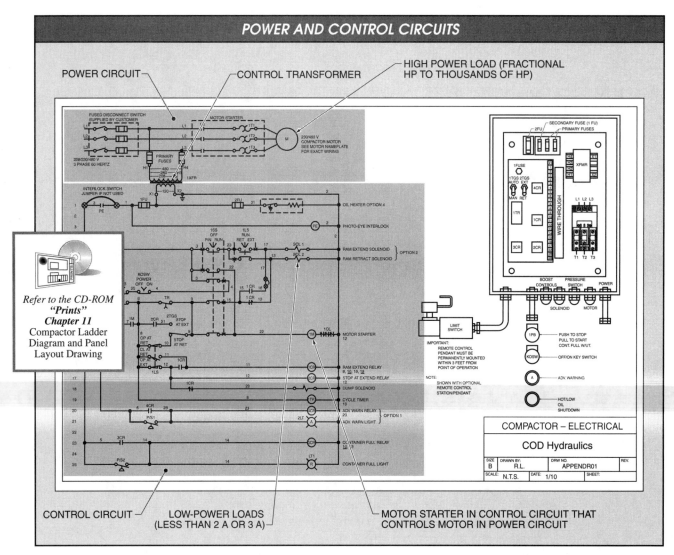

Figure 11-1. *Industrial electrical prints have two parts: the power circuit and the control circuit. The control circuit typically receives power from a step-down transformer.*

In the circuit, two pumps are used to maintain the water level in the tank. Motor starter M-1 in the control circuit controls pump 1 in the power circuit, and motor starter M-2 in the control circuit controls pump 2 in the power circuit.

One pump is turned on any time the level in the tank falls below the low-level switch (LS-1) and remains on until the level reaches the high-level switch (LS-2). LS-1 is normally closed and opens as the water rises above it. LS-2 is also normally closed and opens when the water level reaches it. LS-1 remains in the held-open position until the water drops below the level of LS-2.

An alternating relay is used to alternate the pumps each time LS-1 calls for a pump to turn on. This allows the work to be divided equally between the two pumps (motors) and ensures that the motors share the workload.

Water towers, hydraulic accumulators, pneumatic receivers, and capacitors are all used for the same reason within a system: to store a charge of energy that can be used any time that demand exceeds supply. These items allow the size of the power supply to be reduced.

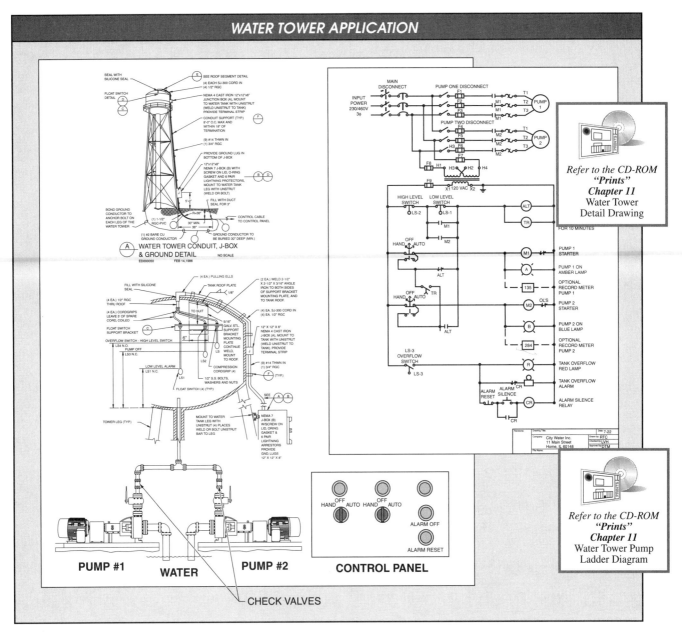

Figure 11-2. Electrical prints use various numbering systems, rules, and circuit logic to create operational electrical systems.

When one pump cannot maintain the water level and has run for 10 min without stopping, the second pump will turn ON. The second pump is turned ON when the on-delay timer (TR) times out and closes its contacts, which closes the contacts on both M-1 and M-2, regardless of which pump was running first. Each pump (motor) is controlled through a three-position (HAND-OFF-AUTO) selector switch that allows each motor to be placed in either the manual position (HAND), automatic position (AUTO), or OFF position.

An overflow alarm is used to signal when the tank has reached an overflow condition (LS-3 closing). Tank overflow can happen during a circuit component malfunction, such as limit switch LS-2 not opening, or when the selector switch is left in the HAND position. The alarm can be silenced after the alarm has been acknowledged, but the overflow condition will still exist until corrected.

An amber lamp is used to indicate when pump #1 is ON, a blue lamp indicates pump #2 is ON, and a red lamp indicates an overflow condition. *Note:* Pressing the "Alarm Silence" pushbutton turns the alarm OFF, but keeps the red lamp ON until the problem is corrected.

A print will typically show where optional time recording meters can be added to log the amount of time (typically in hours) that each pump has run. A run log helps when scheduling preventive maintenance.

BASIC RULES OF LADDER (LINE) DIAGRAMS

Control circuits are typically drawn in ladder (line) diagram format. The electrical industry has established a universal set of symbols and basic rules on how line diagrams are laid out and drawn. Although there are some variations between different companies, there are far more similarities than differences. Learning and applying these standards establishes a common understanding of how circuits operate and makes troubleshooting and circuit modifications easier. An understanding of line diagrams is also important during programmable logic controller (PLC) programming because most PLC software uses the ladder diagram format for circuit design.

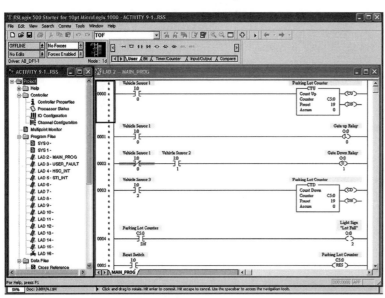

PLC programming software uses ladder diagrams to create PLC programs.

LOAD VOLTAGE DROPS

Voltage drop is the amount of voltage consumed by a component (load) as current passes through it.

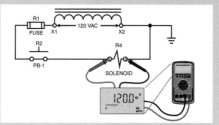

Series Loads. The total voltage applied across loads connected in series is divided by the individual loads. Each load drops a set percentage of the applied voltage, no matter how low the resistance of the load.

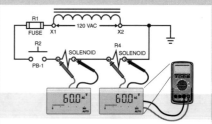

Parallel Loads. The voltage drop across each load is the same when loads are connected in parallel. The voltage drop across each load remains the same when parallel loads are added or removed.

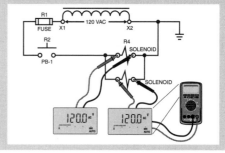

One Load Per Line

No more than one load can ever be placed in any one line between the control circuit power lines (L1-L2, X1-X2, L1-N). The "no more than one load per line" rule exists because, when loads are connected in series, the voltage will divide between the loads. This is the law of voltage in a series circuit. Loads must be connected in parallel when more than one load is connected to the same control section of a ladder diagram. **See Figure 11-3.**

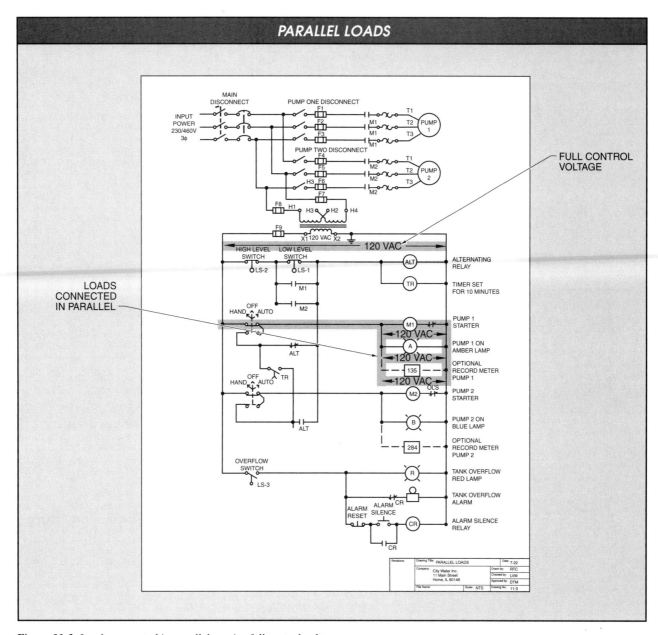

Figure 11-3. Loads connected in parallel receive full control voltage.

For example, when motor starter #1 (M1), the amber lamp, and the hour time recorder are connected in parallel, each of the loads will have the full 120 VAC applied across them when the selector switch is placed in the "HAND" position. When motor starter #1 (M1), the amber lamp, and the hour time recorder are connected in series, each one would only receive a percentage of the 120 V. The exact amount of applied voltage across each component would depend upon each component's resistance (impedance) — the higher the resistance, the higher the voltage drop created by the component.

Load (Component) Connections

A *load* is any electrical component in a line diagram that consumes electrical power from L1. All loads, except magnetic motor starter coils, are connected directly to L2 (N). Control relay coils, solenoids, alarms, and pilot lights are examples of loads that are connected directly to L2. **See Figure 11-4.**

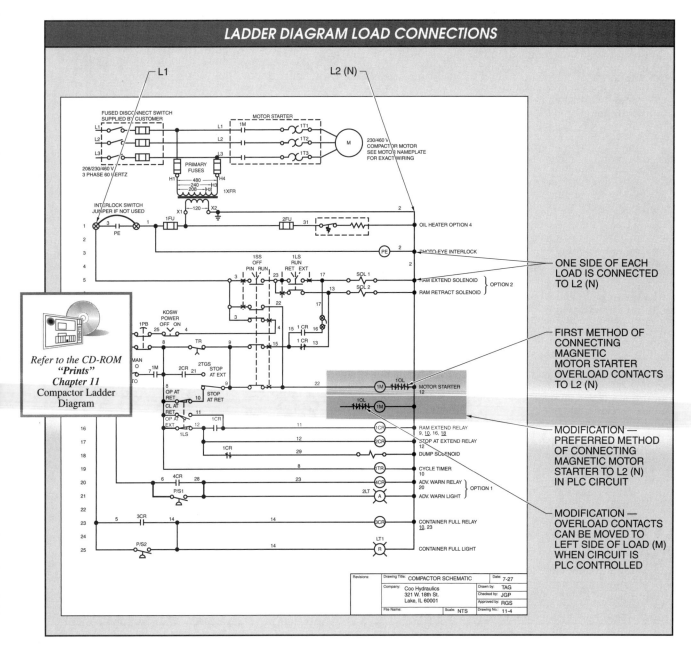

Figure 11-4. All loads must be connected to line two (L2) of a control circuit diagram.

Magnetic motor starter coils are connected to L2 indirectly through normally closed (NC) thermal overload contacts (OLs). A thermal overload contact opens only when an overload condition exists in the motor. In the past, overload contacts were placed between the motor starter coil and L2 (N).

Although this practice is still used, newer circuits place the overload contact to the left of the motor starter coil and connect the output side of the motor starter coil directly to L2, like all other loads. The second method is often used with a PLC. In PLC applications, the overload contacts are an input to the PLC and therefore can be monitored by the PLC.

Although a circuit can have any number of loads, the total number of loads determines the required wire size and the rating of the power supply (which is typically a transformer). As loads are added to a circuit, total current demand increases.

Control Device Connections

Control devices are connected between L1 and a load. Operating coils of contactors and motor starters are activated by control devices such as pushbuttons, limit switches, and pressure switches. **See Figure 11-5.**

Each line of a ladder diagram includes at least one control device. A circuit may contain as many control devices as is required to make the operating loads function as specified. The control devices in a circuit may be normally open (NO) or normally closed (NC) and can be connected in either series or parallel when controlling loads. NO devices that are connected in series develop AND logic and NC devices connected in series develop NOR logic. Also, NO devices connected in parallel develop OR logic and NC devices connected in parallel develop NAND logic.

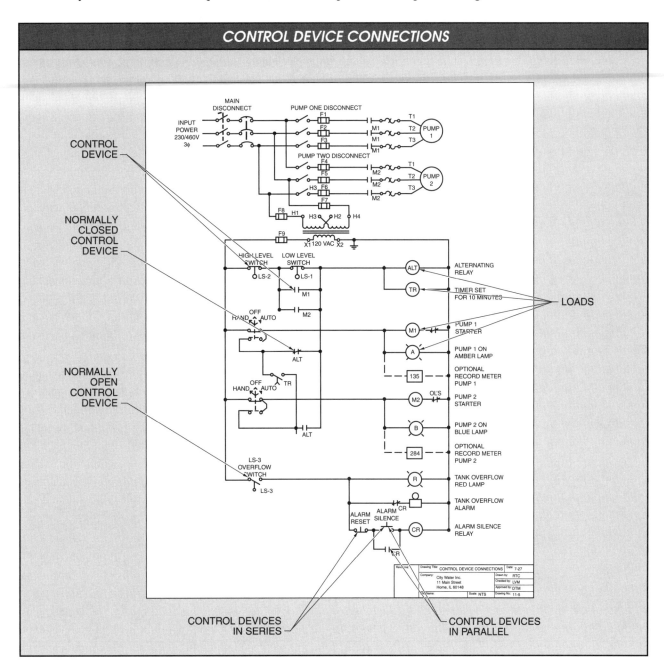

Figure 11-5. Circuit control devices are either normally open (NO) or normally closed (NC) and can be connected in series or parallel.

CONTROL CIRCUIT NUMBERING SYSTEMS

Electrical control circuits are drawn in ladder diagram format to make the control circuit easier to understand, modify, and/or program using software. Small control circuits that have only a few lines can be read and understood without much trouble. Large control circuits are made up of several small control circuits and can be difficult to follow unless there is a method to help tie the smaller sections and parts together. To do this, several numbering systems can be used.

Line Reference Numbers

Each line in a line diagram is numbered starting with the top line and reading down. Numbering each line in a control circuit allows for faster identification of the location of devices and components in smaller individual circuits. **See Figure 11-6.**

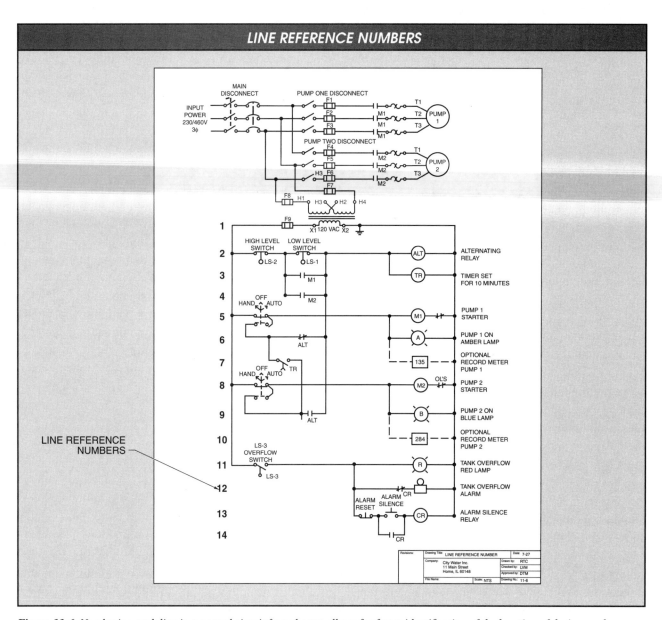

Figure 11-6. *Numbering each line in a control circuit from the top allows for faster identification of the location of devices and components in smaller branch circuits.*

For example, in the water tower pump ladder diagram, an overflow circuit is used to signal when the tank is about to overflow. The overflow circuit is not necessary for normal operation when no problems exist. Overflow and other types of circuits are added as safety options when a circuit is first designed or installed. Overflow and other circuits can also be added after the circuit has been installed. It is easier to identify the overflow circuit as line numbers 11 through 14 when discussing the circuit or showing the circuit as an installation option or add-on.

In the water tower control circuit, line numbers start at the top with the secondary transformer coil (X1 and X2). In other ladder diagrams, the transformer line may not be numbered. The numbering may start with the first part of the control circuit, which would include switches and loads. In general, each company designing and/or installing control circuits will have a preference as to where to start the numbering system and exactly which lines will be numbered. Not all manufacturers follow the basic "starting at the top and numbering down" method.

> When making circuit changes, such as adding a switch into a circuit between wire references #6 and #7, it is acceptable to use letters along with numbers for new wires, such as 6A, 6B, or 6X, so that wire reference numbers downstream do not have to change.

Numerical Cross-References

One reason for adding line reference numbers to each line of a line diagram is so a second numbering system, called the numerical cross-reference numbering system, can be used. Numerical cross-reference systems are necessary to trace the action of a circuit in complex line diagrams. Numerical cross-reference rules simplify the operation of complex circuits and allow for quicker troubleshooting.

Contact Location and Identification. Relays, timers, contactors (heating and lighting), and magnetic motor starters typically have more than one set of contacts or auxiliary contacts that are used in a control circuit. The contacts can appear at several locations in a line diagram, and in larger control circuits the contacts may be hundreds of lines apart.

Numerical cross-reference systems quickly identify the locations and types of contacts controlled by a given component. A numerical cross-reference system consists of numbers in parentheses at the end of a line (to the right of L2). NO and NC contacts are represented by the line number the contact appears in. To differentiate between NO and NC contacts, NC contacts are indicated by a number that is underlined. **See Figure 11-7.**

For example, in the water tower pump ladder diagram, the alternating relay, timer, control relay, and two magnetic motor starters all have contacts used in the control circuit. The two magnetic motor starters also have contacts used in the power circuit, but only contacts used in the control circuit are referenced with a numerical cross-reference number.

In the water tower control circuit, the alternating relay coil found in line 2 has two contacts in the control circuit that are found in lines 6 and 9. Because the contact used in line 6 is NC, line number 6 in the cross-reference bracket at the end of line 2 is underlined. The contact found in line 9 is included but not underlined because the contact is NO.

Wire Reference Numbers

There are several methods of wiring a control circuit. Smaller control circuits can use the direct wire method in which each wire in the control circuit runs from one device to the next device as shown in a line diagram. Another method that is used, typically with larger control circuits or circuits that require faster troubleshooting and/or circuit modifications, connects each component in the control circuit to a terminal strip.

334 Printreading for Installing and Troubleshooting Electrical Systems

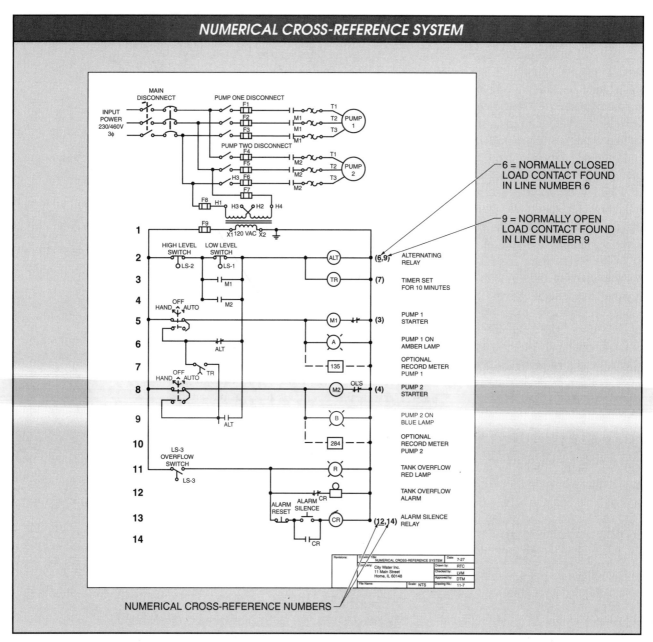

Figure 11-7. A numerical cross-reference system consists of line numbers in parentheses at the end of a line (to the right of L2) that identify the line a component contact can be found on.

The base 10 (decimal) numbering system is the most common numbering system used today. Uncommon numbering systems include the base 12 (dozenal) and base 60 (sexagesimal) numbering systems. The base 2 (binary) numbering system (ON and OFF) is used in computers and electronic systems.

When a terminal strip is used, each wire in the control circuit is assigned a reference point (terminal number) on the ladder diagram and each reference point is assigned a reference number. Wire reference numbers are typically assigned from the top left of the ladder diagram to the bottom right. The reference point numbering system applies to any control circuit regardless of the circuit's size and complexity. **See Figure 11-8.**

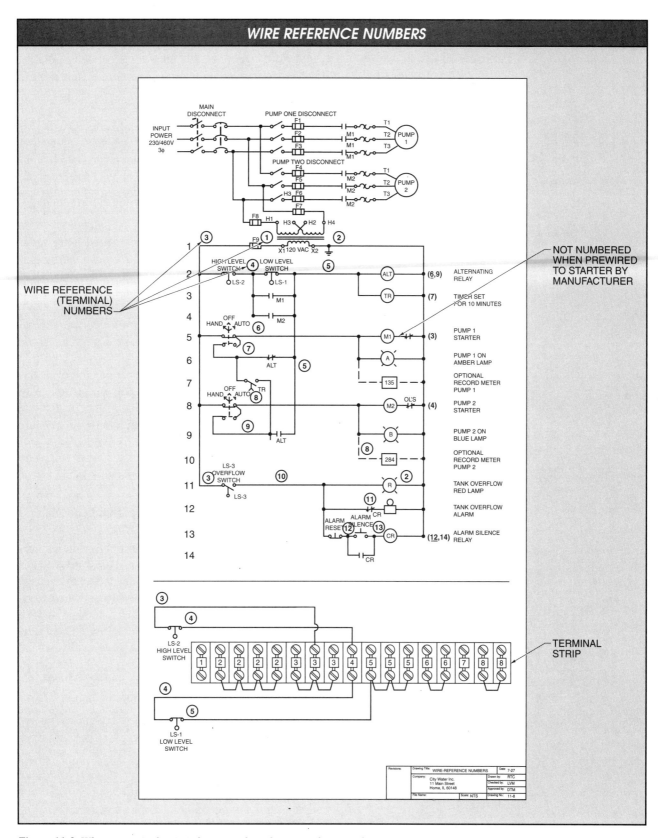

Figure 11-8. When a terminal strip is being used, each wire in the control circuit is assigned a reference number. This helps keep track of the various wires that connect devices and components in a circuit.

Any wire that is connected to a common point is the same electrically as other wires connected to that point and is assigned the same reference number. The wires that are assigned a number vary from two wires to the number required by the circuit. Any wire that is prewired by the manufacturer when the component is purchased is not typically assigned a wire reference number.

The exact numbering system used varies for each manufacturer or design engineer. One common method used is to circle the wire reference numbers. Circling the wire reference numbers helps separate the reference numbers from other types of numbers used on the diagram.

Most newer wire reference numbering systems start with X1 of the control transformer (or the control transformer fuse) as wire (terminal) number 1 and X2 as wire (terminal) 2. All other numbers are then assigned following the top-to-bottom and left-to-right pattern. By assigning the main control circuit power (X1 and X2) terminals 1 and 2, an electrician always knows where to test first for control circuit power during troubleshooting.

General Electric Company
A transformer nameplate identifies which output terminal is X1 and which output terminal is X2.

Manufacturer Terminal Numbers

Manufacturers of electrical relays, timers, counters and other electrical equipment include numbers on the terminal connection points of devices and components. The terminal numbers are used to identify and separate the different parts (coil, NC or NO contacts) on individual pieces of equipment. All control circuits that are designed as "generic circuits" can be used with almost any manufacturer's equipment. Once a piece of equipment has been identified for use in a control circuit, manufacturer terminal numbers are then added to the line diagram. **See Figure 11-9.**

For example, any type of relay can be used in the water tower control circuit as long as the relay coil and contacts meet the electrical and mechanical requirements of the application. Once a model has been selected, the terminal numbers provided by the manufacturer can be added to the control circuit print.

There must be a clear method of distinguishing manufacturer numbers from wire reference numbers. One method is to circle the wire reference numbers but not the manufacturer numbers. Another common method is to make the manufacturer numbers a different font and/or smaller size.

Cross-Referencing Mechanically Connected Contacts

Control devices such as limit, flow, temperature, liquid level, and pressure switches typically have more than one set of contacts operating when the device is activated. Control devices typically have at least one set of NO contacts and one set of NC contacts that operate simultaneously. For all practical purposes, the contacts of multiple contact devices typically do not control other devices in the same line of a control circuit.

> Mechanical contacts are damaged far more often by arcing and burning when contacts open than when contacts close, because the breaking of current causes an arc across opening contacts. Good mechanical contacts have fast-opening contacts ("snap-action" or "fast-break").

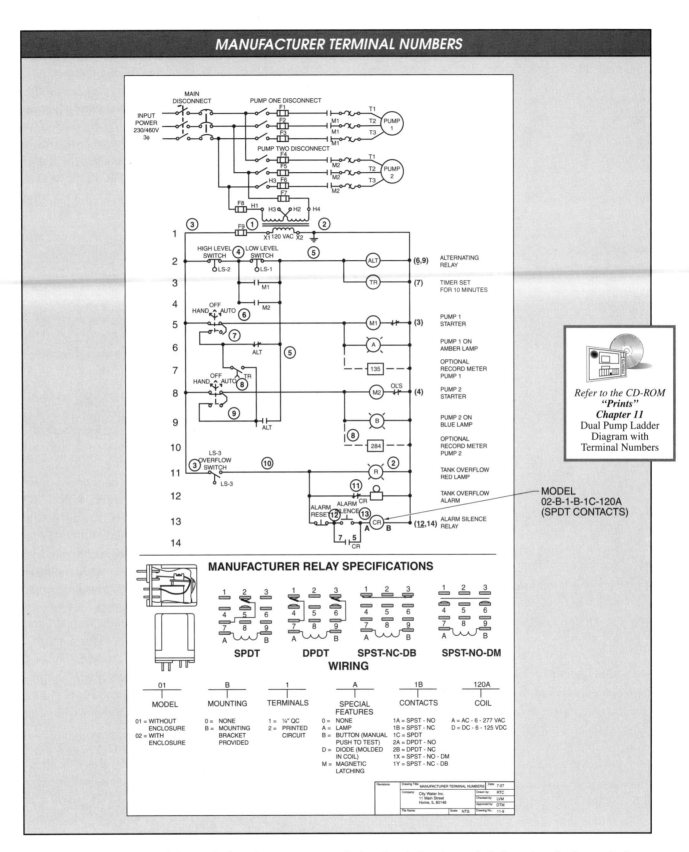

Figure 11-9. Manufacturers of electrical relays, timers, counters and other electrical equipment include numbers for the terminal connection points of devices and components to identify and separate the different parts on the individual pieces of equipment.

However, in control circuits where multiple mechanical contacts are used and separated by more then a couple of lines, some prints use a method of simplification that shows the mechanical link between the contacts. The two methods used to illustrate which mechanically connected contacts found in different control lines belong to the same control switch are the dashed line method and the numerical cross-reference method. **See Figure 11-10.**

Dashed Line Cross-Reference Method. The *dashed line cross-reference method* is a method of identifying the mechanically connected contacts that are on different lines of a ladder diagram by placing a dashed line between the contacts. This indicates that both contacts will move from the normal position when the arm of the limit switch is moved. In the circuit, pilot light PL1 is ON, and motor starter coil M1 is OFF. After the limit switch is actuated, pilot light PL1 turns OFF, and the motor starter coil M1 turns ON.

The dashed line method works well when the mechanically connected contacts are closed together and the circuit is relatively simple. However, when a dashed line needs to cut across many lines of a ladder diagram, the circuit becomes difficult to follow.

Numerical Cross-Reference Method. The *numerical cross-reference method* is a method of identifying various mechanically connected contacts of a device that are on different lines of a complex ladder diagram by using special symbols. For example, a pressure switch with a NO contact in line 1 and a NC contact in line 5 is used to control motor starter #1 and a solenoid. The NO and NC contacts of the pressure switch are simultaneously actuated when a predetermined pressure is reached. A solid arrow pointing down is drawn near the NO contact in line 1 and is marked with a 5 to show the mechanical link with the contact in line 5. A solid arrow pointing up is drawn near the NC contact in line 5 and is marked with a numeral 1 to show the mechanical linkage with the contact in line 1.

The cross-reference method of identifying mechanically connected contacts eliminates the need for a dashed line cutting across lines 2, 3, and 4. The numerical cross-reference method makes the circuit easier to follow and understand and can be used on any switch with mechanically connected contacts found in a control circuit.

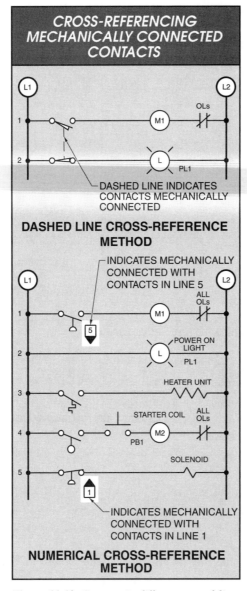

Figure 11-10. Contacts in different control lines that belong to the same control switch are cross-referenced using the dashed line method or the numerical cross-reference method.

CONTROL CIRCUIT LOGIC FUNCTIONS

A circuit must respond as designed, without any changes. Control devices such as pushbuttons, limit switches, and pressure switches are connected into a circuit so that the circuit can function in a predetermined manner. Most control circuits use basic logic functions or combinations of logic functions. Logic functions are common to many trades and disciplines, such as electrical, electronics, hydraulics, pneumatics, and math. The most common circuit logic functions are AND, OR, NOT, NOR, and NAND. **See Figure 11-11.**

AND Circuit Logic

AND circuit logic is a control function that requires two or more switches to be closed before a load can be energized (ON). The switches may be electromechanical, such as pushbuttons, limit switches, and pressure switches, or they can be the contacts of relays, timers, counters, or any other type of mechanical or solid-state device. AND circuit logic is accomplished by hardwiring two normally open switches in series or by using a digital AND gate. **See Figure 11-12.**

For example, an operator cannot turn ON a pump and valve solenoid or a vacuum pump and valve solenoid unless the ON/OFF switch is closed and the foot switch is closed. Using a foot switch in the application allows the hands of the operator to be free for moving an object in or out of the machine.

An electrical circuit that connects the two switches in series can be used to control the 120 VAC pump. The disadvantage of a full voltage control circuit is that the control switches operate at the same voltage level (120 VAC) as the load. In addition to being at the same voltage level, the control switches must also be rated to carry the full current of the load.

CONTROL CIRCUIT LOGIC

Device	Hardwired	Statement	Boolean Expression	Read As	Symbol	Function/Notes
AND GATE		Allows electricity to flow if Switches 1 and 2 are pressed	$Y = A \cdot B$ ($\cdot$ = AND)	Y = A AND B		To provide logic level 1 only if all inputs are at logic level 1 HIGH = 1 LOW = 0
OR GATE		Allows electricity to flow if Switch 1 or 2 is pressed	$Y = A + B$ (+ = OR)	Y = A OR B		To provide logic level 1 if one or more inputs are at logic level 1 HIGH = 1 LOW = 0
NOT (INVERTER) GATE		Allows electricity to flow if Switch 1 is not pressed	$Y = \overline{A}$ (− = NOT)	Y = NOT A		To provide an ouput that is the opposite of the input HIGH = 1 LOW = 0
NOR GATE		Allows electricity to flow if neither Switch 1 nor Switch 2 is pressed	$Y = \overline{A + B}$	Y = NOT (A OR B)		To provide logic level 0 if one or more inputs are at logic level 1 HIGH = 1 LOW = 0
NAND GATE		Allows electricity to flow if both Switch 1 and Switch 2 are not pressed	$Y = \overline{A \cdot B}$	Y = NOT (A AND B)		To provide logic level 0 only if all inputs are at logic level 1 HIGH = 1 LOW = 0

Figure 11-11. Most control circuits use the basic logic functions (AND, OR, NOT, NOR, and NAND) or any combination of the logic functions.

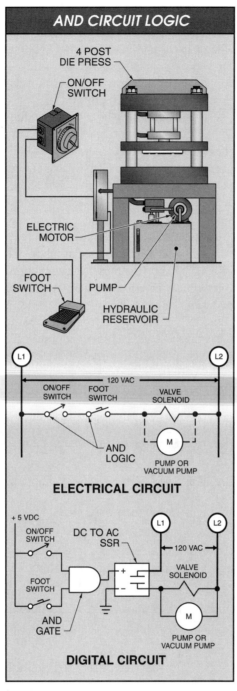

Figure 11-12. AND circuit logic requires that all switches and contacts be closed to energize a load.

be used because they only have to be rated for the voltage of the logic gate. The voltage rating of a logic gate is typically 5 VDC.

The low voltage also allows smaller gauge wires (18 AWG or smaller) to be used to connect all of the control switches. A DC to AC solid-state relay (SSR) is used as the interface between the DC control circuit and the 120 VAC power circuit containing the pump motor and valve solenoid.

OR Circuit Logic

OR circuit logic is a control function that requires any one or more switches to be closed before a load can be energized (ON). The switches can be any type of mechanical or solid-state control switches or contacts. **See Figure 11-13.**

In the circuit, moving any one of the limit switches energizes the hydraulic pump. An SSR is used as the interface between the digital circuit and the 24 VDC motor circuit because the motor draws more current and operates at a much higher current level than the digital integrated circuit (IC) gate can deliver.

NOT Circuit Logic

NOT circuit logic is a control function that allows a load to be energized (ON) as long as the control switch or contact is not activated. NOT circuit logic is accomplished in a hardwired circuit by using a switching device that is normally closed, or in a digital circuit by using an inverter (NOT) digital logic gate to convert a normally open switch into a normally closed output. **See Figure 11-14.**

For example, an NC liquid level switch that is made in a hardwired circuit operates the same way as an NO switch with an inverter in the digital circuit. Without inverters, NO and NC control switches are required to perform circuit operations. NO switches are typically used to start, or allow, the flow of current in a hardwired circuit. NC switches are typically used to stop, or remove, the flow of current in a hardwired circuit. Any NO or NC switch can be changed to the opposite operating function by adding an inverter gate.

The pump and solenoid valve control circuit will operate in the same manner when a digital AND logic gate is used. The advantage of a digital logic gate control circuit is that lower-rated control switches can

Chapter 11—Industrial Control Systems 341

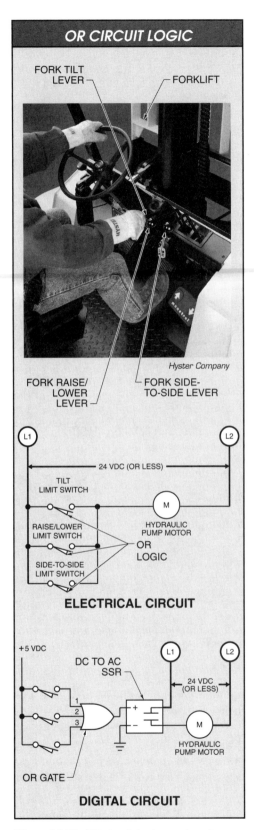

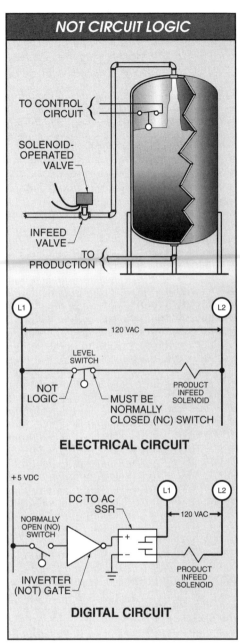

Figure 11-13. OR circuit logic requires that one or more switches or contacts be closed to energize a load.

Figure 11-14. NOT circuit logic energizes a load as long as the input is not activated.

In 1949, the first commercial computer using vacuum tubes was placed into service (UNIVAC 1). The need for improvements in this technology lead to the development of the first transistor in 1954, the first IC (integrated circuit) in 1958, the first microprocessor chip in 1970, and the first IBM PC in 1981.

SERIES AND PARALLEL CONNECTED SOLID-STATE SENSORS

As a rule of thumb, a maximum of three sensors can be connected in series to provide AND logic. Factors that limit the number of AC 2-wire sensors that can be wired in series to provide AND logic include the following:
- **AC Supply Voltage.** Typically, the higher the AC supply voltage, the higher the number of sensors that may be wired in series.
- **Voltage Drop across the Sensor.** Voltage drop varies for different sensors. The lower the voltage drop, the higher the number of sensors that may be connected in series.
- **Minimum Operating Load Voltage.** The minimum operating load voltage varies depending on the load that is being controlled. For every proximity sensor added in series with the load, less supply voltage is available across the load.

As a rule of thumb, a maximum of three sensors can be connected in parallel to provide OR logic. Factors that limit the number of AC 2-wire sensors that can be wired in parallel to provide OR logic can include the following:
- **Photoelectric and Proximity Switch Operating Current.** The total operating current flowing through a load is equal to the sum of each sensor's operating current. The total operating current must be less than the minimum current required to energize the load.
- **Amount of Current a Load Draws When Energized.** The total amount of current a load draws must be less than the maximum current rating of the lowest-rated sensor. For example, when 3 sensors that are rated 125 mA, 250 mA, and 275 mA are connected in parallel, the maximum rating of the load cannot exceed 125 mA.

NOR Circuit Logic

NOR circuit logic is a control function that allows a load to be energized (ON) as long as no control switches or contacts are activated. NOR logic is an expansion of NOT logic where two or more normally closed control switches (or contacts) are used or a NOR digital gate is used.

NOR logic is used in control circuits to detect conditions above the set limit of an input switch setting, such as excessive temperature, pressure, liquid level, current level, voltage level, or count. An example of NOR circuit logic is when a heater is made to turn OFF when the temperature or pressure in a system is above the set limits. **See Figure 11-15.**

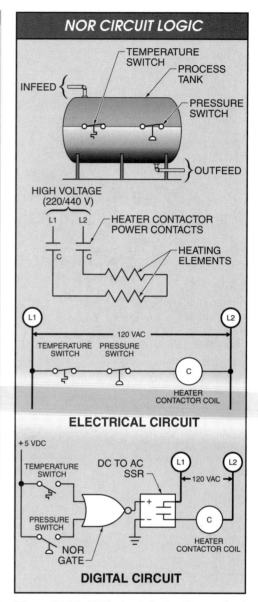

Figure 11-15. NOR circuit logic allows a load to be energized as long as none of the inputs are activated.

In the circuit, a NOR logic function is accomplished by using two NC switches in the electrical circuit or two NO switches and a NOR gate in the digital circuit. In the electrical circuit, the temperature and pressure switches are used to control the heater contactor coil. The heater contactor coil is used to control the heating elements. In the digital circuit, the temperature and pressure switches are used to control an SSR. The SSR is used to control the heater contactor coil.

NAND Circuit Logic

NAND circuit logic is a control function that allows a load to be energized (ON) until all the control switches or contacts are activated. An example of NAND logic can be seen in a solenoid-operated valve that controls the flow of product critical to a process. **See Figure 11-16.** For example, a solenoid-operated valve controls the flow of water though a pipe. The water can flow to a fire sprinkler system, cooling system, process system, or other vital system.

Dangerous problems develop when the flow of water is stopped. However, at times the flow must be stopped. By using NAND circuit logic, more than one switch must be activated to stop the flow. When two key-operated switches are used, both switches must be activated before the flow is stopped. An operator may have one key and a supervisor may have the other key.

NAND circuit logic is applied to other circuits that require more than one switch to be activated. Solenoids can also be used to control an electrically operated prison door lock or any other electrically operated lock.

Combination Circuit Logic

Control circuits are made up of smaller logic circuits combined into a larger interconnected system. Most control circuits will include many or all of the basic circuit logic functions (AND, OR, NOT, NAND, and NOR). When reading an electrical print, it helps to understand and identify the individual logic sections in order to understand the whole circuit operation. This is especially true when working with manufacturer prints because these prints vary somewhat depending upon the manufacturer and the information the manufacturer is willing to release. **See Figure 11-17.**

For example, some circuits provided by manufacturers are part of service bulletins. The control circuit part of the schematic diagram is not drawn in standard ladder diagram format; however, the drawing follows all the rules of ladder diagrams. Three loads, the indoor fan relay (IFR), the contactor (CONT), and the compressor time-delay relay (CTD), are all connected directly to the 24 V power supply to form line 2 (the neutral line).

Also if the jumper wire connecting terminal Y to the CONT (as noted by NOTE #14) was removed, three control switches, the high-pressure switch (HPS), the discharge-temperature switch (DTS), and the low-pressure switch (LPS), would control the timer coil and the timer's normally open time-delayed contact, which controls the coil of the contactor.

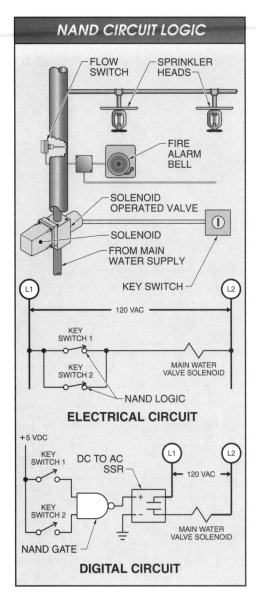

Figure 11-16. NAND circuit logic keeps a load energized until all inputs are activated.

Refer to the CD-ROM "Quick Quizzes" Chapter 11

The four switches (the HPS, DTS, LPS, and the timer contact) controlling the CONT coil use a combination of AND circuit logic and NOR circuit logic. To energize (turn ON) the CONT coil, the normally open timer contact AND the normally open (held-closed if pressure is reached) LPS must be closed, but neither the HPS nor the DTS can be energized. The HPS and DTS switches are used to create the NOR circuit logic function.

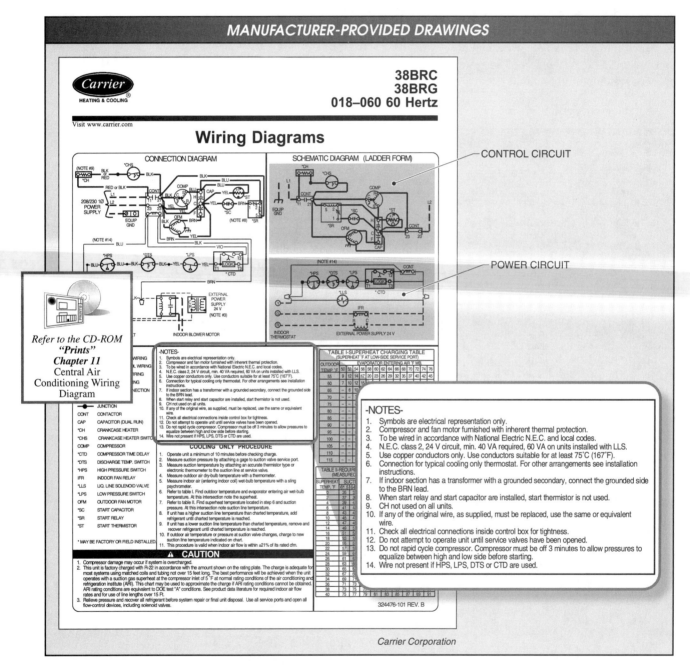

Figure 11-17. When reading manufacturer electrical prints, it helps to understand and identify the individual logics used in order to understand the operation of the whole circuit.

Name _____   Date _____

Activity—Installing an Industrial HVAC Compressor 11-1

Scenario:
Activity 11-1 is an installation activity requiring that a 30 ton compressor and water-cooled condenser be installed for the refrigerated processing room of a breakfast sausage facility.

Task:
Using the Component Location drawing, the Power Circuit schematic, and the Control Circuit ladder diagram, wire the high-pressure switch and the low-pressure switch to the proper terminals of the terminal strip in the control panel. Also, wire the primary side of the control transformer to the main disconnect (120/208 V, 3φ, 4-wire service).

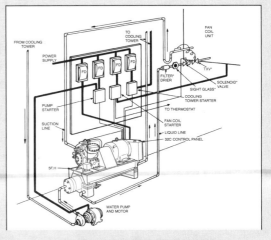

30 TON COMPRESSOR WITH WATER-COOLED CONDENSER

Reference Prints:

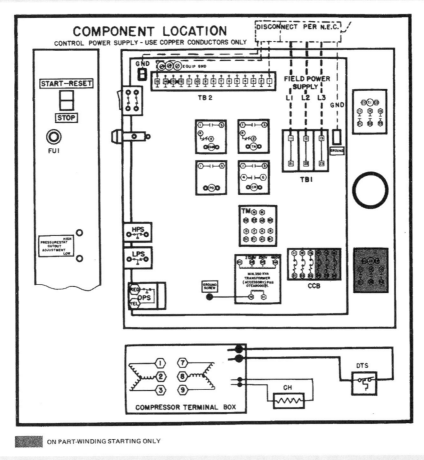

Refer to the CD-ROM "Prints" Chapter 11 HVAC Electrical Component Location Drawing

Carrier Corporation

347

Activity—Installing an Industrial HVAC Compressor

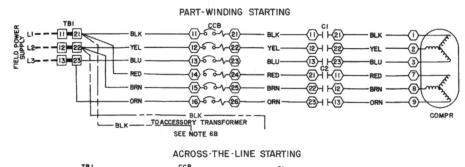

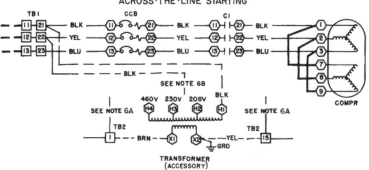

POWER CIRCUIT

Refer to the CD-ROM
"Prints"
Chapter 11
HVAC Compressor –
Power and Control
Circuits

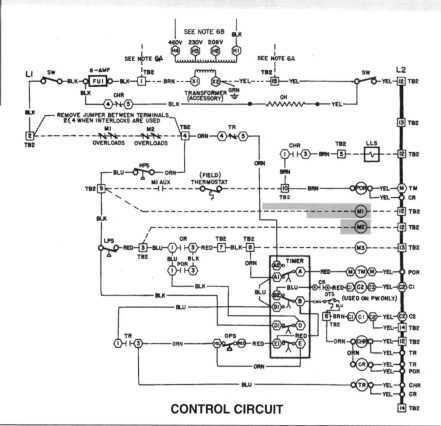

CONTROL CIRCUIT

Carrier Corporation

Name _____ Date _____

Activity—Troubleshooting a Motor Control Circuit 11-2

Scenario:
Activity 11-2 is a troubleshooting activity requiring that a motor control circuit for a corn oil deodorizer feed pump be investigated for problems. The motor control circuit uses an electric motor drive with a bypass contactor (for drive or circuit failure) to control the pump motor.

Task:
To troubleshoot the motor control circuit, the prints must first be found. To make the prints easier to understand, fill in the control circuit line reference numbers, wire reference numbers, and numerical cross-reference numbers for the control circuit prior to troubleshooting the wired circuit.

Square D Company

DRIVE AND BYPASS CONTACTOR MOTOR CONTROL ENCLOSURE

Reference Print:

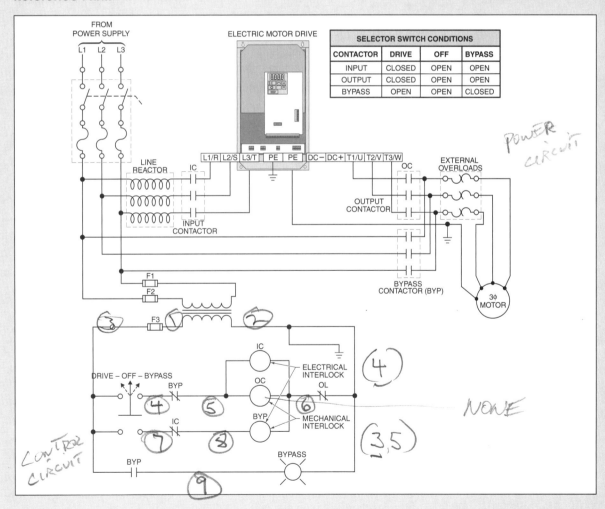

Activity—Troubleshooting a Motor Control Circuit

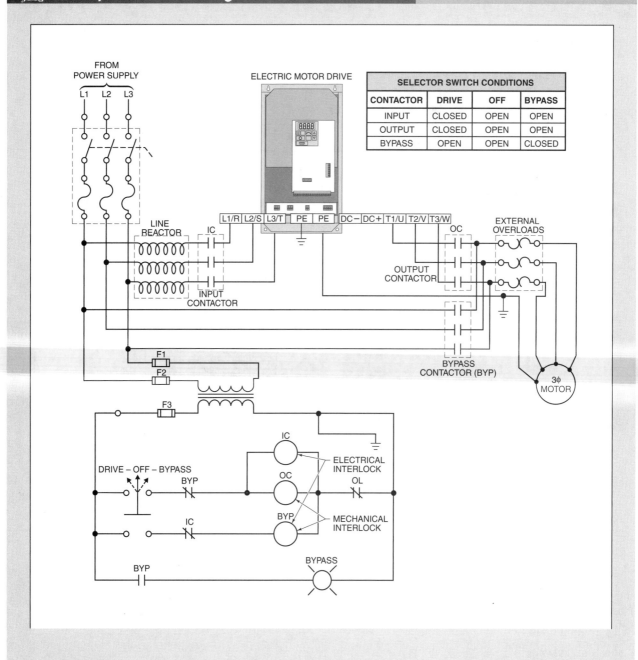

Industrial Power Systems

Printreading for Installing and Troubleshooting Electrical Systems

Electricity can be easily transmitted, controlled, and changed into another form of energy, such as heat, light, sound, or mechanical energy. The production of electricity begins with a generator, battery, photocell, or thermocouple, depending on the type of electricity desired (AC or DC) and the amount of watts required. The produced electricity ends at a heating element, lamp, speaker, motor, solenoid, visual display such as a monitor screen, or any other component that consumes electrical energy.

POWER DISTRIBUTION

Power distribution refers to everything between the generator that produces the electricity and the components that consume it. Automobiles have small power distribution systems that exist between the battery and the loads (solenoids and motors). Utility companies have large power distribution systems between the generating plant and the customer (residential, commercial, and industrial). **See Figure 12-1.** Power distribution from a utility company begins at the generator and ends at the service entrance of a building or facility.

Power distribution for the customer starts with the service entrance of a facility and ends with the loads (whether lamps, motors, computers, and heating elements, for example). **See Figure 12-2.** Power distribution for almost everyone that works in the electrical field, other than utility workers, is a system that takes the power from the service entrance of a building and delivers it throughout the building to the loads.

TYPES OF POWER DISTRIBUTION

Electrical utility companies distribute high-voltage 3φ power produced by AC generators. The voltage created by these generators is stepped up to 245,000 V, with the amount of current reduced to 1 A by transformers, which allow the size of transmission line wires to be smaller. As a transmitted high voltage nears a customer, the voltage is stepped down to a safer voltage level, such as 12,500 V, for delivery to a building. **See Figure 12-3.** Residential customers use low-voltage 1φ power, with commercial and industrial customers using both low-voltage 1φ power and high-voltage 3φ power.

354 Printreading for Installing and Troubleshooting Electrical Systems

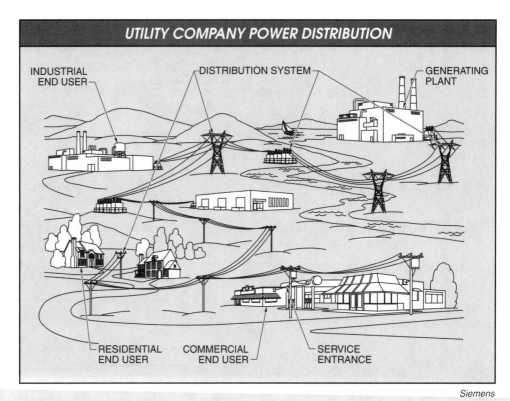

Siemens

Figure 12-1. *Power distribution from a utility company begins at the generator and ends at the service entrance of a building or facility.*

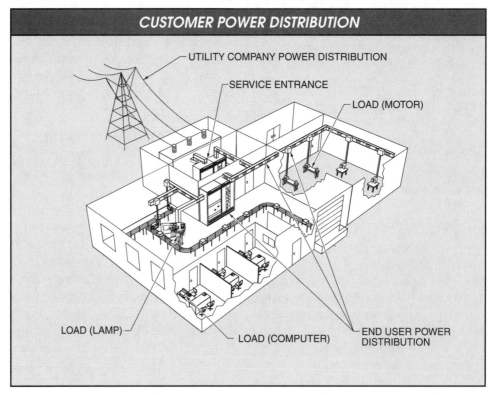

Figure 12-2. *Customer power distribution starts at the service entrance and ends at the loads.*

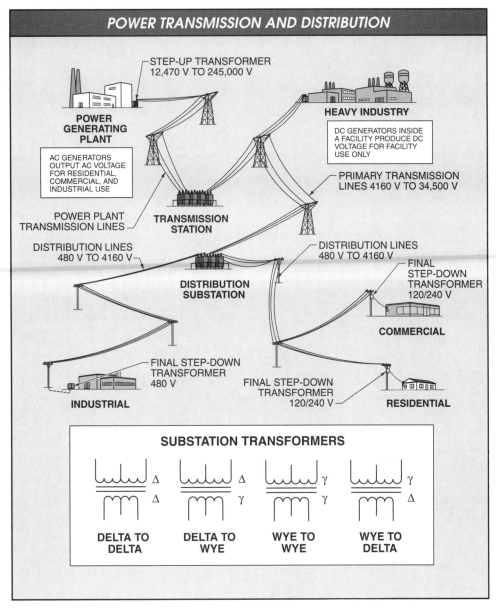

Figure 12-3. AC voltage from utility company generators is stepped up to 245,000 V with current reduced to 1 A. As transmitted voltages near the customer, the voltage is stepped down to 12,500 V.

On November 16, 1896, the first large-scale generation of AC power by the Niagara Falls generation facility was started. It was thought that the Niagara Falls facility would supply all the electrical power ever needed by the entire east coast of the U.S. The first generators operated at a slow 25 Hz, but were later converted to 60 Hz (the U.S. frequency standard).

Power companies primarily use 1ϕ transformers connected in several different configurations to deliver the correct amount of power to customers. **See Figure 12-4.** Three transformers (except in an open-delta system) are typically interconnected to deliver 3ϕ voltage to commercial and industrial customers. The exact voltage level delivered is determined by customer needs and the distribution system of the utility company.

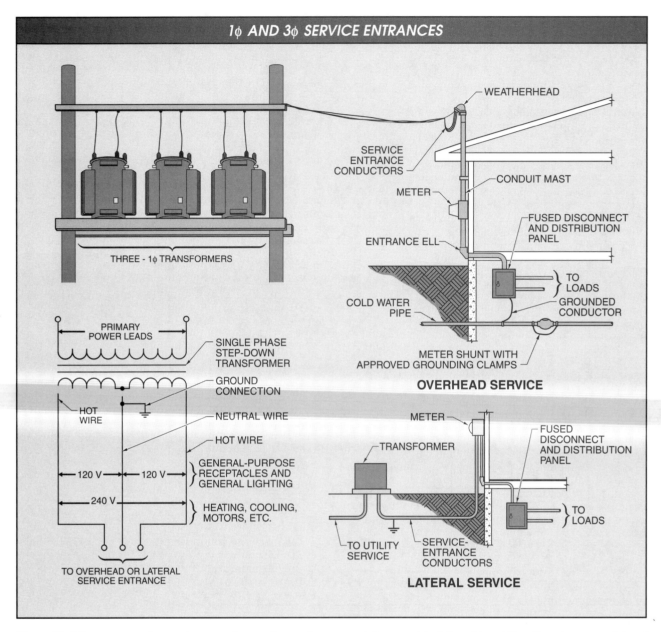

Figure 12-4. *Power companies use three transformers (except in an open-delta system) to deliver 1ϕ and 3ϕ voltage to commercial and industrial customers.*

For example, residential buildings receive 120/240 V, 1ϕ, 3-wire service as standard service, while schools and office buildings typically receive 120/208 V, 3ϕ, 4-wire service. The service to a building is accomplished by using an overhead service or lateral (underground) service.

Industrial and commercial power requires the same 1ϕ voltage need as residential customers in addition to requiring 3ϕ voltage for high-power loads. Similar to residential buildings, apartment buildings and office buildings require lighting and low-voltage 1ϕ for small appliance circuits, except on a much larger scale. Other commercial buildings, such as bakeries and restaurants, require about the same 1ϕ lighting circuits and 3ϕ high-voltage appliance circuits as some industrial systems, except on a smaller scale. Commercial systems such as outdoor sports stadiums, require many large-scale lighting circuits.

The distribution system used varies based on the application, but a few common systems are used for industrial applications. **See Figure 12-5.** The most common distribution systems include the following services:

- 120/240 V, 1ϕ, 3-wire
- 120/208 V, 3ϕ, 4-wire
- 120/240 V, 3ϕ, 4-wire
- 277/480 V, 3ϕ, 4-wire

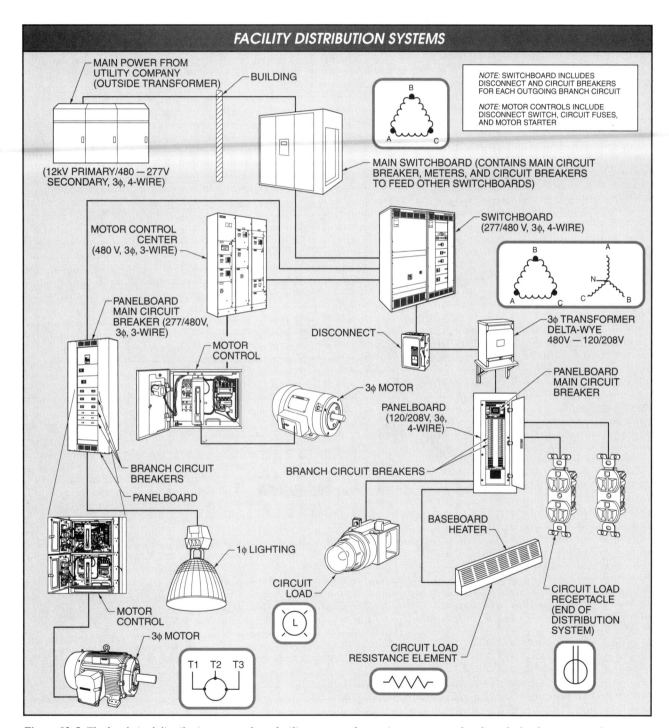

Figure 12-5. The hardwired distribution system for a facility starts at the service entrance and ends at the loads or receptacles.

120/240 V, 1ϕ, 3-Wire Service

A 120/240 V, 1ϕ, 3-wire service is used to supply power to customers that require 120 V and 240 V 1ϕ power. This service provides all of its power to 120/240 V 1ϕ circuits. The neutral wire is grounded, therefore, the neutral wire cannot be fused or have a switch at any point. **See Figure 12-6.**

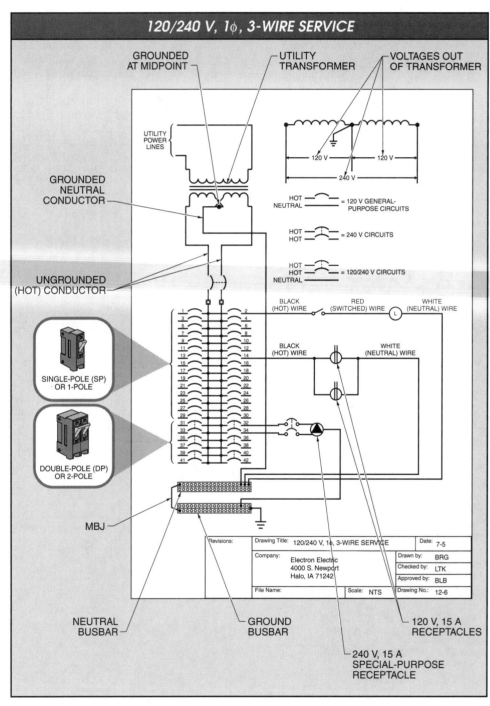

Figure 12-6. As part of all residential, commercial, and industrial power distribution systems, 120/240 V, 1ϕ, 3-wire service supplies power for 120 V and 240 V loads.

A 120/240 V, 1ϕ, 3-wire service is typically used for interior wiring for lighting and small appliances. For this reason, it is the primary service used to supply most residential buildings. It is also used for commercial and industrial applications to supply power to 1ϕ loads and receptacles, although in commercial and industrial applications, large power panels or a large number of panels are used.

In commercial and industrial wiring, small 1ϕ loads include more than the typical small appliances found in a residence. This is because commercial and industrial appliance circuits typically include large lighting loads; high-power, 1ϕ equipment such as copying machines, commercial refrigerators, ovens, and washers; and various appliances such as hot tubs and saunas. In hotels, automatic ice machines and heating elements embedded in the concrete of entranceways are included. Other commercial and industrial small appliance circuits include office computers, printers, motors less than 5 HP, cooking equipment, security equipment, large entertainment systems, and hospital equipment.

In addition to the various appliances, commercial and industrial circuits often require duplication because there may be hundreds of rooms or offices at one location. Panels for 120/240 V 1ϕ service are included in commercial and industrial systems even when the main utility power delivered is 3ϕ. The 1ϕ power panels are tapped off of the 3ϕ power system.

Receptacle and Plug Configurations

A hardwired distribution system in a facility delivers power directly to loads without any switching. When the loads are portable, or not known, power is delivered to convenient locations by receptacles. Receptacles have various configurations so only matching plugs can be connected to a specific receptacle. The plug-to-receptacle matching system prevents the connection of equipment to the wrong power source. For example, a plug rated 125 V, 20 A cannot be connected to a receptacle rated 125 V, 15 A.

TESTING RECEPTACLES

Test Light Application

When a receptacle is properly wired, a test light bulb will illuminate when the test light leads are connected from the neutral slot to the hot slot. A test light bulb will also illuminate when the test leads are connected from the ground slot to the hot slot. If the test light illuminates when the test leads are connected from the neutral slot to the ground slot, the ground (green) and neutral (white) wires are reverse wired. The condition of having the hot and neutral wires reversed is a safety hazard and must be corrected.

If a test light illuminates when the test leads are connected to the neutral slot and hot slot but does not light when connected to the ground slot and the hot slot, the receptacle is not grounded. When a test light illuminates but is dimmer than when connected between the neutral slot and hot slot, the receptacle has an improper ground. An improper ground is also a safety hazard and must be corrected.

Receptacle Tester Application

Testing receptacles is also possible with a receptacle tester. A receptacle tester is a device that is plugged into a standard receptacle to determine if the receptacle is properly wired. The three indicator lights indicate if a receptacle is wired correctly or if there is a fault (any of five faults identified) with the receptacle.

The prong configuration of a receptacle determines the available voltage and current at that location. By looking at the configuration of the receptacle, the type of building power distribution system for that circuit can better be understood. **See Figure 12-7.**

When electrical power is constant, changes in voltage will produce corresponding changes in current. For example, electrical power (P) is applied voltage (E) multiplied by current (I), thus the higher the voltage, the lower the current. This is an application of the power formula ($P = E \times I$). The power formula is why high-power electrical loads are designed for higher voltages whenever possible.

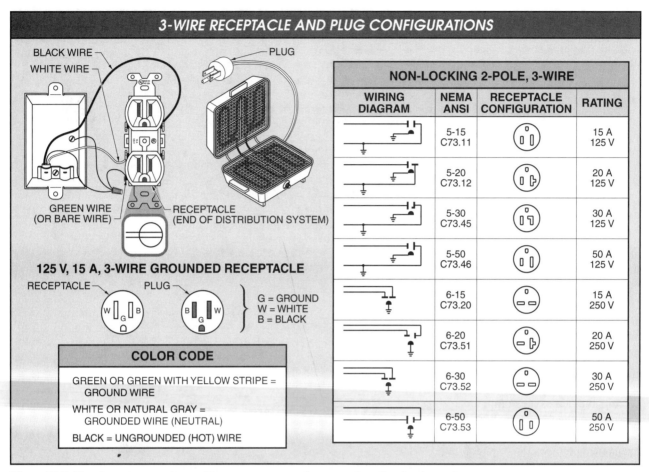

Figure 12-7. The plug-to-receptacle matching system prevents the connection of equipment to the wrong power source.

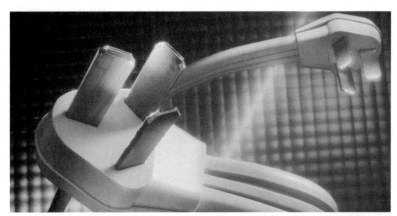

Cooper Wiring Devices

Some plug and receptacle configurations, such as the standard appliance plug and receptacle, are very common and can be found in various amperage ratings: NEMA 10-20P (20 A), 10-30P (30 A), and 10-50P (50 A).

Single-phase high-power loads such as electric heating elements, furnaces, and water heaters are designed to operate at 240 V instead of 120 V. Like 120 V plugs (which actually have a 125 V rating because 125 V is the maximum voltage allowed), 240 V receptacles also have a configuration that matches a 250 V current rating and are designed so smaller- and larger-current-rated plugs will not fit into the 240 V configuration. Some 240 V receptacles are wired to provide only 240 V and a ground, and other receptacles are wired to provide both 240 V and 120 V. **See Figure 12-8.**

Receptacles and plugs may be nonlocking or locking. Nonlocking plugs fit into nonlocking receptacles in a firm but nonlocking manner. Locking plugs fit into locking receptacles, with the plug inserted and twisted to lock it into place. Once locked, locking plugs cannot be pulled out without first untwisting the plug and receptacle. Because appliances and machinery come in various voltage and current sizes, plugs and receptacles also come in various sizes. **See Appendix.**

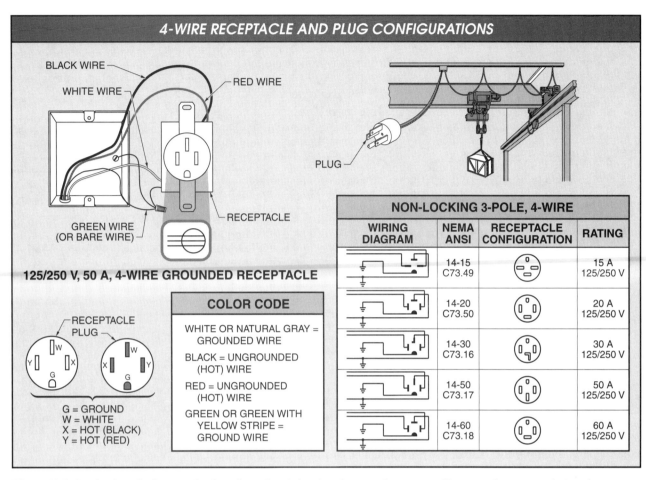

Figure 12-8. Single-phase high-power loads such as electric heating elements, furnaces, and hot water heaters are designed to operate only on 240 V, while some loads can operate on either 240 V or 120 V.

In a global world, equipment prints from various countries are common. In the U.S., "ground" is a common term, but in much of the world "earth" is used. The terms "grounding" and "earth connection" on electrical prints mean exactly the same thing.

Grounding

A power distribution system typically includes a conductor for grounding the system and any equipment connected to the system in addition to the normal current-carrying conductors (hot/ungrounded, hot/switched, and neutral wires). Electrical circuits are grounded to help prevent fires and electrical shock, and to ensure fast operation of circuit overcurrent protection devices. A *ground* is the connection of all exposed non-current-carrying metal parts of an electrical system to earth. The purpose of grounding is to provide a safe path for a fault current to flow. A complete grounding path must be maintained when installing electrical equipment in any facility.

When using metal conduit (rigid or EMT), the conduit can be used to provide a grounded metal pathway back to the service panel. **See Figure 12-9.** When using nonmetallic cable such as Romex, a bare or green insulated copper wire serves as the grounding path. Grounding wires are connected to all receptacles and metal boxes to provide a continuous pathway for short-circuit current. A bare or green ground wire may also be included even when conduit is being used.

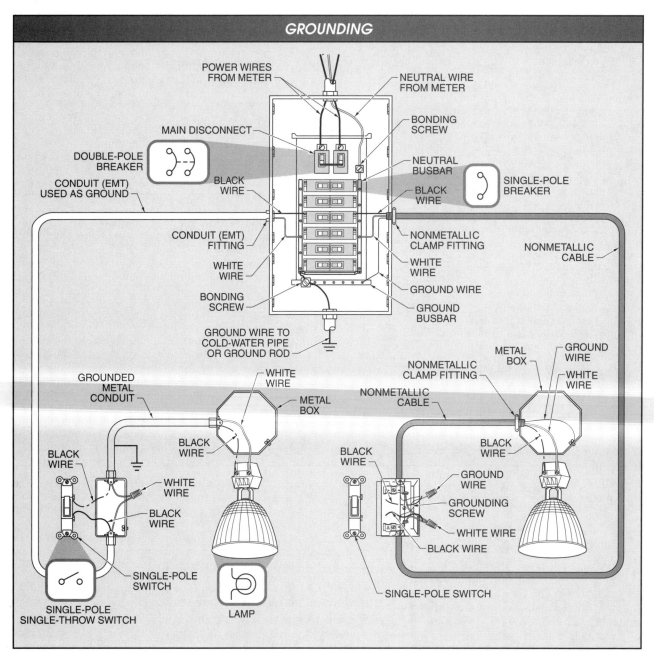

Figure 12-9. *A ground wire is part of all power distribution systems and is needed to prevent fires and also to prevent electrical shock to personnel.*

NEC® Phase Arrangement and High-Phase Markings

Large commercial facilities and all industrial buildings have service entrances that are supplied 3φ power from the utility company. Three-phase circuits include three individual ungrounded (hot) power lines. The power lines are referred to as phases A (L1), B (L2), and C (L3). Phases A, B, and C must be connected to a panelboard or switchboard according to NEC® subsection 408.3(E). The phases must be arranged A, B, C from front to back, top to bottom, or left to right as viewed from the front of the panelboard or switchboard. **See Figure 12-10.**

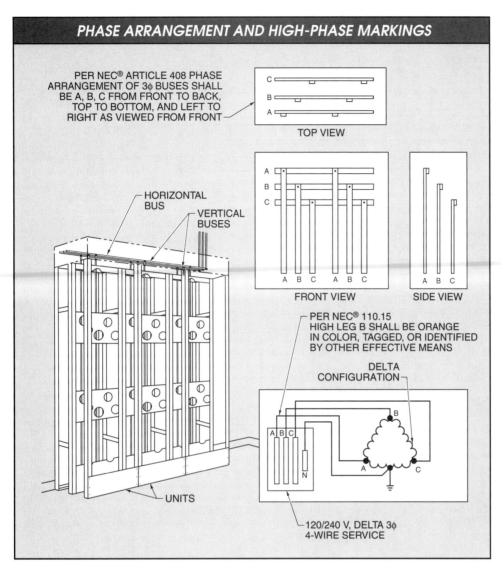

Figure 12-10. NEC® Article 408 requires that the phases in a panelboard or switchboard be arranged A, B, C from front to back, top to bottom, or left to right as viewed from the front of the panelboard or switchboard.

A delta high-voltage leg must be used as phase B and must be colored orange (or clearly marked) when the panelboard or switchboard is fed from a 120/240 V, 3ϕ, 4-wire service. The high-voltage leg of a delta service must be marked because there are 195 V between phase B and the neutral (grounded conductor) conductor. This amount of voltage is considered a source of power because although 195 V is too high for 120 V loads, it can be used for some 1ϕ, 240 V loads.

120/208 V, 3ϕ, 4-Wire Service

A 120/208 V, 3ϕ, 4-wire service is used to supply commercial customers (such as schools and office buildings) and industrial customers (such as warehouses and factories) that require a large amount of 120 V and 208 V 1ϕ power and low-voltage 3ϕ power. A 120/208 V, 3ϕ, 4-wire service includes three ungrounded (hot) lines and one grounded (neutral) line. Each hot line has 120 V to ground when used with the neutral line. **See Figure 12-11.**

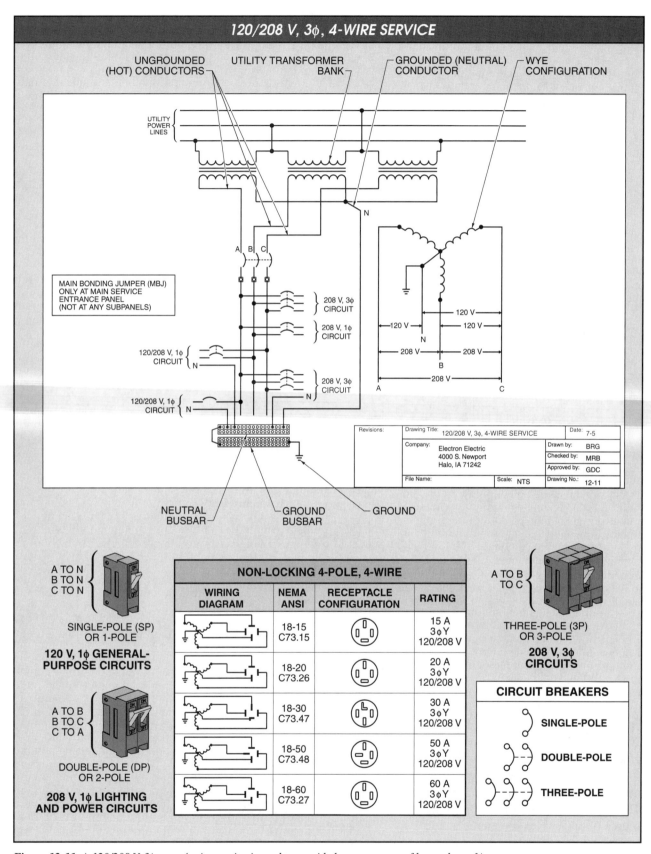

Figure 12-11. A 120/208 V, 3ϕ, wye 4-wire service is used to provide large amounts of low-voltage 1ϕ power.

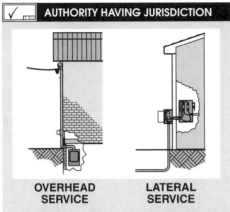

AUTHORITY HAVING JURISDICTION

OVERHEAD SERVICE LATERAL SERVICE

The AHJ has the final word regarding electrical installations no matter what the NEC® states. Because local jurisdiction may require service installation practices that are different from NEC® practices, be sure which practices meet AHJ and NEC® requirements.

For example, many local AHJs require that all electrical services be installed in metal conduit despite the wiring methods the NEC® recognizes. Communication with the local AHJ must be maintained so there are no surprises at the end of a service installation job.

A 120/208 V, 3ϕ, 4-wire service is used to provide large amounts of low-voltage (120 V, 1ϕ) power. All 120 V circuits must be balanced by equally distributing power consumption among the three hot lines (L1, L2, and L3). Phase balancing is accomplished by alternately connecting the 120 V circuits to power panels so each phase, A to N, B to N, and C to N, is divided among the loads (lamps and receptacles). Likewise, 208 V, 1ϕ loads such as 208 V lamps and heating appliances must also be balanced between the three phases (A to B, B to C, and C to A).

Three-phase loads, such as heating elements designed to be connected to 3ϕ power, can be connected to phases A, B, and C. Three-phase motors can also be connected to 3ϕ power lines. However, the applied 3ϕ voltage to the motor will only be 208 V and may present a problem when using some 3ϕ motors. The problem occurs because 208 V is at the lowest end of the 3ϕ voltage range. Therefore, any voltage variation in an AC induction motor will affect motor operation.

277/480 V, 3ϕ, 4-Wire Service

A 277/480 V, 3ϕ, 4-wire service is the same as a 120/208 V, 3ϕ, 4-wire service except that the voltage levels are higher. A 277/480 V, 3ϕ, 4-wire service includes three ungrounded (hot) lines and one grounded (neutral) line. Each hot line has 277 V to ground when measured to the neutral, or 480 V when measured between any two hot lines (A to B, B to C, or C to A). **See Figure 12-12.**

A 277/480 V, 3ϕ, 4-wire service provides 277 V, 1ϕ or 480 V, 1ϕ power, but not 120 V, 1ϕ power. For this reason, 277/480 V, 3ϕ, 4-wire service is not used to supply 120 V, 1ϕ power for general lighting and appliance circuits. However, the service can be used to supply 277 V, and 480 V, 1ϕ lighting circuits. Such high-voltage lighting is used in commercial and industrial fluorescent and HID lighting circuits.

A system that cannot deliver 120 V, 1ϕ power appears to have limited use. However, in many commercial applications (sport complexes, schools, offices, and parking lots), lighting is the major part of the electrical system. Because large commercial applications include several sets of transformer banks, 120 V, 1ϕ power is available through other transformers. Additional transformers can also be connected to a 277/480 V, 3ϕ, 4-wire service to reduce the voltage to 120 V, 1ϕ. **See Figure 12-13.**

For example, a 120 V, 250 W high-pressure sodium lamp has a typical operating current of 2.70 A; a 240 V, 250 W lamp draws 1.36 A; and a 480 V, 250 W lamp draws 0.65 A. When the numbers are substituted into the power formula, the result is that the higher the voltage rating of a lamp, the less operating current the lamp requires. A number of 250 W lamps can be connected with other lamps of the same wattage rating to a 15 A circuit. The number of lamps will vary based on the actual applied voltage.

366 Printreading for Installing and Troubleshooting Electrical Systems

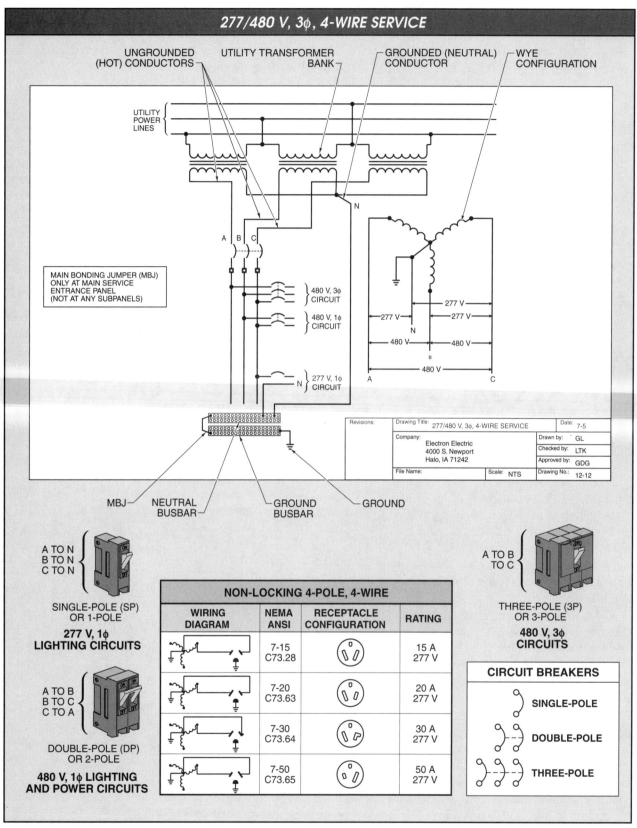

Figure 12-12. Each hot line of a 277/480, 3ϕ, wye 4-wire service has 277 V to ground when measured to the neutral or 480 V when measured between any two hot lines (A to B, B to C, or C to A).

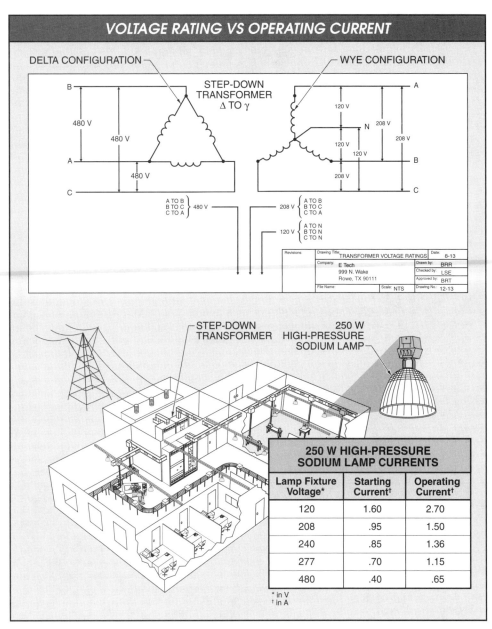

Figure 12-13. When component numbers are placed into the power formula, the higher the voltage rating of a component, the less operating current the component requires when wattage (power) remains the same.

120/240 V, 3ϕ, 4-Wire Service

A 120/240 V, 3ϕ, 4-wire service is used to supply customers with large amounts of 3ϕ power and some 120 V and 240 V, 1ϕ power. A 120/240 V, 3ϕ, 4-wire service supplies 1ϕ power delivered by one of the three transformers and 3ϕ power delivered by all three transformers. Single-phase power is provided by center-tapping any one of the transformers. **See Figure 12-14.**

Typically, only one transformer delivers all of the 1ϕ power. A 120/240 V, 3ϕ, 4-wire service is used in applications that require mostly 240 V, 3ϕ power and some 120 V, 1ϕ power. However, large power-consuming electric loads such as air compressors and pumps with motors over 1 HP, or any machine with motors over 1 HP, and electric heating elements are typically connected to 3ϕ power.

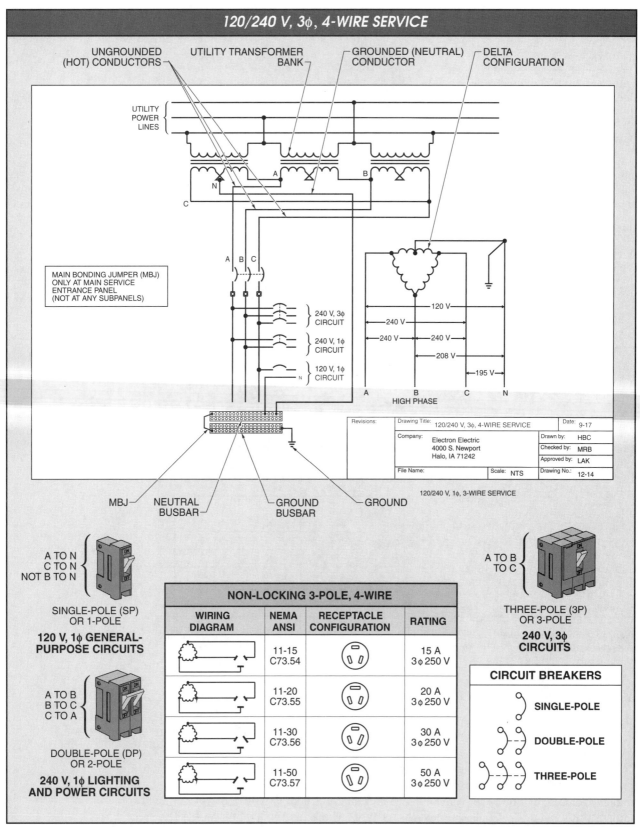

Figure 12-14. 120/240 V, 3φ, 4-wire service is used by customers that require large amounts of 3φ power and some 120 V and 240 V 1φ power.

In many commercial applications and all industrial applications, the total amount of 1ϕ power used is small when compared to the total amount of 3ϕ power used. When large amounts of 1ϕ power are needed, the size of the transformer that supplies the 1ϕ loads can be increased.

Power Distribution System Conductor Color-Coding

Conductors (wires) are covered with an insulating material that is available in various colors. The advantage of using conductors that have different colors is that the function of each conductor can be easily determined. Some colors have precise meanings; for example, the color green always indicates a conductor used for grounding. Other colors may have more than one meaning depending on the circuit; for example, a red conductor may be used to indicate a hot wire in a 240 V circuit or a switched wire in a 120 V circuit.

Conductor color-coding makes balancing loads between the three phases easier and helps with troubleshooting. For this reason, conductor color-coding is also used in applications that do not require every conductor to be color-coded. Standard colors (green, white, black, and red) should always be used where applicable. **See Figure 12-15.** Green (or green with a yellow stripe) and white (or gray) are required conductor colors per Article 210.5 of the NEC®.

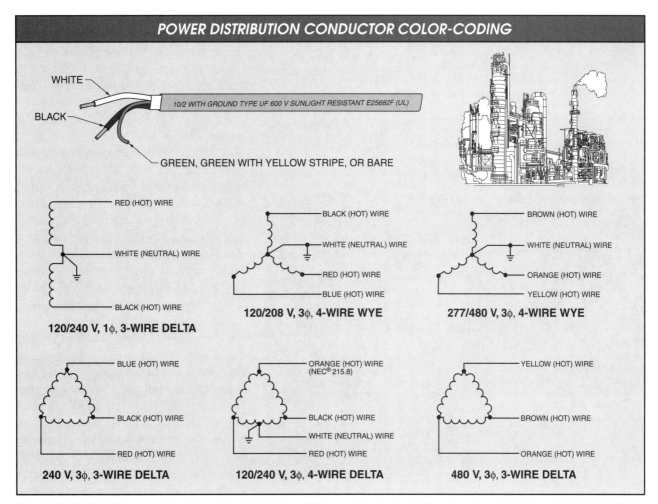

Figure 12-15. The advantages of using conductors that have different colors is that the function of each conductor can be easily determined, balancing loads between the three phases is easier, and troubleshooting is easier.

Green or green with a yellow stripe is the standard color of a grounding conductor in the electrical field. Solid green is the most common color used. Green is used to indicate a grounding conductor regardless of the voltage level or whether the circuit is 1ϕ or 3ϕ. A grounding conductor is a conductor that only carries current during a fault such as a short circuit.

A neutral conductor is a current-carrying conductor that is intentionally grounded. White or gray is used for the neutral conductor. Neutral conductors carry current from one side of a load to ground. Neutral conductors are connected directly to loads and the grounded power source and are never connected through fuses, circuit breakers, or switches.

Electrical circuits include ungrounded (hot) conductors in addition to grounding and neutral conductors. An ungrounded conductor is a current-carrying conductor that is connected to loads through fuses, circuit breakers, and switches. Ungrounded conductors can be any color other than white, gray, green, or green with a yellow stripe. Black is the most common color used for ungrounded conductors.

Red, blue, brown, orange, and yellow are also used for ungrounded conductors. Although these colors are used to indicate hot conductors, the exact color used to indicate specific hot conductors (L1, L2, and L3) will vary. The one exception to the color-coding rule is listed in Article 215.8 of the NEC®. Article 215.8 states that in a 4-wire delta-connected secondary system, the high-voltage phase must be colored orange (or clearly marked) because the 195 V leg is too high for 120 V, 1ϕ power.

Busways

An electrical distribution system must deliver power from the electrical service or the power source to the loads. In residential and most commercial systems this power distribution is accomplished using conductors. Industrial systems also use conductors for power distribution, unless the distribution system needs to be flexible due to the need to relocate production machinery. A flexible manufacturing facility must have a power distribution system that can be easily changed to accommodate a changing machinery floor plan.

A *busway* is an electrical distribution system made up of busbars inside metal-enclosed boxes (busducts) that are available in prefabricated sections. Prefabricated fittings such as tees, elbows, and crosses simplify the connecting and reconnecting of power distribution systems. By bolting busway sections together, electrical power can be made available at many locations throughout a system. **See Figure 12-16.**

A busway does not have exposed conductors because the power in a plant distribution system is at a high level. To offer protection from the high voltage, the conductors of a busway power distribution system are supported with insulating blocks and covered with an enclosure to prevent accidental contact. A typical busway power distribution system provides for the quick connection and disconnection of machinery. Facilities can be retooled or re-engineered without major changes in the electrical distribution system.

Ruud Lighting, Inc.

Busways can be assembled, disassembled, and reassembled as many times as required to meet the requirements of an ever-changing production floor.

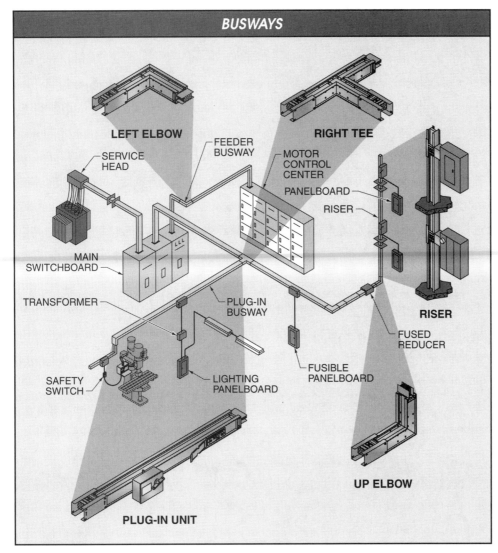

Figure 12-16. *Busway power distribution systems use fittings such as tees and elbows to provide flexibility in the system.*

The most common length of busway is 10′. Shorter lengths are typically purchased and used as required. Prefabricated elbows, tees, and crosses make it possible for electrical power to run up, down, around corners, and to be tapped off the distribution system at any point. Prefabricated busways allow simple and easy connections to be made when new systems are installed and during system modifications.

The two basic types of busways are feeder and plug-in busways. **See Figure 12-17.** Feeder busways deliver the power from the power source to a load-consuming component. Plug-in busways serve the same function but also allow load-consuming components to be conveniently added anywhere along the bus structure. A plug-in power module is used on a plug-in busway system.

The three general types of plug-in power panels used with busways are fusible switch, circuit breaker, and specialty plug. The fusible switches and circuit breaker plug-in panels allow the system to be tapped at almost any point so that conduit and wire can be run to a machine or load.

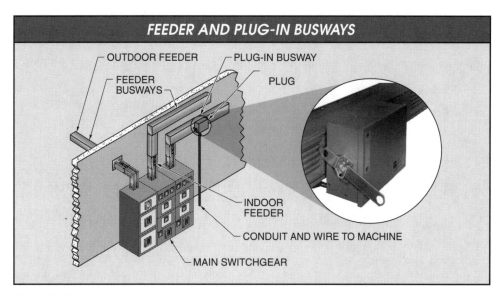

Figure 12-17. Feeder busways deliver power from the power source to load-consuming components, and plug-in busways allow load-consuming components to be conveniently added anywhere along the bus structure.

INDUSTRIAL POWER CIRCUIT APPLICATION

Most industrial circuits include a system for moving product along a line, such as a conveyor. Conveyor motion is created by motors, which are controlled by motor control circuits that include magnetic motor starters or drives. The product is modified or changed at various points along the line. For example, a product can be sent through a shrink-wrap process before being sent to a caser and palletizer. **See Figure 12-18.**

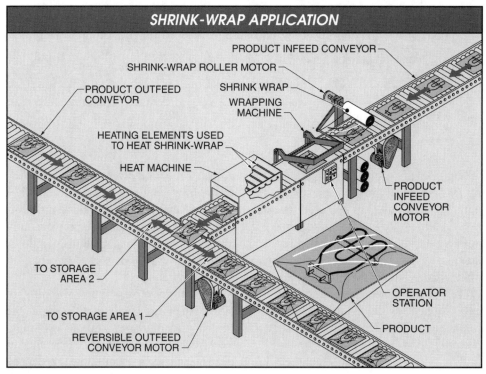

Figure 12-18. In a shrink-wrap application, conveyors move product to a wrapping machine that applies a shrink-wrap to the product.

In a shrink-wrap application, conveyors move the product to a wrapping machine that applies a shrink-wrap to the product. The shrink-wrap process includes numerous controls for controlling the required motors, heating elements, and assorted loads. Motor controls for the infeed conveyor, shrink-wrap motor, heating elements, and outfeed conveyor are typically grouped together and connected to a common operator station. **See Figure 12-19.**

POWER SYSTEM TESTS

There are several tests that can be conducted on an electrical power system. Loose connections cause high resistance, which results in electrical equipment overheating. Infrared thermography or temperature measurements are used to identify hot spots in electrical equipment.

Unbalanced loads cause early failure in electrical power system equipment. An ammeter measurement on each phase will identify any significant unbalance. Harmonic currents cause overheating and damage to power system devices. A power quality analyzer can be used to identify the presence and amount of harmonics. In addition, a reading from a true-rms ammeter can be compared to the reading from an analog ammeter. If there is a large difference between the two readings, there are probably significant harmonics in the system.

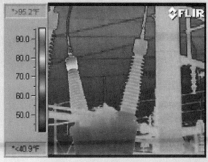
FLIR Systems

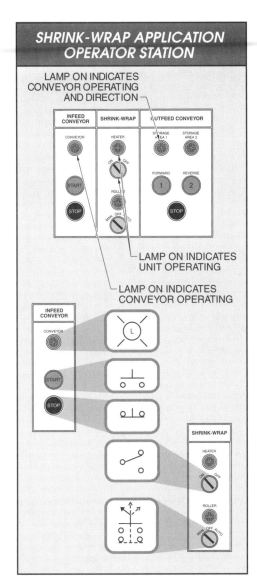

Figure 12-19. The infeed conveyor, shrink-wrap motor, heating elements, and outfeed conveyor are typically grouped together and connected to a common operator station.

The operator station is typically designed to control the loads with many different levels of control. In the shrink-wrap example, the infeed conveyor motor is controlled by a standard start/stop pushbutton. Included with the pushbutton is a lamp that indicates when the motor is running. The heating elements used to heat the shrink-wrap after it is applied are controlled by a selector switch and a temperature switch. The setting of the selector switch determines when the heating elements are to be operating, and the temperature switch automatically maintains a set temperature. Also included at the operator station is a lamp that indicates when the heating elements are energized.

The shrink-wrap roller motor is controlled by a selector switch and a limit switch. The limit switch automatically controls the motor when product is detected and the selector switch is in the automatic position. The manual position of the selector switch allows the motor to be operated manually. A lamp is included that indicates when the shrink-wrap roller motor is operating. **See Figure 12-20.**

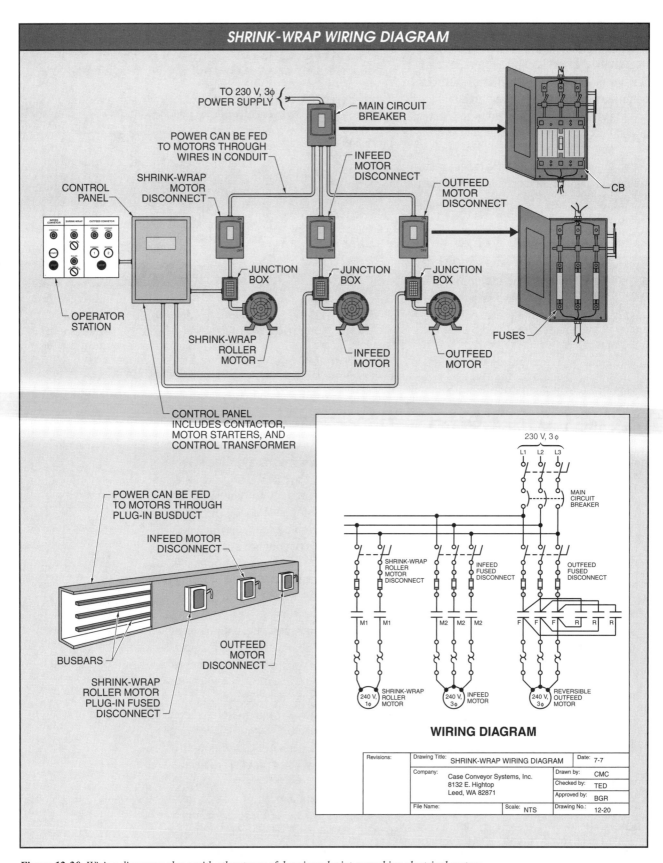

Figure 12-20. Wiring diagrams along with other types of drawings depict a machine electrical system.

The outfeed conveyor motor is controlled by a standard forward/reverse/stop pushbutton. The forward motor direction sends the product to Storage Area 1. The reverse motor direction sends product to Storage Area 2. Lamps are used on the control panel to indicate the direction of the motor.

A power circuit and control circuit are required to operate the shrink-wrap system. The power circuit is connected to the building's power distribution system. The loads in the power circuit are selected (single-voltage loads) or wired (dual-voltage loads) to be compatible with the supply voltage. **See Figure 12-21.**

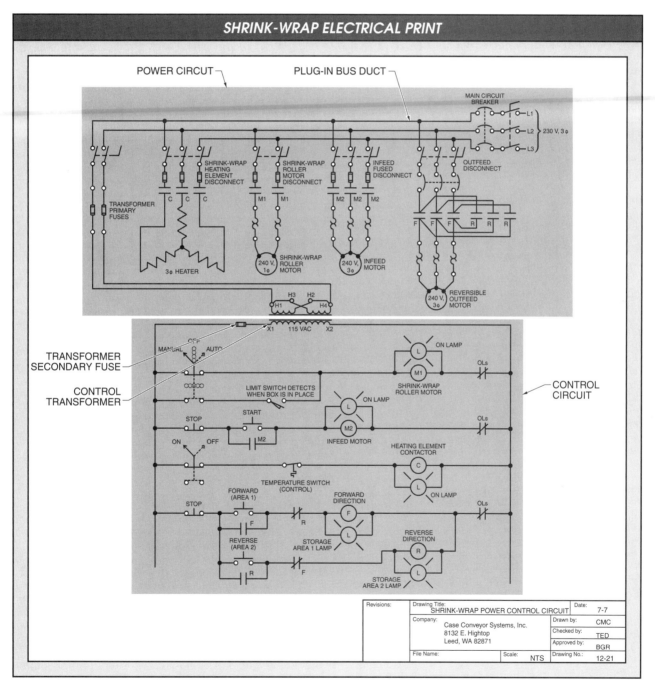

Figure 12-21. Machine electrical prints require a power circuit and a control circuit.

ASI Robicon
All the power circuit equipment for a specific process or part of a system is typically enclosed together in a section of an electrical room.

In the shrink-wrap application, power is connected to a 240 V, 3ϕ power distribution system and includes a 240 V, 3ϕ heating element, a 240 V, 1ϕ shrink-wrap roller motor, a 240 V, 3ϕ infeed motor, a 240 V, 3ϕ outfeed motor, and 240 V to the primary (high-voltage) side of the control transformer. Also included in the power circuit are the required disconnects, protection devices, contactor, and motor starters for each motor.

The control circuit includes the contactor coil, motor starter coils, lamps, selector switches, pushbuttons, a temperature switch, and a limit switch. Also included in the control circuit is the secondary (low-voltage) side of the transformer, a fuse, and motor starter overload contacts (OLs).

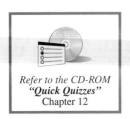
Refer to the CD-ROM "Quick Quizzes" Chapter 12

Name _____ Date _____

Activity—Installing an Electrical Service for Air Conditioning 12-1

Scenario:
Activity 12-1 is an installation activity requiring that an electrical service be purchased for a new air conditioning system that will be installed in the office of a factory.

Task:
Determine which electrical service should be used for supplying power to the office air conditioning system.

Possible Services:
120/240 V, 1φ, 3-wire service   277/480 V, 3φ, 4-wire service
120/208 V, 3φ, 4-wire service   120/240 V, 3φ, 4-wire service

Reference Prints:
Notes:
1. Factory wiring is in accordance with the NEC®. Field modifications or additions must be in compliance with all applicable codes.
2. Use copper conductors only.
3. **CAUTION:** Not suitable for use on systems exceeding 150 V to ground.
4. Manual reset circuit breaker is integral to the transformer.

ASI Robicon

SMALL 3φ FILTERED POWER SUPPLY

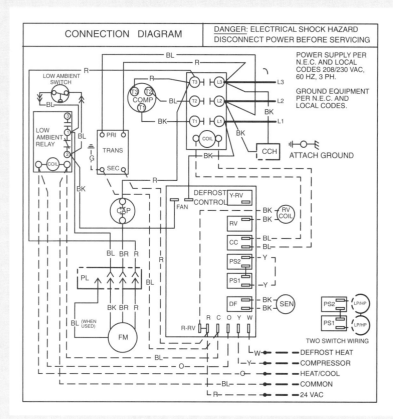

LEGEND

C	—	Contractor
CAP	—	Capacitor
CCH	—	Crankcase Heater
COMP	—	Compressor
DFC	—	Defrost Control
DFR	—	Defrost Relay
FM	—	Fan Motor
HP	—	High-Pressure
LAS	—	Low Ambient S
LAR	—	Low Ambient R
LP	—	Low-Pressure S
PL	—	Plug
PS	—	Pressure Switc
RV	—	Reversing Valv
TRANS	—	Transformer
		Factory Power
----		Factory Contro
━━━		Field Control W
▬▬▬		Field Power Wi

Refer to the CD-ROM "Prints" Chapter 12 Air Conditioning System Legend and Connection Diagram

NOTES:
1. Factory wiring is in accordance with National Electrical Code (NEC). Field modifications or additions must be in compliance with all applicable codes.
2. USe copper conductors only.
3. CAUTION: Not suitable for use on systems exceeding 150 volts to ground.
4. Manual rest circuit breaker is integral to the transformer.

Carrier Corporation

✍ ACTIVITY—Installing an Electrical Service for Air Conditioning

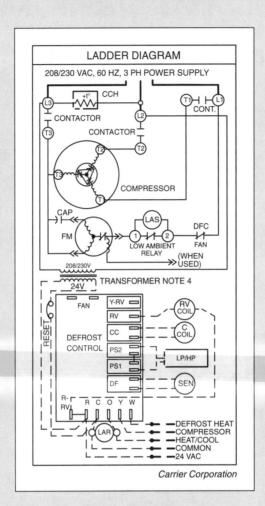

Refer to the CD-ROM
"Prints"
Chapter 12
Air Conditioning System
Ladder Diagram

Name _____   Date _____

Activity—Inspecting an Electrical Service

12-2

Scenario:
Activity 12-2 is a troubleshooting activity requiring that the senior electrician on an industrial job site act as an inspector. A new employee with the electrical contractor has wired the service panel and a subpanel per a provided print. As the senior electrician, inspect the work of the new employee.

Task:
To inspect the electrical work, the electrician must understand how the service and subpanel should be wired. Wire the utility transformer to the service panel and wire a subpanel from the service panel. Ground the transformer and panel(s) and use the main bonding jumper (MBJ) appropriately.

THREE 1ϕ TRANSFORMERS

Reference Print

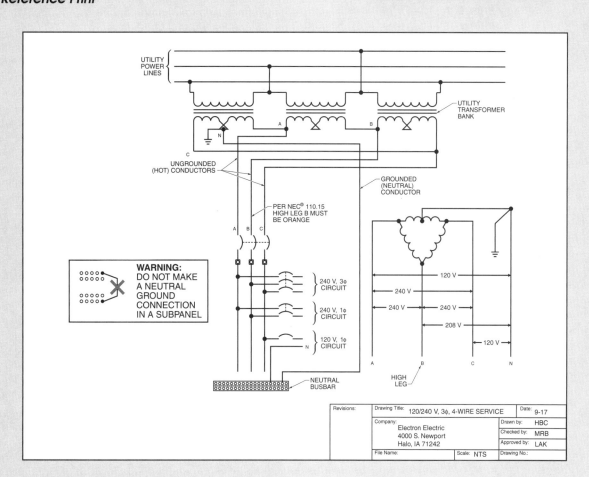

Activity—Inspecting an Electrical Service

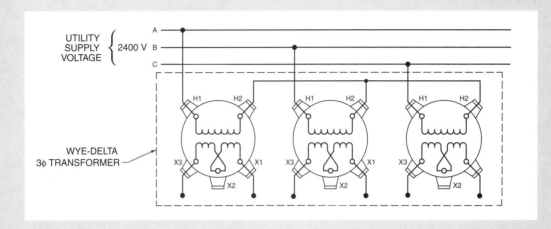

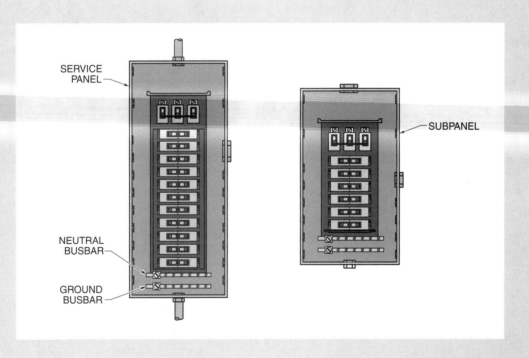

Industrial Equipment

Printreading for Installing and Troubleshooting Electrical Systems

Electrical circuits on various types of drawings and prints are shown using symbols. An electrical device or component may be represented by several symbols depending on the type of drawing or print the symbol is used on. For example, the symbol for a 3-way switch is different on an architectural plan (S_3) than it is on a line diagram, where it is represented by a graphic symbol.

Regardless of the variations of a symbol and how and where it is used, the symbol will still represent the same physical piece of industrial equipment. A person reading an industrial print must recognize what a symbol represents for a specific type of drawing on a specific type of print.

CIRCUIT WIRING

Just as an electrical device or component can be represented using various symbols, the actual device or component can be connected into a circuit using one of several wiring methods. Each wiring method is different and has advantages and disadvantages, but the devices and components will still perform the same function in the system and will control the circuit in the same manner.

For example, a temperature switch can be directly wired to a small motor or wired to a magnetic motor starter coil to control a large motor, wired to the control input terminals of an electric motor drive, or wired to the input terminals of a PLC. **See Figure 13-1.** In each case, the temperature switch symbol and the actual device remain the same, but the manner in which the temperature switch is wired into the circuit changes.

All electrical circuits are designed to perform some type of work. In order to perform work, circuit output components such as motors, solenoids, lamps, and heating elements are used. To control the work, input devices such as temperature switches, pushbuttons, and limit switches are used to turn the loads ON and OFF. **See Figure 13-2.**

> Circuits accomplish work by controlling matter through a loop. The matter is electrons in electrical systems, atmospheric air in pneumatic systems, and incompressible oil in hydraulic systems.

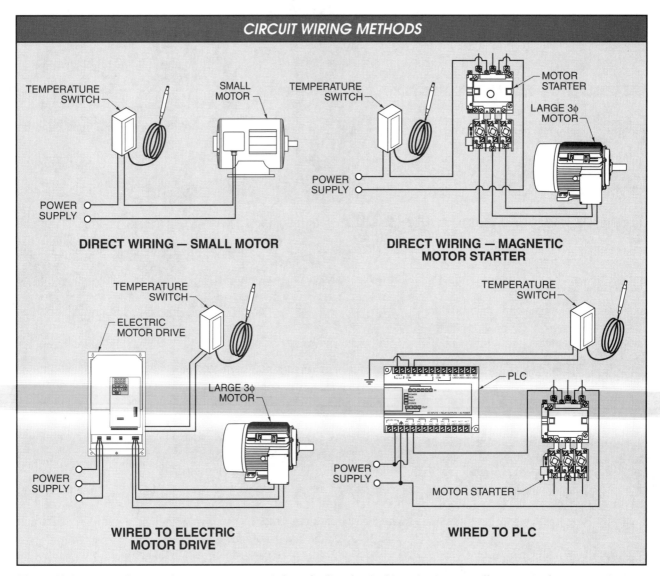

Figure 13-1. *An input device such as a temperature switch can be directly wired in a circuit to a small motor, wired to a magnetic motor starter coil for control of a large motor, wired to the control input terminals of an electric motor drive, or wired to the input terminals of a PLC.*

When designing an electrical system and preparing the required prints, consideration is given to how the physical pieces of equipment (power supply, switches, motors, and lights) are to be wired together and which specific type of equipment will work best for an application. Once a wiring method is known, electrical prints can be prepared that clearly show how a circuit operates. The electrical prints are then used for circuit installation, circuit troubleshooting, and when making circuit modifications.

Temperature switches have a set temperature or adjustable differential setting. The differential setting determines the temperature range within which the temperature switch turns ON and OFF. A temperature switch that turns on the heating element at 72°F and OFF at 75°F has a 3° differential setting. Setting the differential too high causes wide temperature fluctuations and setting it too low can burn out the heating coil due to short cycling times.

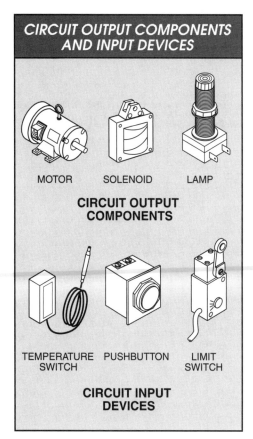

Figure 13-2. To perform work, circuit output components such as motors, solenoids, lamps, and heating elements consume energy. To control work, input devices such as temperature switches, pushbuttons, and limit switches turn loads ON and OFF.

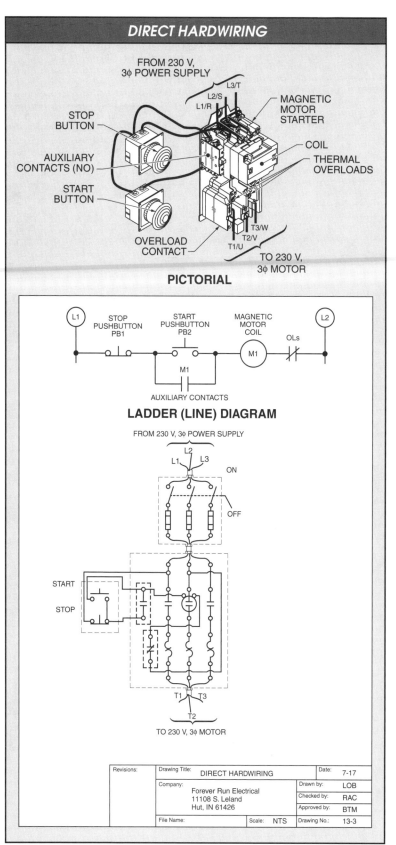

Wiring Methods

Understanding the different methods used to wire an electrical circuit is important when working with electrical prints. For example, a basic motor control circuit can be hardwired using a magnetic motor starter, terminal strips, an electric motor drive, a PLC, or a combination of these.

Direct hardwiring is the oldest and most straightforward wiring method. In direct hardwiring, each device and component is wired directly to the next device or component as specified by the line diagram. **See Figure 13-3.** Direct hardwiring is used for wiring receptacles, lamps, doorbell circuits, entertainment systems, and basic industrial motor control circuits.

Figure 13-3. Basic motor control circuits are hardwired using a magnetic motor starter.

Direct hardwiring using a step-down control transformer is the second oldest wiring method. **See Figure 13-4.** A low control voltage (typically 24 V) allows for greater safety when working inside the motor control enclosure.

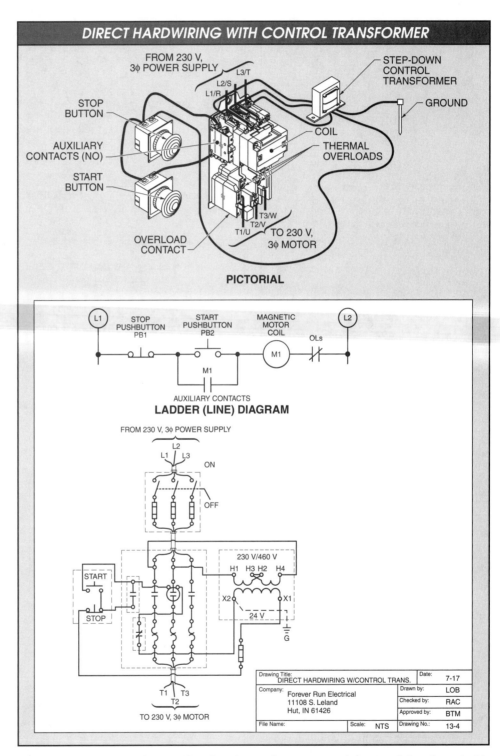

Figure 13-4. Control transformers allow the voltages of a motor control circuit and the voltages inside a motor control enclosure to be lower (typically 24 V).

Direct hardwiring to terminal strips is a way of organizing and simplifying the wiring for an electrical control circuit (or any circuit). Direct hardwiring to terminal strips allows for easy circuit changes and simplifies the troubleshooting process. Using terminal strips to organize circuit wiring is the standard for most control circuits having any complexity or with more than a few control circuit rungs. **See Figure 13-5.** Direct hardwiring and wiring to terminal strips were the two standard methods of wiring control circuits prior to the development of electric motor drives and PLCs.

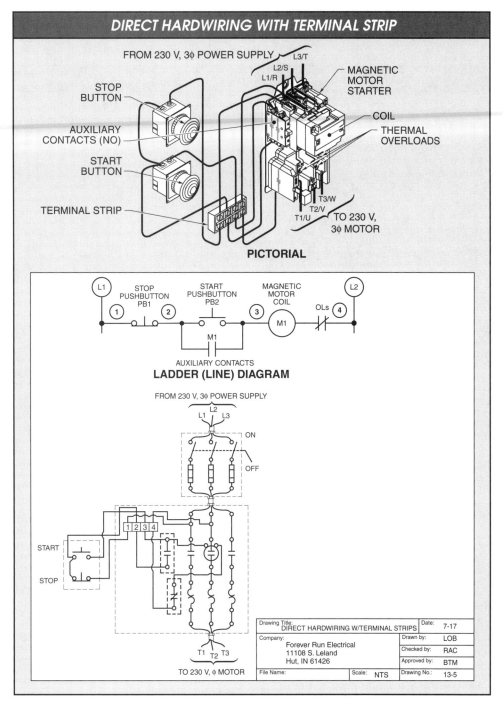

Refer to the CD-ROM "Prints" Chapter 13 Direct Hardwiring with Terminal Strips

Figure 13-5. Direct hardwiring to terminal strips simplifies the wiring for any electrical circuit.

An advantage of wiring to an electric motor drive as compared to a motor starter is that an electric motor drive allows for direct control of voltage, current, frequency, and power to the motor in addition to providing the basic control functions of starting, stopping, and jogging. Most electric motor drives internally reduce the voltage supplied to the input devices to 12 VDC or 24 VDC.

Another advantage to using an electric motor drive instead of a magnetic motor starter is that most drives allow an analog signal as an input signal. When an analog input signal such as 0 VDC to 10 VDC is used, a 2 VDC signal typically equals a motor speed of 20%, and a 5 VDC signal equals a motor speed of 50%. **See Figure 13-6.** Combining a PLC and an electric motor drive allows for maximum control and monitoring of a motor and circuit.

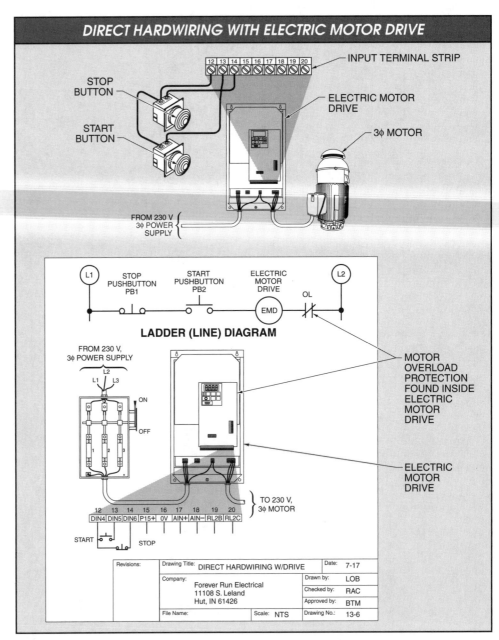

Figure 13-6. Electric motor drives allow for direct control of voltage, current, frequency, and power to a motor in addition to providing the control functions of starting, stopping, and jogging.

Using a PLC for circuit control gives greater flexibility as it allows for monitoring of circuit input devices, output components, and the program being run. The main advantage of using PLCs lies in the flexibility of system control and not in the delivery of power to motors, lamps, heating elements, or other loads on the power circuit. A PLC can monitor and control all systems that it is connected to, but it cannot directly monitor and display motor parameters such as voltage, current, frequency, and power. **See Figure 13-7.**

The advantage to wiring input devices to a PLC is that a step-down transformer is not necessarily required to reduce voltage. Many PLCs internally reduce voltage to 12 VDC or 24 VDC, which is then used to supply power to the input and/or output terminals.

PLC PROGRAMMING DIAGRAMS

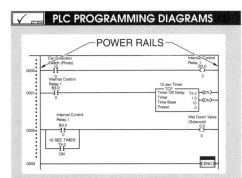

A PLC programming diagram appears much like a ladder diagram, except power does not actually flow through the outermost vertical rails (called power rails). A PLC programming diagram contains the logic for the PLC application and is also used to test and troubleshoot the application. The PLC programming instructions take the place of control wiring and control devices such as relays, timers, and counters. The PLC programming diagram resides in the memory of the PLC and various files within the memory relate to the running of the program.

PLC programming is similar regardless of the manufacturer. However, specific programming rules and procedures must be followed for each manufacturer and model of PLC.

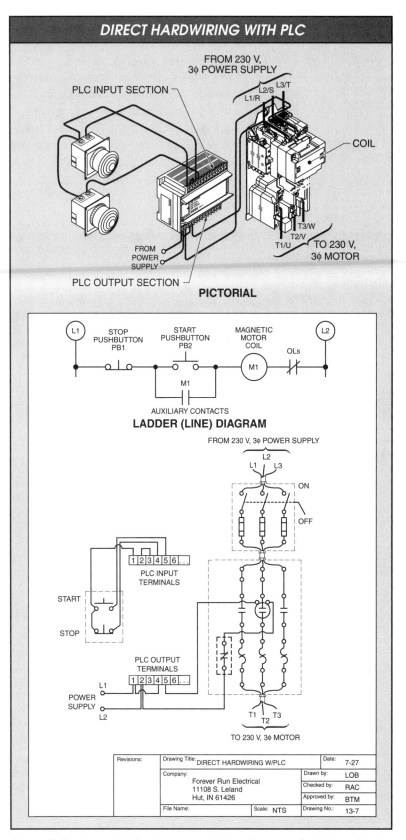

Figure 13-7. *Using a PLC for circuit control allows for greater flexibility and for the monitoring of circuit input devices, output components, and the program being run.*

Component Layout and Location

Electrical prints such as line and wiring diagrams use symbols to show devices and components and their interconnections. However, wiring diagrams do not show the actual physical location of each device and component in the system. To understand where each piece of electrical equipment is physically located, a component layout and location drawing is required. **See Figure 13-8.**

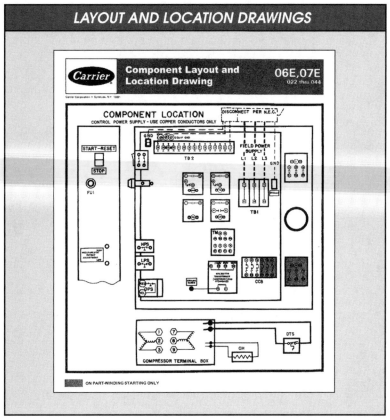

Carrier Corporation

Figure 13-8. A device and component layout and location drawing is a drawing that is helpful in locating devices and components when troubleshooting or modifying a system.

For example, in the Carrier® compressor-unit application, the component location for the control circuit enclosure is shown. The device and component layout and location drawing shows the location of relays, timers, the control transformer, and the other devices and components used in the system.

A *layout and location drawing* is a drawing that helps locate devices and components for testing during troubleshooting and for system modification. Component layout and location drawings are used in the assembly of control panels and may be available from the manufacturer. Some manufacturers even make their equipment prints available on-line.

A power circuit and control circuit for an air conditioning compressor unit is shown on the application. **See Figure 13-9.** In the power circuit drawing, the manufacturer shows two ways the motor can be started. The part-winding method is preferred on larger units because it reduces the amount of starting current drawn during the first few seconds of motor startup. High-current draw during starting can cause a voltage dip in the power lines and additional problems for the end user, such as computers resetting. When a small horsepower motor is used, the motor can be started using the "across-the-line" starting method.

Most sets of electrical prints do not include component layout and location drawings. Typically, only the power circuits and control circuits have these drawings. A basic knowledge of electrical circuits and systems is necessary in order to install electrical devices and components when a component layout and location drawing is not provided. The following general statements apply to most electrical circuits and systems:

- Electrical output components (loads) such as lamps, motors, heating elements, and solenoids are placed where loads are required. Load location is typically a matter of necessity, not convenience.

- Electrical input devices (switches) such as pushbuttons, pressure switches, and limit switches are placed where switches are required. Switch location is typically a matter of necessity, not convenience.

- Electrical control devices are typically placed in an enclosure and can be located almost anywhere as long as the location is convenient for servicing. Electrical devices and components such as relays, timers, counters, motor starters, heating or lighting contactors, and control transformers can all be located together in one central enclosure.

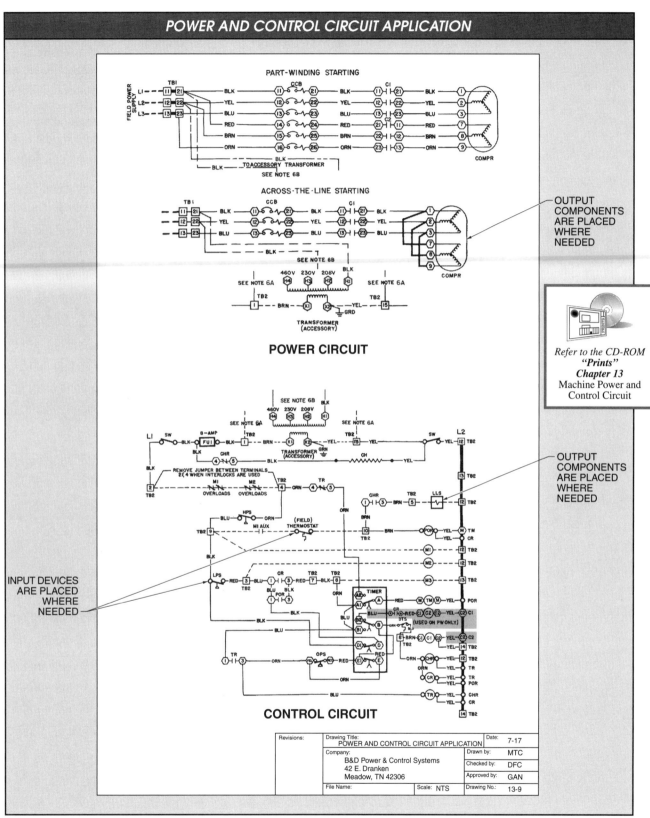

Figure 13-9. A basic knowledge of electrical circuits and systems is necessary in order to install electrical devices and components when a component layout and location drawing is not provided.

DIRECT HARDWIRING

Direct hardwiring is the way all electrical circuits were wired in the past and is still a common method of wiring today. Fewer components means there is less of a need for change or for troubleshooting; thus, the simpler the circuit, the more likely it is to be hardwired. For example, lamp circuits controlled by a switch, whether manual or automatic, are always hardwired. **See Figure 13-10.**

In the lamp circuit application, the 277 VAC HID lamps are automatically controlled by a photoelectric day/night switch. Photoelectric switch control is typical for outdoor lighting in which lamps turn ON when it becomes dark. The circuit is simple and the wiring is straightforward. The only disadvantage is that troubleshooting and circuit modifications take longer because it takes time to determine the exact location and layout of each wire, device, and component. This is typical in small systems but becomes a problem with larger systems that include many components.

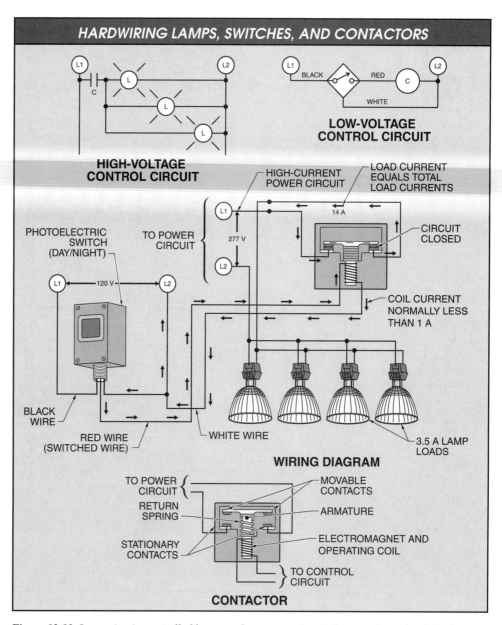

Figure 13-10. Lamp circuits controlled by manual or automatic switches are always hardwired.

Electrical noise is typically the result of unwanted electrical signals entering a control circuit of a system and disrupting the actual signals in the circuits. Noise occurs in power circuits, but becomes a real problem when entering signal (control) circuits. Signal and data circuits are especially vulnerable to noise because of the low-voltage levels used in these circuits.

Wiring Variations

Troubleshooting or making circuit modifications to direct hardwired circuits can sometimes be difficult. Although line diagrams show how components are to be interconnected, the actual wiring diagram for the circuit can and will vary from location to location. For example, the wiring of two 3-way switches to control a lamp can change based on where the main power comes into the circuit. **See Figure 13-11.**

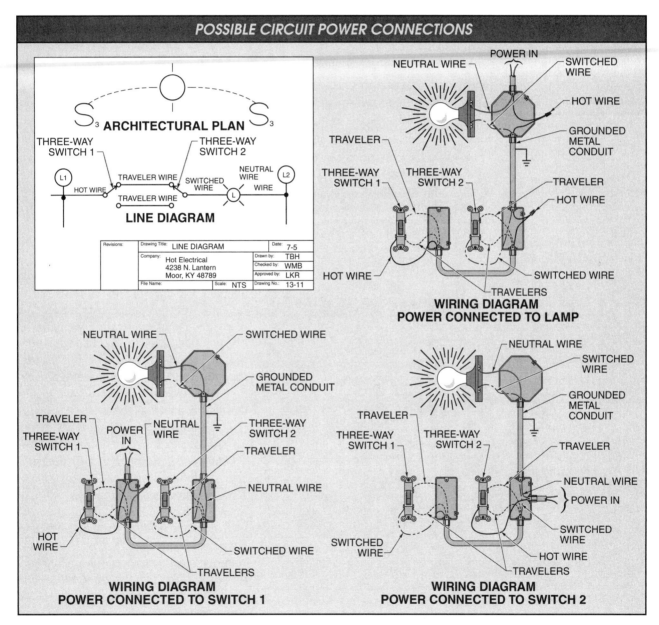

Figure 13-11. Problems with direct hardwired circuits occur when troubleshooting or making circuit modifications. Line diagrams show how components are to be interconnected, but the actual wiring diagram for the circuit can and will vary from location to location.

In the lamp circuit, the architectural plan symbol remains the same and the floor plan symbols remain the same regardless of how the lamp and switches are to be wired. Likewise, the line diagram showing how the switches and lamp are interconnected remains the same regardless of how the circuit is wired. What does change is the actual hardwired circuit. The actual hardwiring of the circuit is not determined until the devices and components are actually wired and consideration is given to the location of the walls and whether circuit power is coming up from the basement or down from the ceiling.

Hardwired Reversing Circuit

When learning and/or comparing the different methods of wiring (direct hardwired, wiring using terminal strips, and wiring using a PLC), it is helpful to see the exact same circuit wired in different ways. A motor reversing circuit is a very common electrical circuit and can be used to show all three of the common wiring methods used today.

> When a control transformer is shown on a print, the grounded side "X2" is always shown on the right side of the circuit. However, on most control transformers the X2 terminal is physically on the left side of the transformer. Care must be taken when wiring a transformer to ensure that the X2 side of the transformer becomes the grounded side and is used as shown on the print, regardless of where it is physically located on the transformer.

A typical motor control circuit for controlling the direction of a motor uses forward, reverse, and stop pushbuttons. **See Figure 13-12.** In the reversing motor control circuit, a magnetic motor starter (FC) is used to control the motor running forward, and a second magnetic motor starter (RC) is used to control the motor running in reverse. The motor in the power circuit is a 3φ motor, but the same control circuit can be used for a 1φ or DC motor.

Each device and component in a hardwired circuit is wired to the next device or component as specified by the line diagram. For example, the X1 terminal of the transformer is connected directly to a fuse, the fuse is connected directly to the stop pushbutton, the stop pushbutton is connected directly to the reverse pushbutton, the reverse pushbutton is connected directly to the forward pushbutton, and so on, until the final connection from the OL contact is made back to the X2 terminal of the transformer.

Although the motor reversing circuit operates as designed when hardwired, troubleshooting and circuit modifications will take time to perform. For example, if a forward indicator lamp and a reversing indicator lamp need to be added to a reversing motor control circuit, the exact connection points of the indicator lamps must be found, and even when the exact connection points are found, problems may arise when making the actual connections, such as a lack of room under the terminal screws.

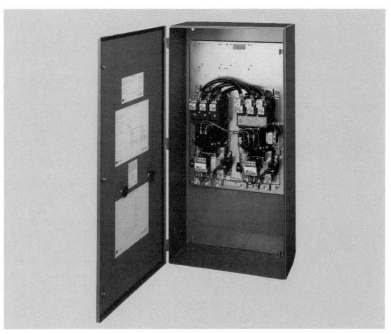

Rockwell Automation, Allen-Bradley Company, Inc.

Hardwired reversing motor control circuits require two motor starters. The left starter is for motor forward and the right starter is for motor reverse.

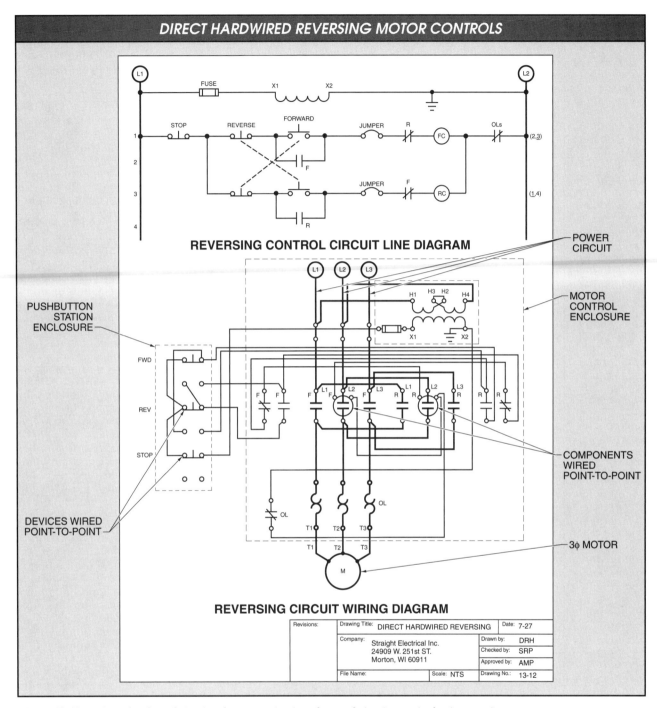

Figure 13-12. In direct hardwired circuits, the power circuit and control circuit are wired point-to-point.

Some circuit modifications may not be difficult, such as adding forward and reverse indicator lamps. This change may only require that new wires be added. In other modifications, old wires do not need to be removed. Direct hardwired circuits typically require the most wires be removed.

Some circuit modifications, such as adding limit switches, are more difficult. For example, when forward and reverse limit switches are to be added to a circuit, some wiring must be removed from the circuit. In addition, new wiring from the limit switches must be added. **See Figure 13-13.**

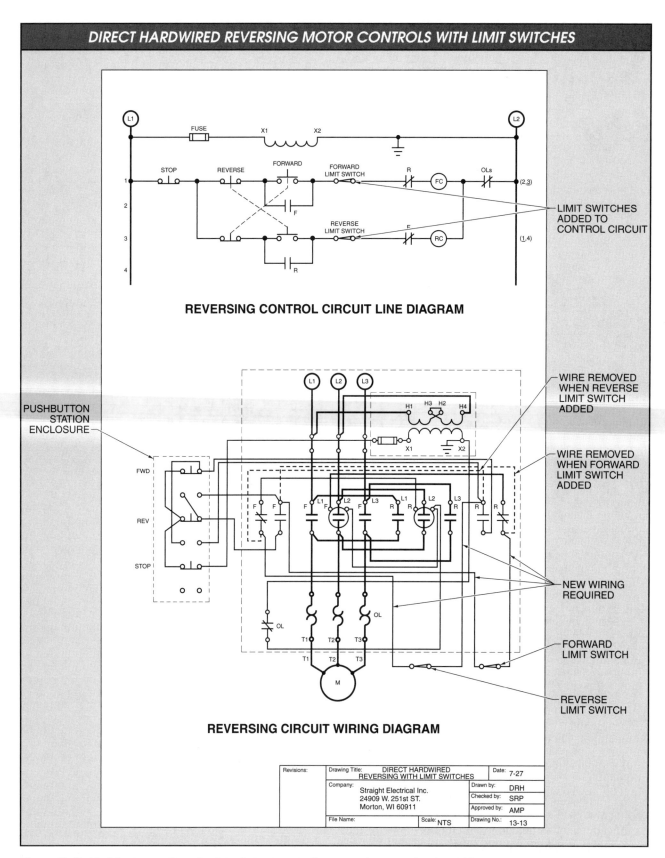

Figure 13-13. Modifications to direct hardwired circuits typically require the removal and/or addition of circuit wiring.

HARDWIRING USING TERMINAL STRIPS

Hardwiring using terminal strips allows for easier circuit modifications and simplifies circuit troubleshooting. When wiring using terminal strips, each wire in the control circuit is assigned a reference point on the line diagram to identify the different wires that connect the devices and components. Each point is assigned a wire reference number. **See Figure 13-14.**

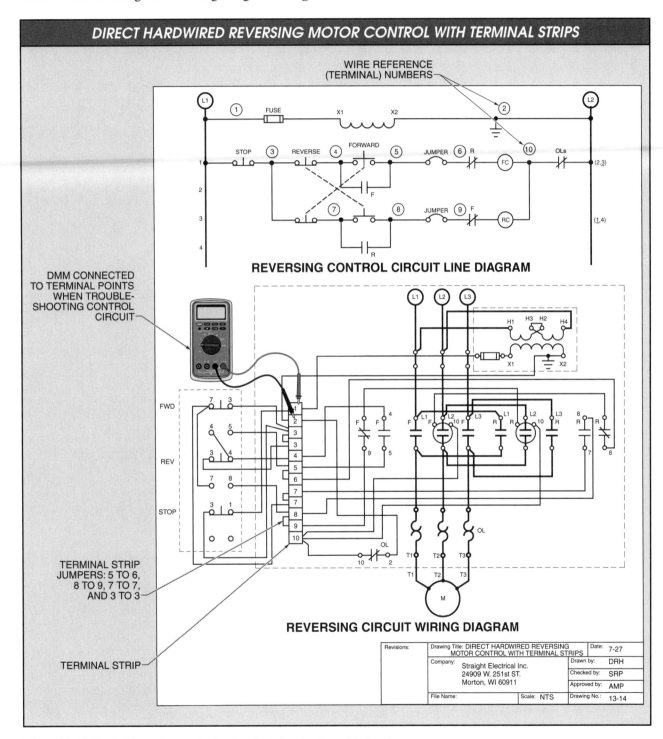

Figure 13-14. Hardwiring using terminal strips simplifies circuit troubleshooting.

When troubleshooting a circuit with a terminal strip, the troubleshooter can go directly to the terminal strip and take measurements to help isolate the problem. Terminal strip and wire reference numbers help when troubleshooting and make circuit modifications easier. Circuit modifications are easier because most, if not all, of the wires required to make the changes to the circuit are disconnected and reconnected at the terminal strip. See Figure 13-15.

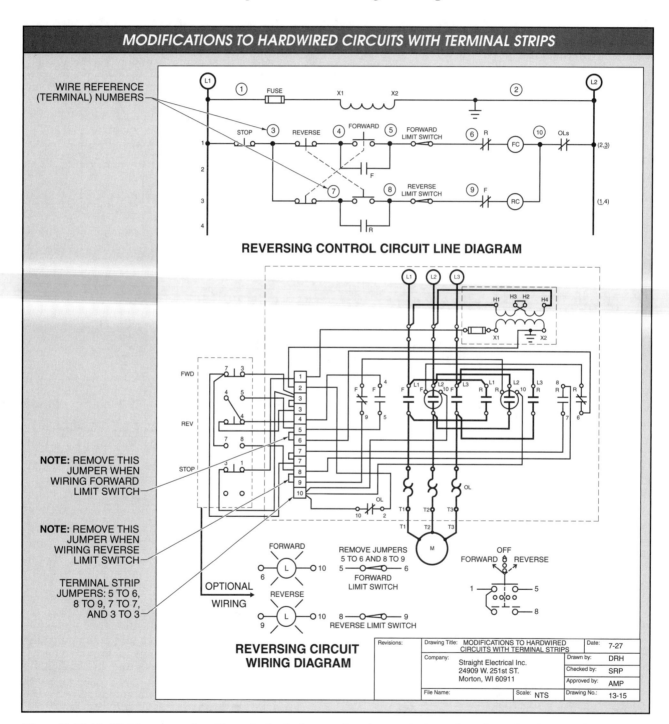

Figure 13-15. Modifications are easier with terminal strips because the wires required to make a change are disconnected and reconnected at the terminal strip.

Dual Compressor Application

In some applications, pumps or electric-motor-driven compressors must continuously operate to maintain water level or air pressure, to supply production with enough volume of a liquid, or to satisfy environmental requirements. To ensure that the loss of a motor does not mean the loss of operations, two motors are often used. If one motor fails, the other motor can operate until the first one is repaired. In such dual motor applications, it is common to use an alternating relay (AR) to alternate the two motors so that only one motor runs at a time. **See Figure 13-16.**

WIRE GAUGE

Wire is sized by using a number, such as #12 or #14. The wire number is based on the American Wire Gauge (AWG) numbering system in which the lower the number, the larger the wire diameter.

AWG	AMPACITY	DIA (Mils)
18	—	40.0
17	—	45.0
16	—	51.0
15	—	57.0
14	20	64.0
12	25	81.0
10	30	102.0
8	40	128.0

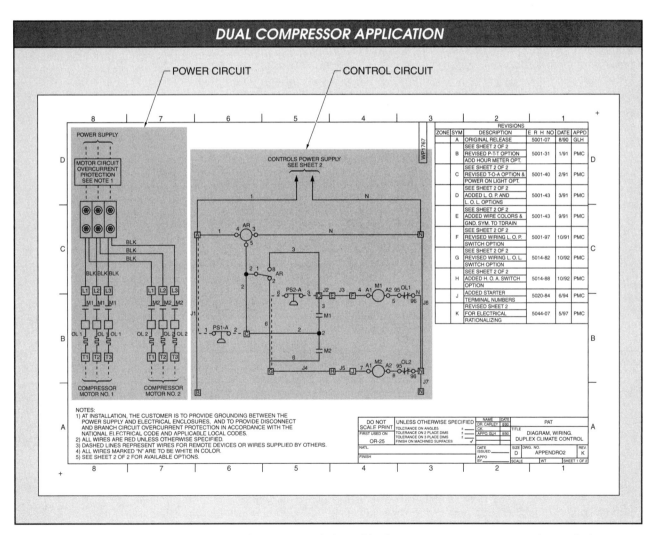

Figure 13-16. To ensure that the loss of a motor does not mean the loss of facility operations, two motors are often used. If one motor is lost, the other can still operate.

The dual compressor print (sheet 1 of 2) shows the power circuit and control circuit used for controlling the motors of the two compressors, which are controlled by magnetic motor starters M1 and M2. The control circuit shows how pressure switch PS1-A is used to energize a compressor motor any time the pressure in the system is below the setting of the pressure switch. The AR determines which motor starter will be energized at startup. The addition of pressure switch PS2-A allows both compressor motors to turn ON if the system pressure drops below the setting of the pressure switch, which may happen during high usage times.

As with most electrical prints, circuit wiring options and/or modifications can be required. The dual compressor print (sheet 2 of 2) shows several circuit changes that can be made easily because terminal strip wiring was used for the control circuit. **See Figure 13-17.**

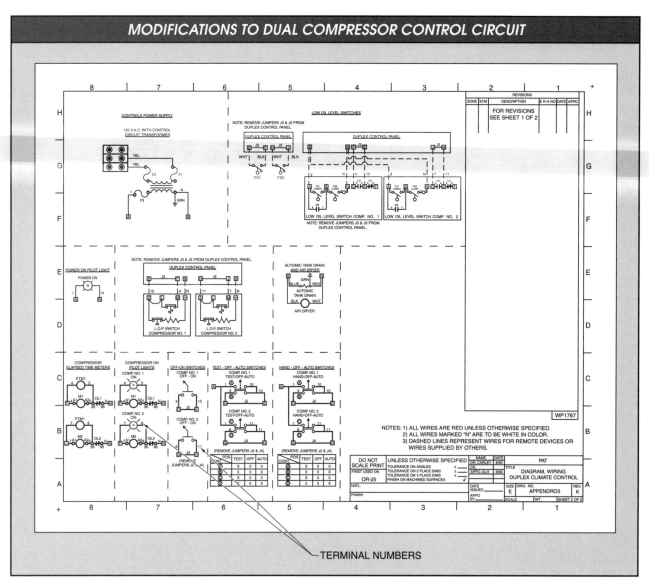

Figure 13-17. The dual compressor print shows several circuit changes that can be made easily because a terminal strip was used to wire the control circuit.

Chapter 13—Industrial Equipment **403**

For example, a green "POWER ON" lamp can be added to terminals A and N to show that the control circuit is powered. Likewise, red lamps can be added to terminals F and N to show when compressor starter M1 is ON and to terminals J and N to show when compressor starter M2 is ON. The wiring options also show variations of how selector switches can also be added to modify the operation of the control circuit.

Pump Application

Pumps are used to move fluids. Motors drive the pumps to maintain fluid flow. Pump motor control circuits can range from a few rungs to hundreds of rungs depending upon how complex the system is and how many pumps and motors are being used. Large ladder diagrams of electrical systems include line numbers, cross-reference numbers, terminal numbers, and as much other information as needed. **See Figure 13-18.**

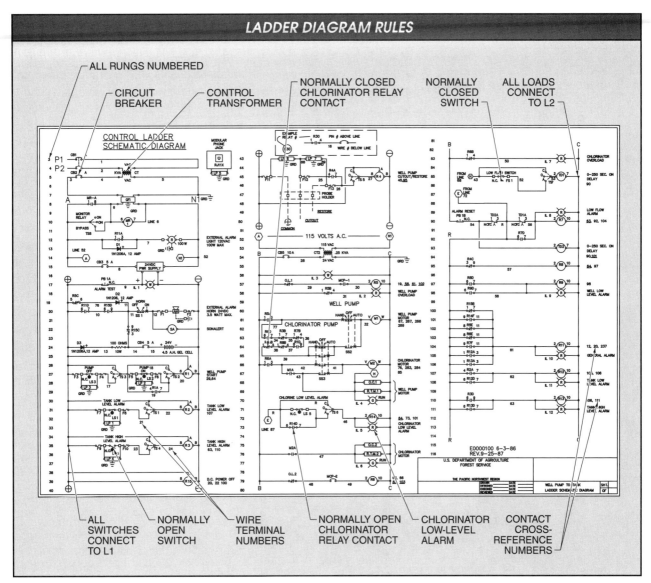

U.S. Department of Agriculture—Forest Service

Figure 13-18. *Large ladder diagrams of electrical systems must always include line numbers, cross-reference numbers, terminal numbers, and as much other information as needed.*

Although the ladder diagram for the well pump application is large, the ladder diagram follows standard numbering systems and rules that help make the print and circuit easy to understand:

- Each individual rung is numbered. Rung numbering allows for easy referencing of devices and components, such as the control transformers on lines 5 and 54 and the red chlorinator low-level alarm lamp on line 73.
- Each control device has contacts in the control circuit that are cross-referenced. The chlorinator low-level relay (R14 on line 72) has a normally closed contact on line 64, and a normally open contact on lines 73 and 101.
- Each wire is assigned a terminal number. Circuit breaker CB2 (on line 5) is connected between terminals 3 and 4. Troubleshooting the circuit using a voltmeter is simple when terminal numbers are used.
- All loads are connected directly to power line 2.
- All switches are connected between the hot conductor and the loads. Switches can be normally open, normally closed, connected in series, and/or connected in parallel.

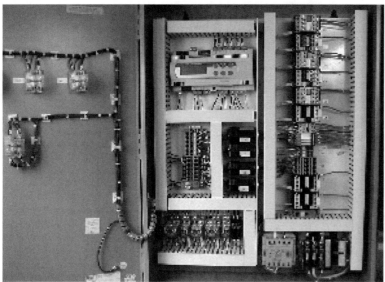

PLC Multipoint

The operation of most PLC-wired circuits can only be understood when looking at ladder diagrams and/or wiring diagrams.

Electric Motor Drive Control Variations

An electric motor drive can be used in place of manual or magnetic motor starters. Using an electric motor drive instead of a magnetic motor starter allows for much greater control of a motor. Electric motor drives can be used to control ON-delay and deceleration times, provide programmable overload protection, and monitor voltage, current, power, frequency, and other control functions.

When wiring an electric motor drive, both the power circuit and control circuit are included as terminals on the drive. The power circuit terminals are large and spaced far apart because of the high voltage and current traveling through them. Control circuit terminals will be smaller and closer together because control circuits typically only use 12 VDC or 24 VDC. An electric motor drive manufacturer will provide alternative ways for a control circuit to be wired based on the motor drive's specific application. **See Figure 13-19.**

In the electric motor drive application, a three-position selector switch (DRIVE/OFF/BYPASS) is used to place the control circuit in the condition of using the electric motor drive to control the motor or of bypassing the drive and starting the motor directly from the power lines. In most electric motor drive circuits, there are several voltages being used. The power circuit supplying power to the motor would typically be 240/480 VAC, the control circuit starting at the transformer would be 120 VAC, and the drive control circuit would typically be 12 VDC or 24 VDC. Understanding how to read an electric motor drive print is important when troubleshooting.

PLC WIRING

In a PLC-wired control circuit, the electrical circuit/system has much greater flexibility than a direct hardwired circuit and has circuit-monitoring capabilities. For example, in a reversing motor control circuit, the PLC is part of the control circuit. The power circuit remains the same in all wiring methods. What changes is that the input devices of the control circuit are wired to the PLC input section or module, and the control circuit output components are wired to the PLC output section or module. **See Figure 13-20.**

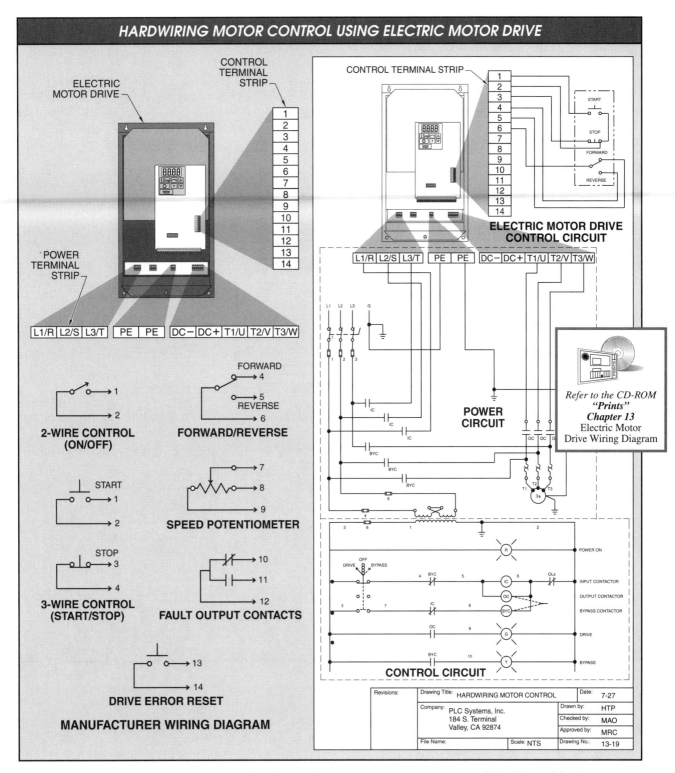

Figure 13-19. Electric motor drives are used in place of manual or magnetic motor starters to control ON-delay and deceleration times and provide programmable overload protection.

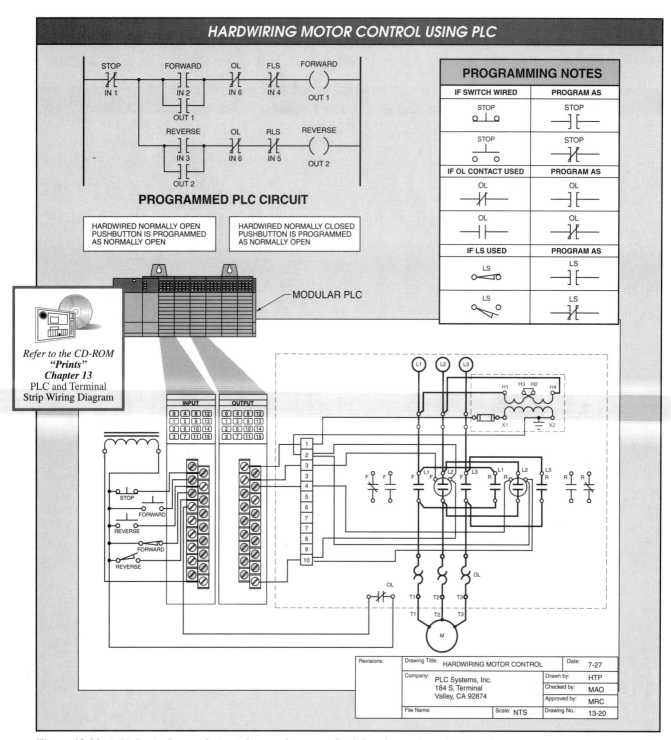

Figure 13-20. A PLC-wired control circuit has much greater flexibility than a direct hardwired circuit and has circuit-monitoring capabilities.

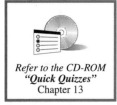

Circuit logic is programmed using PLC software on a PC. The program is then downloaded to the PLC. The PLC programming software can be used to monitor and display the condition (ON or OFF) of the circuit input devices and output components. When changes in the control circuit are required, the program can be modified and downloaded to the PLC. The hardwiring of the circuit does not change.

Name _____ Date _____

Activity—Installing Power to Two Milling Machines 13-1

Scenario:
Activity 13-1 is an installation activity requiring that two milling machines be wired up on a machine shop floor. The first vertical milling machine is older and is direct hardwired. The second vertical milling machine is a little newer and has terminal strip wiring.

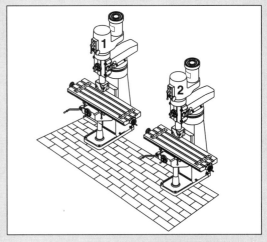

VERTICAL MILLING MACHINES

Task:
Using the correct wiring diagram for the application, add numerical cross-reference numbers to both of the line diagrams. Finish wiring the first milling machine, then finish wiring the second milling machine. Wire the pushbutton stations using the least amount of wires in the conduit running between the pushbutton stations and the motor control enclosures. Also, no wire splices are allowed in the control station box or the motor control enclosure.

Reference Prints:

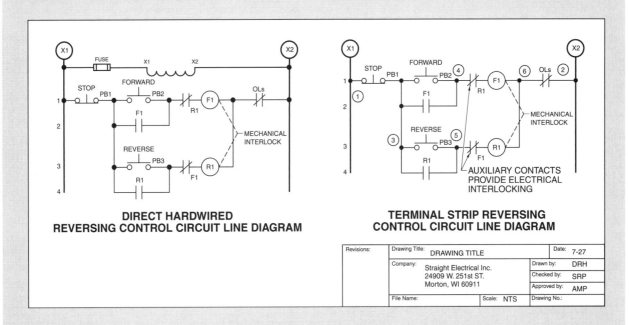

DIRECT HARDWIRED REVERSING CONTROL CIRCUIT LINE DIAGRAM

TERMINAL STRIP REVERSING CONTROL CIRCUIT LINE DIAGRAM

Activity—Installing Power to Two Milling Machines

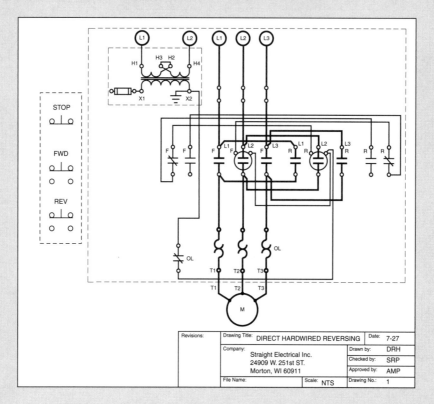

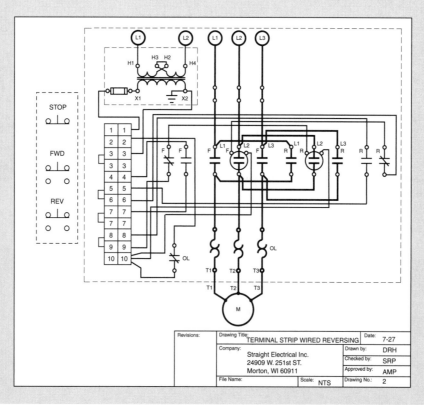

Name _____ Date _____

Activity—Wiring a PLC into a Milling Machine Control Circuit 13-2

Scenario:
Activity 13-2 is a troubleshooting activity that requires an understanding of the PLC of an automated machining cell. The PLC controls the operation of the vertical milling machine, product placement robot, feed conveyor, and take-away conveyor.

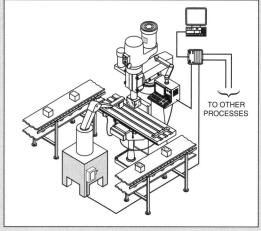

AUTOMATED MACHINING CELL

Task:
Using the PLC programming diagram and the PLC circuit line diagram, finish wiring the vertical milling machine control box to the input module of the machining cell PLC. No wire splices are allowed in the milling machine control box or the motor control enclosure. Also, a second milling machine must have the control box wired to an electric motor drive to control speed and motor torque. When using an electric motor drive, motor starter overloads are included within the drive. The drive also supplies 24 VDC to the control terminals, so no step-down control transformer is required.

Reference Print

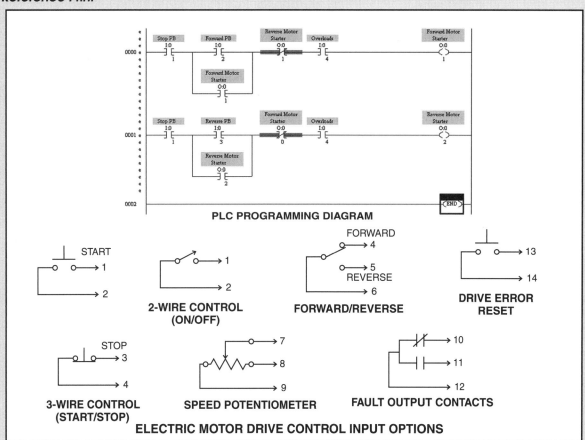

411

☑ Activity—Wiring a PLC into a Milling Machine Control Circuit

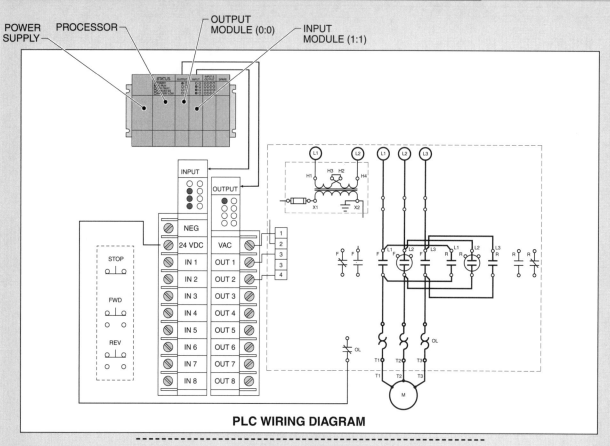

PLC WIRING DIAGRAM

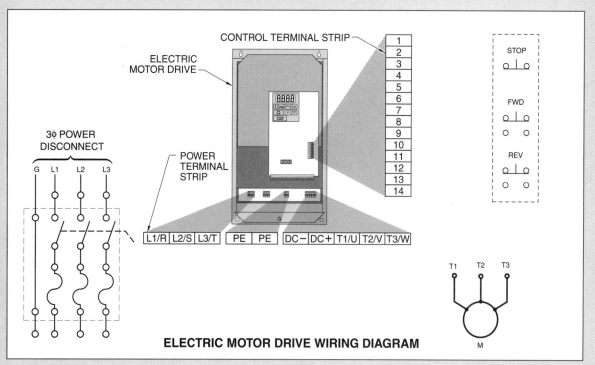

ELECTRIC MOTOR DRIVE WIRING DIAGRAM

Fluid Power Systems

Printreading for Installing and Troubleshooting Electrical Systems

At first glance, fluid power symbols seem to be more complex and harder to understand than electrical symbols. Many fluid power symbols are more detailed than electrical symbols and that makes fluid power prints seem more difficult to understand. Once fluid power symbols are understood, however, it becomes clear that fluid power prints tell much more about the devices and components the symbols represent than do electrical prints.

FLUID POWER SYMBOLS

The symbol for a fluid power valve is very specific as far as how a device or component operates. This is not true for electrical switch symbols. Electrical switch symbols that control current flow are more basic than symbols for fluid power valves. **See Figure 14-1.**

The electrical symbol for a 2-way, manually operated, spring-return, normally open pushbutton shows the manual operator for a normally open switch, but does not show that a spring is used to automatically open the switch when the manual operator is not being pressed. The manually operated position (closed) is not evident, neither is the fact that a spring automatically returns the switch to the open position.

A fluid power symbol for a 2-position, 2-way, spring-offset, normally open, manually actuated, directional control valve shows what the valve looks like in both the open and closed positions, how the valve gets into the closed position (manually actuated), and how the valve returns to the open position (spring offset).

It is important for anyone working on a fluid power system to understand both electrical and fluid power symbols because most fluid power circuits are electrically powered and controlled. For example, airplane wheels and wing parts, earth-moving equipment, and most manufacturing operations use fluid power to develop high mechanical forces and include electrical controls to ensure precision. **See Figure 14-2.**

——	Solid line is a main fluid power line (flow and pressure)
----	Short dashed line is a pilot line (pressure)
— —	Long dashed line is a drain line (flow)
+	Lines crossing but not connected
⊥	Lines connected

415

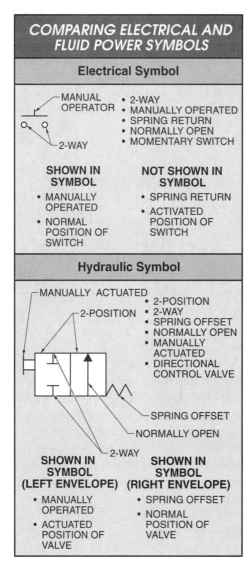

Figure 14-1. Fluid power symbols are more descriptive than electrical symbols in explaining how a device or component operates.

Hydraulic systems provide the force required for the stamping and fabrication of most steel products.

Atlas Technologies, Inc.

Figure 14-2. Hydraulic and pneumatic symbols are very similar in design, with a few special symbols used specifically only for hydraulic or pneumatic components.

FLUID POWER

Fluid power is the transmission and control of energy by means of a pressurized fluid. **See Figure 14-3.** A *fluid* is a liquid or a gas that can move and change shape without separating when under pressure. The main advantage of a fluid power system is that pressurized fluids multiply forces simply and efficiently. Fluid power is used to move, cut, bend, punch, and align various materials with forces ranging from several pounds to several hundred thousand pounds. Fluid power uses liquid (oil for hydraulics) or gas (air for pneumatics) to produce work.

> **⚠ WARNING**
> Industrial hydraulic systems operate at thousands of psi and extremely high temperatures. To help prevent severe injuries, follow all safety rules when performing any maintenance work on hydraulic equipment.

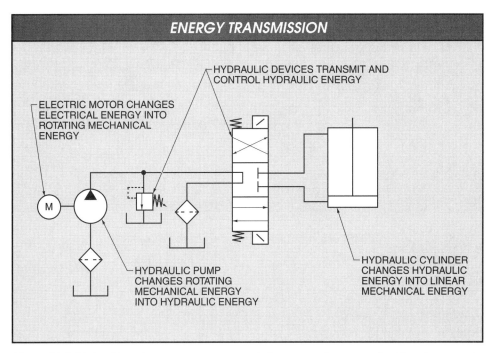

Figure 14-3. Hydraulics refers to an energy transmission system that uses a liquid, and pneumatics refers to an energy transmission system that uses a gas.

Hydraulic Systems

A *hydraulic system* is a system that transmits and controls energy using a liquid (typically oil). A *liquid* is a fluid substance that can flow readily and assume the shape of its container. The flow rate in a hydraulic system is easily regulated because liquid is considered noncompressible. All hydraulic systems are closed systems because the fluid is returned, filtered, and reused. Hydraulic systems are preferred for applications requiring large forces as forces of any magnitude can be delivered by hydraulic cylinders, motors, and oscillators. Hydraulic systems are standard for most applications requiring a high force from a small area or footprint. **See Figure 14-4.**

Applications for hydraulic systems include forklifts, forming and stamping presses, clamping, pile driving, industrial robots, farm equipment, mining equipment, most high-force industrial tool machinery, and power steering systems for automobiles, trains, ships, and airplanes. In industrial tool machinery, hydraulic force is directly used to perform work on a product. Such work includes cutting, clamping, bending, forming, punching, feeding, and ejecting parts, as well as moving worktables in and out of machines and clamping tools.

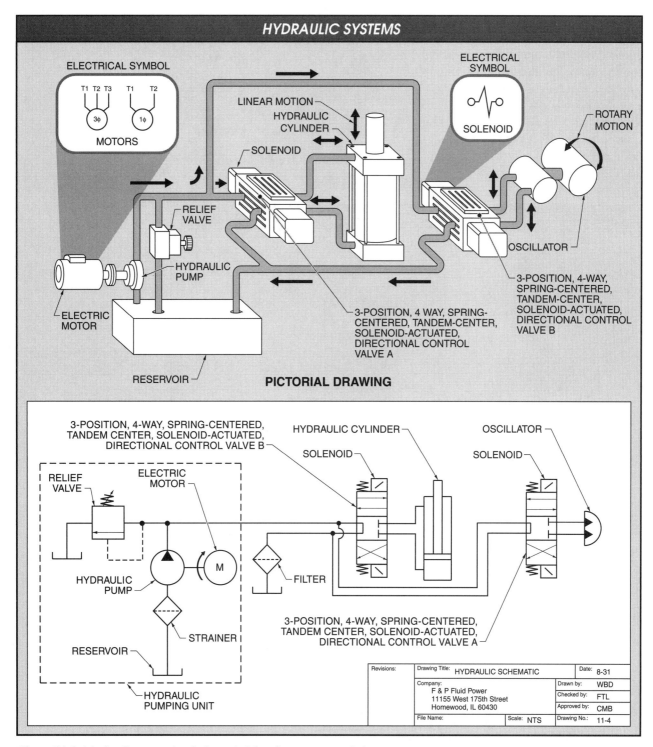

Figure 14-4. A hydraulic system is a hydrostatic (closed) energy transmission system.

Pneumatic Systems

A *pneumatic system* is a system that transmits and controls energy using a gas such as air. *Air* is the mixture of gases that are present in the atmosphere. Air has the properties of a fluid. However, unlike the molecules of a solid or liquid, air molecules are not freely attracted to each other and are easily compressible. Pneumatic systems are open systems because air is exhausted to the atmosphere after it performs its work. **See Figure 14-5.**

Pneumatic systems are preferred for applications requiring low forces. The advantages of using pneumatic systems in low-force applications is that pneumatic systems have a relatively low cost, are capable of high speeds, and can have leaks that do not shut down the process or operation. Examples of pneumatic applications include pneumatic wrenches, air brakes, dental drills, air clamps, injection molding machines, automatic gates and doors, impact guns, and any food-processing machinery.

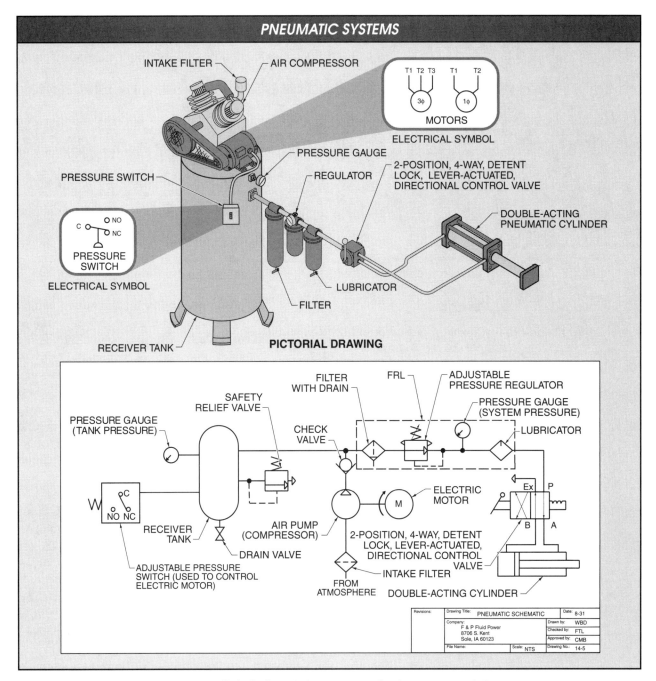

Figure 14-5. *A pneumatic system is a partially hydrodynamic (open to atmosphere) energy transmission system.*

Pneumatic systems are also used in wet locations or in areas that may contain an explosive atmosphere. For example, pneumatic systems are used in many mining operations as control systems to reduce or eliminate the hazards of sparks and shock that are present with electrical control systems.

Eaton Corporation

Hydraulic gear pumps can have pressure ratings of 1600 psi, vane pumps have pressure ratings of 3200 psi, and piston pumps have pressure ratings of 6400 psi or more.

PUMPS			
Device	Type	Symbol	
		Hydraulic	Pneumatic
FIXED DISPLACEMENT PUMPS	Unidirectional		
	Bidirectional		
VARIABLE DISPLACEMENT PUMPS	Manually Compensated		
	Pressure Compensated		

Figure 14-6. Fluid power systems use positive displacement pumps to create flow in the system.

Pumps

A *pump* is a fluid power component that converts mechanical energy into hydraulic or pneumatic energy by pushing fluid into a system. Pumps are classified as positive-displacement or non-positive-displacement. A *positive displacement pump* is a pump that delivers a finite quantity of fluid for each revolution of the shaft. A *non-positive-displacement (centrifugal) pump* is a pump that circulates or transfers fluid using rotational speed.

Hydraulic systems and most pneumatic systems use positive displacement pumps. **See Figure 14-6.** A *fixed displacement pump* is a positive displacement pump that develops a fixed amount of flow for each revolution of the shaft. Flow rate can only be changed by changing the drive speed of the pump. A *variable displacement pump* is a positive displacement pump in which the amount of flow can be manually changed or automatically changed for each revolution of the shaft.

Electric motors are the most common prime movers used to drive hydraulic pumps and pneumatic pumps. The shaft of an electric motor is connected directly to the shaft of a hydraulic or pneumatic pump using a coupling, or indirectly to some pneumatic pumps (or compressors) using belts. Different sized pumps or variable displacement pumps are used to deliver different flow rates to various parts of a fluid power circuit.

In a hydraulic pump and electric motor control application, an electric motor is turned ON and OFF using a standard electrical two-way switch for small motors (typically ¼ HP or less) and a motor starter for large motors. **See Figure 14-7.** A motor starter is typically controlled by a standard two-way (ON/OFF) selector switch or a start/stop pushbutton station. Three-phase motors are preferred in high-horsepower applications because 3ϕ motors are more energy-efficient than 1ϕ motors.

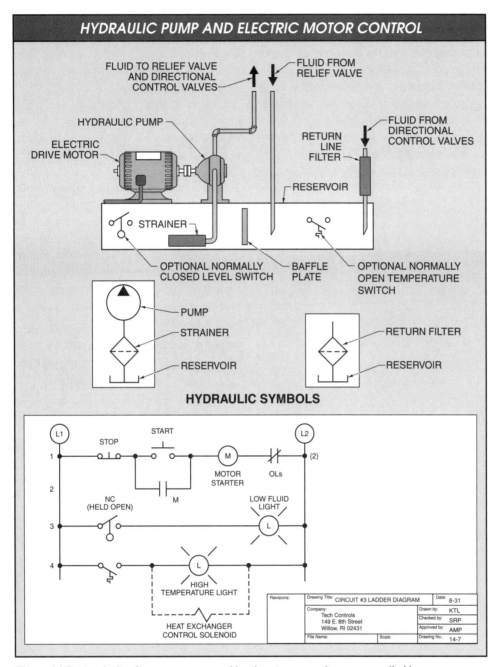

Figure 14-7. Most hydraulic pumps are rotated by electric motors that are controlled by motor starters and start/stop control stations.

Fluid Conditioners

A *fluid conditioner* is a device that maintains clean fluid in a system. Fluid systems, devices, and components require clean fluid, whether air or oil, to operate properly. Fluid conditioners include filters, filters with drains, lubricators, lubricators with drains, and heat exchangers. A *filter* is a device that removes solid contaminants from a fluid power system. A filter with drain is a device that removes contaminants and moisture from a pneumatic system. A *lubricator* is a device that adds lubrication (oil) to a pneumatic system. A *heat exchanger* is a device that removes heat from fluid in a hydraulic or pneumatic system.

Hydraulic Fluid Conditioners. Fluid conditioners must keep the hydraulic oil clean and at the proper temperature. An oil temperature in a hydraulic system that is too low causes pressure valve problems. To eliminate low oil temperatures, hydraulic systems use oil heaters to raise the oil to a specific temperature. As with any filtering system, hydraulic filters must be replaced before they become clogged and cause resistance to flow. **See Figure 14-8.**

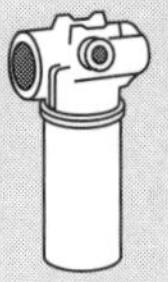

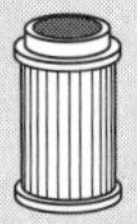

Pneumatic Fluid Conditioners. Air in a pneumatic system is not recirculated as oil in a hydraulic system is. A pneumatic fluid conditioner eliminates or reduces dirt and moisture in atmospheric air before the air is allowed to enter a pneumatic system. A large amount of dirt can be removed by using a filter on the inlet side of the compressor. Air filters are typically located in the pressure lines before the components and immediately preceding power tools. Filters protect valves and actuators against failure caused by large particles as well as gradual failure caused by small particles that increase wear in pneumatic equipment. Filters must be replaced before a filter becomes clogged. A clogged filter reduces compressor and system efficiency. **See Figure 14-9.**

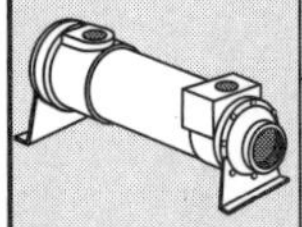

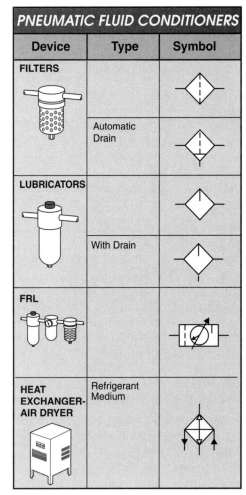

Figure 14-8. Hydraulic conditioners must keep hydraulic oil clean and at the proper temperature.

Figure 14-9. Pneumatic conditioners must keep the air as clean, cool, and dry as possible.

Air compressed by a compressor contains moisture as it air enters the piping of a pneumatic system. The presence of moisture in a pneumatic system causes problems such as rust, pressure drops, and lubricant removal from surfaces. To properly condition compressed air, air must pass through an air filter, a pressure regulator, and a lubricator before it is delivered to pneumatic equipment.

A *pressure regulator* is a pressure valve that controls the pressure in one leg of a system. Pressure regulators control the pressure in a leg of a system downstream from the valve. Some pressure regulators (vented pressure regulators) also can act as relief valves. A *lubricator* is a device that is used to add small droplets of oil to compressed air. Lubrication increases efficiency and helps maintain the sliding surfaces inside valves, cylinders, and air tools.

> A fluid power color code can be found on fluid power prints and drawings:
> red = supply pressure
> violet = intensified pressure
> orange = reduced or pilot pressure
> blue = drain
> green = intake or exhaust
> yellow = metered flow

FRLs. A *filter/regulator/lubricator (FRL)* is a special unit that is typically located at each workstation to condition, regulate air pressure, and lubricate air from the main supply line. **See Figure 14-10.** Porous metal filters combined with deflectors use centrifugal force to separate moisture and solid particles out of the air. The moisture is periodically removed from the filter by means of a drain cock located at the bottom of the filter.

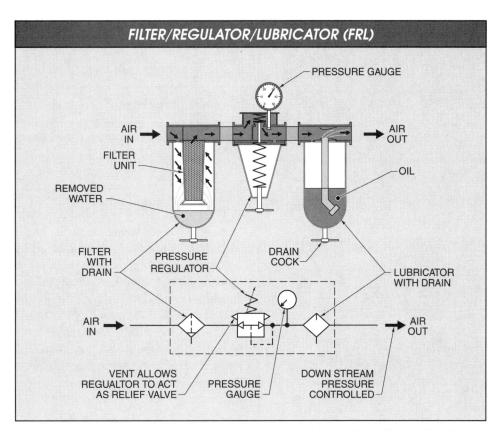

Figure 14-10. Compressed air must be cleaned, regulated, and lubricated by a filter/regulator/lubricator (FRL) before being used in pneumatic machinery or equipment.

Actuators

The pressurized fluid in a hydraulic or pneumatic system must be converted back to mechanical power after the fluid is transmitted to the point of use. An actuator is a fluid power component that converts fluid energy into mechanical energy. Fluid power actuators include cylinders, motors, and oscillators.

Cylinders. Cylinders are the most common type of actuator. A *cylinder* is a component that converts fluid energy into linear mechanical force. Cylinders include rams and single acting, double-acting, and double-rod cylinders. Rams and single-acting cylinders provide a force in one direction only. A *double-acting cylinder* is a type of cylinder that provides a high force in both directions of movement and requires fluid pressure to extend and fluid pressure to retract. Double-rod cylinders provide pushing and pulling forces at the same time. **See Figure 14-11.**

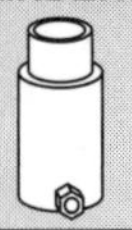

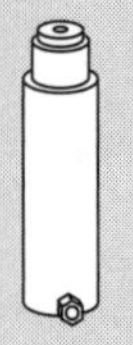

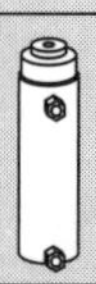

Figure 14-11. Cylinders convert fluid power energy into linear mechanical force.

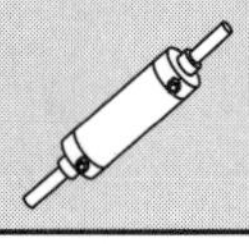

Motors. A *motor* is a hydraulic or pneumatic component that converts fluid energy into rotary mechanical energy. Fluid power motors include fixed displacement, variable displacement, and bidirectional motors. A *fixed displacement motor* is a fluid power motor that provides rotary motion that is unidirectional or bidirectional and a constant torque and speed output. A *variable displacement motor* is a fluid power motor that is unidirectional and provides a variable torque and speed output. A *bidirectional fluid power motor* is a fluid power motor that can rotate a load in both directions. **See Figure 14-12.**

Figure 14-12. Motors convert fluid power energy into rotary mechanical force.

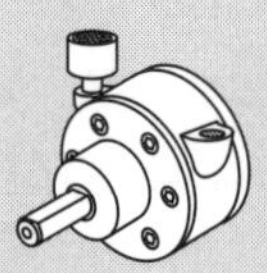

Fluid power motor applications include operating in explosive environments or high ambient temperatures, totally submerged applications, and applications in which a rotating shaft can often become stalled. No damage is done to the source of power or control devices when a fluid power motor is stalled. The speed at which a fluid power motor rotates depends on the volume of fluid delivered to the motor. The volume of fluid is adjusted by increasing or decreasing the flow rate through a flow control valve. The higher the flow rate, the faster the motor rotates. The lower the flow rate, the slower the motor rotates. **See Figure 14-13.**

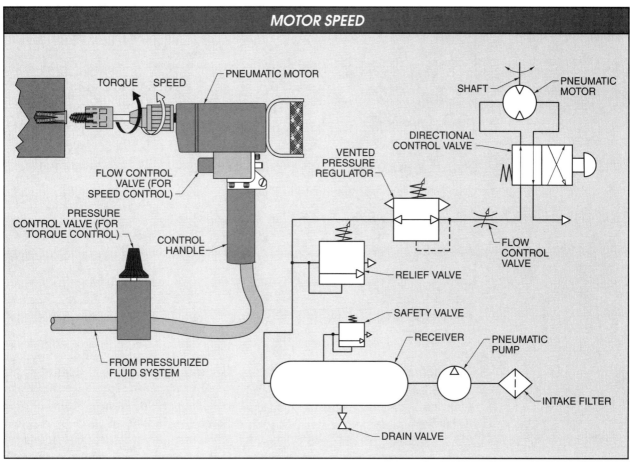

Figure 14-13. A flow control valve placed on the output side of a motor is used to change the flow rate through the motor, which changes the speed of the motor.

There are several fluid power laws. Charles' law states that a volume of air that is heated expands, and a volume of air that is cooled contracts. Boyle's law states that if a volume of a gas is decreased, pressure increases. Pascal's law states that pressure in a liquid or gas is exerted with equal force in all directions.

The amount of fluid pressure determines the amount of torque a motor has. The pressure of the fluid is set by pressure control valves. The higher the pressure, the higher the torque. The lower the pressure, the lower the torque. Most fluid power motors are designed to be bidirectional. The direction of rotation of a fluid power motor is reversed by reversing the direction of fluid flow through the motor.

Oscillators. An *oscillator* is a type of motor actuator that moves in fixed rotational increments each time fluid pressure is applied, then automatically reverses in direction. Typical increments are 15°, 30°, 45°, 90°, 120°, and 180°. Oscillators are used for clamping, opening and closing gates, loading parts into a machine, transferring parts, unloading or dumping parts, and applications that require an object to be turned over. **See Figure 14-14.** The advantage of using an oscillator is that oscillators can move very lightweight and/or very heavy loads without being adjusted.

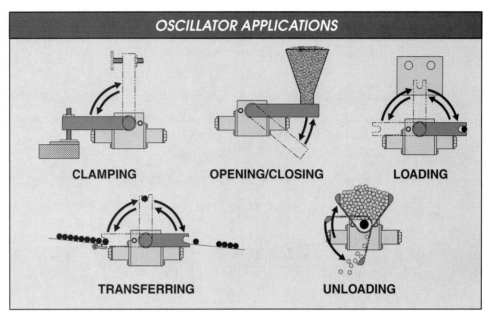

Figure 14-14. Oscillators rotate in fixed increments each time fluid pressure is applied, and then automatically reverse in direction.

Directional Control Valves

A *directional control valve* is a fluid power valve that connects, disconnects, or directs fluid flow from one part of a circuit to another. A directional control valve is connected before an actuator (sometimes immediately before) to control the movement of the actuator. The number of positions, ways, and types of actuators a valve contains identifies a directional control valve.

Positions. A directional control valve spool is placed in specific positions to start, stop, or change the direction of fluid flow. A *position* is an envelope within a valve in which the spool can be placed to direct fluid flow through the valve. A directional control valve typically has two or three positions. **See Figure 14-15.** Some directional control valves have four or more positions.

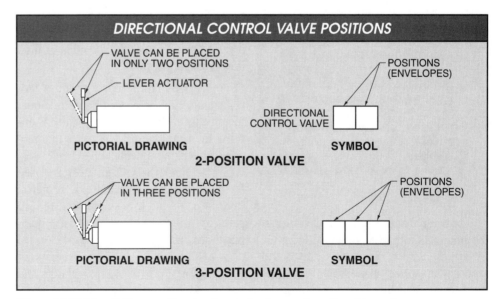

Figure 14-15. Most fluid power directional control valves are 2- or 3-position valves.

Hydraulic fluids are customized for specific applications. Hydraulic fluids are formulated to withstand extremely high or low temperatures, and they have excellent lubricating qualities. Except for biodegradable (vegetable-based) fluids, hydraulic fluids are highly toxic.

To place a hydraulic or pneumatic cylinder in the fully extended or fully retracted position or to start and stop the flow of fluid, 2-position directional valves are used. To place a hydraulic or pneumatic cylinder in the fully extended position, fully retracted position, or any position between fully extended and fully retracted, 3-position valves are used. Special applications use 4-position valves and up.

Ways. A *way* is a flow path through a valve. Most directional control valves are 2-way, 3-way, or 4-way valves. The number of ways required depends on the specific application. A 2-way directional control valve has two main ports that allow or stop the flow of fluid. Shutoff and quick-exhaust valves are 2-way valves.

A 3-way valve allows flow, stops flow, or allows oil back to the tank in hydraulic applications or exhausts air in pneumatic applications. To control single-acting cylinders, fill and drain tanks, and control nonreversible fluid power motors, 3-way valves are used.

A 4-way directional control valve has four main ports in hydraulic applications and typically has five ports in pneumatic applications. To control the direction of double-acting cylinders and bidirectional motors, 4-way valves are used. **See Figure 14-16.**

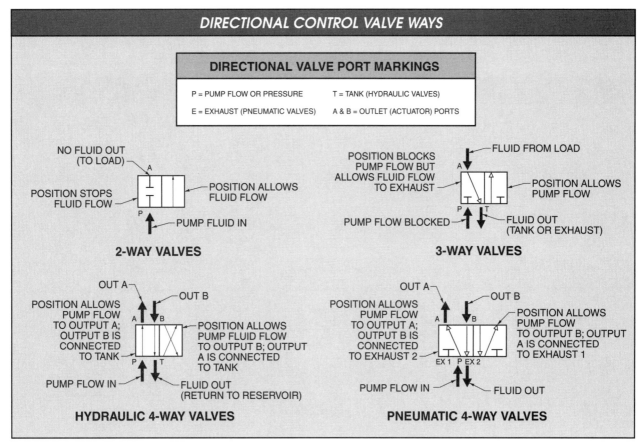

Figure 14-16. The number of ways in a hydraulic valve is determined by counting the number of ports (3 ports = 3 way, 4 ports = 4 way). However, in a pneumatic valve, 3 ports = 3 way, but both 4 ports and 5 ports = 4 way.

The most popular type of hydraulic or pneumatic directional valve actuator in use today is the solenoid actuator.

Pushbutton, lever, and foot pedal actuators are used to manually operate valves. Solenoids are used to electrically operate valves from a remote location. Pilot pressure actuation also allows a valve to be actuated from a remote location. Springs are used to set a valve in the "normal" position.

Actuators are used individually or in combination. For example, many solenoid-actuated valves also include a manual operator to serve as an override. The manual operator is typically included so that the valve can be manually tested when troubleshooting a circuit and to allow the valve to be actuated when there is no electrical power available.

Valve Actuators. Directional control valves must have a means to change the position of the valve spool. An *actuator* is the part of the directional control valve that changes the position of the spool in the valve. Directional control valve actuators include pushbuttons, levers, foot pedals, springs, and solenoids. **See Figure 14-17.**

Normally Closed and Normally Open Directional Valves

A *normally closed (NC) valve* is a valve that does not allow fluid to flow in the spring-actuated (normal) position. A *normally open (NO) valve* is a valve that

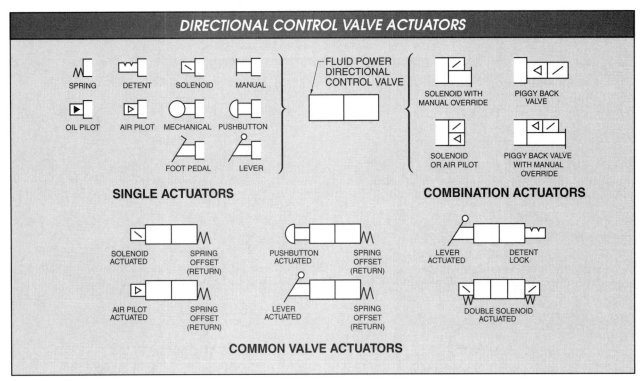

Figure 14-17. Directional control valve actuators (manual or automatic) change the position of the valve spool, which changes the flow path through the valve.

allows fluid to flow in the spring-actuated (normal) position. Directional control valves that are 2-way and 3-way valves are typically either normally open or normally closed. Typically, a valve must have a spring to be considered normally open or normally closed. **See Figure 14-18.**

> In fluid power terms, "normally open" means there is fluid flow, and "normally closed" means there is no fluid flow. In electrical terms the opposite is true: "normally closed" means there is current flow, and "normally open" means there is no current flow.

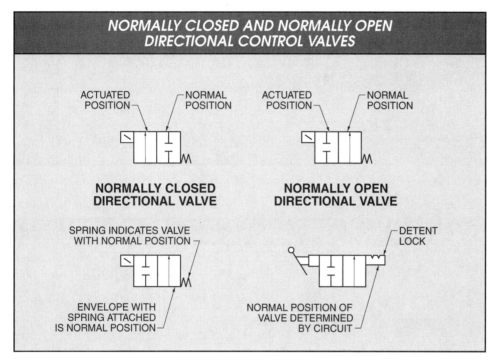

Figure 14-18. A normally closed (NC) valve does not allow fluid to flow out of the valve in the normal position. A normally open (NO) valve allows fluid to flow out of the valve in the normal position.

Two-Way Directional Control Valves. A *2-way directional control valve* is a fluid power valve used to start and stop the flow of a fluid (liquid or gas) and has two ports located on the valve. One port (the inlet port) is connected to pump flow or pressurized fluid and the second port is the outlet port. Pump flow or pressurized fluid may be air, oil, or another type of product. For example, 2-way plumbing valves are used to control the flow of pressurized paint, glue, food products, and gas in industrial facilities.

A normally closed valve does not allow fluid to flow through it in the normal (spring) position. A very common valve is the 2-position, 2-way, spring-offset, normally closed, solenoid actuated, directional control valve. **See Figure 14-19.** The envelope that the spring is attached to is the normal position, and the valve is in the normal (spring) position any time the solenoid is not energized. The valve changes position and allows fluid to flow when the solenoid is energized.

A 2-position, 2-way directional control valve can also be normally open. A normally open directional control valve allows fluid to flow through it in the normal (spring) position. **See Figure 14-20.**

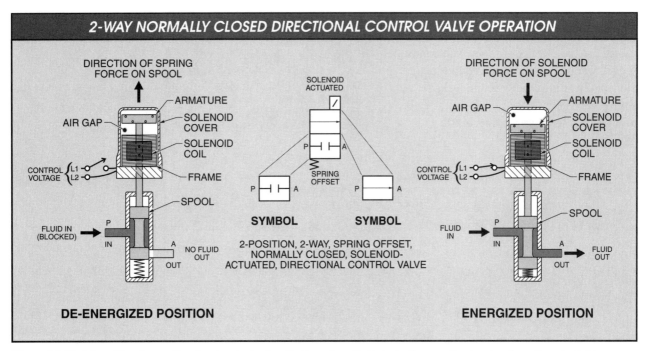

Figure 14-19. A 2 position, 2-way, directional control valves that is spring offset is typically normally closed and must be actuated to open.

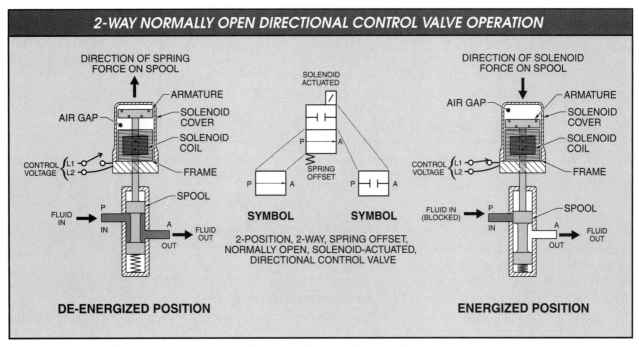

Figure 14-20. Normally open directional control valves are used as part of system safety.

A 2-position, 2-way, spring-offset, normally open, solenoid-actuated directional control valve is a commonly used directional control valve in hydraulics and pneumatics. The spring position indicates the normal position, and the valve is in the spring position any time the solenoid is not energized. The valve changes position and does not allow fluid to flow when the solenoid is energized. The valve changes back to the spring position when the solenoid is not energized, allowing fluid to flow through the valve. Two-position, 2-way, solenoid-actuated directional control valves are typically used in applications that require the starting and stopping of product flow, such as in air conditioning units to start and stop the flow of refrigerant. **See Figure 14-21.**

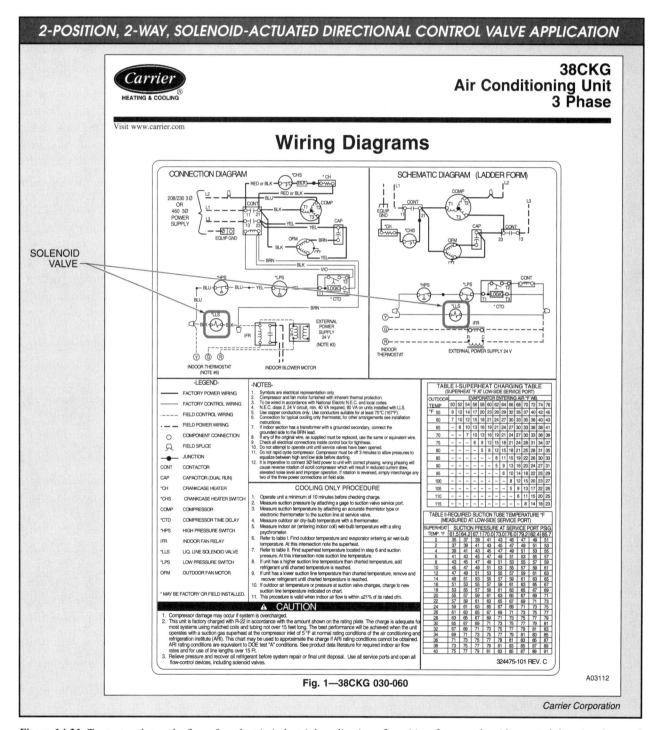

Figure 14-21. To start and stop the flow of product in industrial applications, 2-position, 2-way, solenoid-actuated directional control valves are typically used.

PIGGYBACK DIRECTIONAL VALVES

Piggyback directional valves are usually large in size. The valve is of such size that solenoids cannot be used to directly shift the spool of the valve. For electrical control of a large directional valve, the main system valve has a small directional slave valve attached to the top. The main system valve is designed to be pilot operated so that fluid pressure can be used to actuate the main valve.

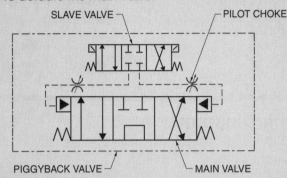

The slave valve is solenoid actuated, allowing electrical control. The slave valve actuator ports are internally connected to the main system valve's pilot-operated ports. This arrangement allows for the use of chokes, which cause the main system valve shifting to be slowed down. The arrangement also allows pilot pistons to be used on the main system valve to increase the valve shifting speed.

Three-Way Directional Control Valves. A *3-way directional control valve* is a fluid power valve used to start and stop the flow of a liquid or gas, allow fluid to return to the reservoir in a hydraulic system, or allow air to exhaust to atmosphere in a pneumatic system. Three-way directional valves have three ports located on the valve. One port is connected to pump flow or pressurized fluid (inlet), another port is the actuator (outlet) port, and the third port is the exhaust (air) or reservoir (oil) port. The fluid is typically oil or air.

Unlike 2-way valves, 3-way directional control valves are generally not used to control any type of product flow. This is because the flow of a product is normally only started or stopped, not returned to a tank. Directional control valves that are 3-way are typically used to control single-acting cylinders and nonreversible fluid power motors.

Like 2-way valves, 3-way valves that include a spring position are either normally closed or normally open. A normally closed valve does not allow fluid to flow through the valve in the normal (spring) position. For example, a 2-position, 3-way, spring-offset, normally closed, solenoid-actuated, directional control valve is typically only used for hydraulic or pneumatic applications. **See Figure 14-22.**

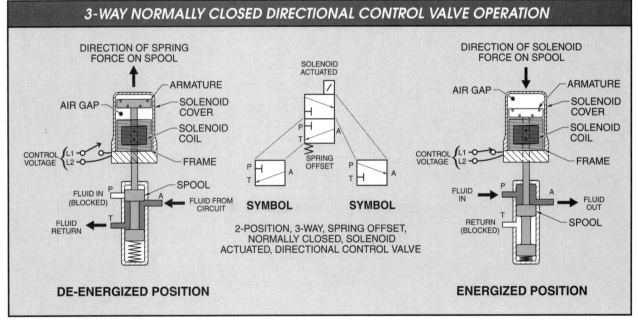

Figure 14-22. Unlike 2-way valves, 3-way directional control valves for hydraulics include a reservoir or tank port (T), and pneumatic valves include an exhaust port (E or EX).

The spring position is the normal position and the valve is in the spring position any time the solenoid is not energized. The valve changes position and allows fluid to flow when the solenoid is energized. The valve returns to the spring position and connects outlet port A to the exhaust or tank port when the solenoid is not energized. The normal position allows any air in the cylinder to exhaust to the atmosphere in a pneumatic system, or any oil in the cylinder to return to the reservoir in a hydraulic system.

A 2-position, 3-way valve can also be normally open. A normally open valve allows fluid to flow through the valve in the normal (spring) position. The 3-way valve is a solenoid-actuated, spring-return valve. **See Figure 14-23.** The spring position is the normal position, and the valve is in the spring position any time the solenoid is not energized. The valve changes position and does not allow pump flow to flow through the valve when the solenoid is energized. The fluid returning from the circuit is connected to the tank, or reservoir, port. The valve returns to the normal position and pump flow is allowed to flow through the valve when the solenoid is de-energized.

Three-way directional control valves are typically used in applications where the valve controls a spring-return cylinder. **See Figure 14-24.** For example, a 3-way valve can be used to control a spring-return cylinder that is used to clamp a part in place. Flow from the pump is used to advance the cylinder and clamp the part using high force. When the 3-way directional valve is de-energized, the spring in the cylinder is used to retract the cylinder. The advantage of using a single-acting cylinder with spring return is that only one fluid line needs to be connected to the cylinder.

In the example, a foot-actuated valve or a solenoid-actuated valve controlled by an electrical foot switch is used. The advantage of using a foot-actuated valve is that no electricity is required to operate the circuit. The advantage of using a solenoid-actuated valve is that additional circuit control can easily be added for safety. In the example, a limit switch can be added to the electrical circuit to prevent the cylinder from being operated unless a guard is in place. The guard activates the limit switch, which is connected in series with the foot switch.

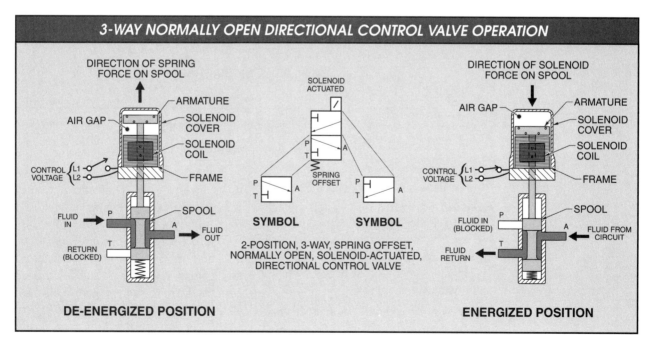

Figure 14-23. Three-way normally open directional control valves are typically used as intermediate valves (directional valves before and after valve) in a fluid power circuit.

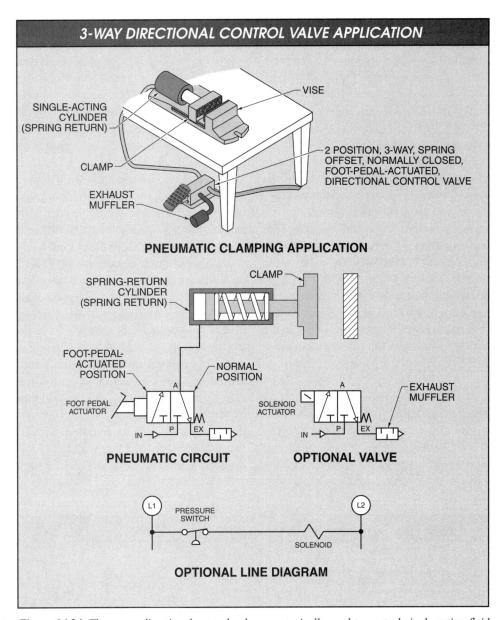

Figure 14-24. Three-way directional control valves are typically used to control single-acting fluid power cylinders and nonreversible motors.

Likewise, a pressure switch can also be added to the electrical circuit to prevent the clamp cylinder from advancing unless the required clamping pressure is present in the system (low system pressure could create a dangerous condition by not clamping the part with enough force).

Four-Way Directional Control Valves. A *4-way directional control valve* is a fluid power valve used to control the movement of a double-acting cylinder and bidirectional motor. A 4-way directional control valve has four ports on hydraulic valves and four or five ports on pneumatic valves.

With a 4-way hydraulic directional control valve, one port (the inlet) is connected to the pump, two ports (A and B) are used as actuator ports, and one port (T) is used as the reservoir, or tank port. **See Figure 14-25.** The two actuator ports are typically connected to a double-acting cylinder or bidirectional motor.

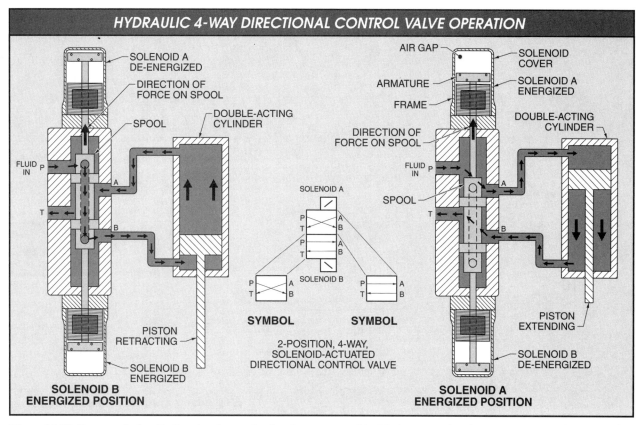

Figure 14-25. Four-way hydraulic directional control valves have one port (the inlet) connected to the pump, two ports that are used as actuator ports (A and B), and one port that is used as the reservoir or tank port (T).

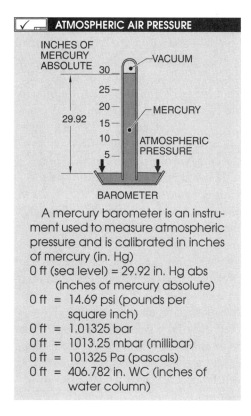

ATMOSPHERIC AIR PRESSURE

A mercury barometer is an instrument used to measure atmospheric pressure and is calibrated in inches of mercury (in. Hg)

0 ft (sea level) = 29.92 in. Hg abs (inches of mercury absolute)
0 ft = 14.69 psi (pounds per square inch)
0 ft = 1.01325 bar
0 ft = 1013.25 mbar (millibar)
0 ft = 101325 Pa (pascals)
0 ft = 406.782 in. WC (inches of water column)

With a 4-way pneumatic directional control valve, one port (the inlet) is connected to the air compressor, two ports (A and B) are used as outlet ports to actuators, and one or two ports (E or EX) are used as exhaust ports. The two actuator ports are typically connected to a double-acting cylinder or bidirectional motor. **See Figure 14-26.**

Unlike 2-position, 2- and 3-way valves, a 2-position, 4-way valve does not typically have a normally open or normally closed position. This is because the pressurized fluid is always flowing out of one of the two outlet ports. The valve changes the outlet port that the fluid from the pump is flowing out of, but does not stop the flow of fluid. A 3-position directional valve is used in applications requiring that fluid not flow out of the valve in one position. In a 3-position valve, the center position is the normal position and is typically used to stop the flow of fluid through the valve when needed. **See Figure 14-27.**

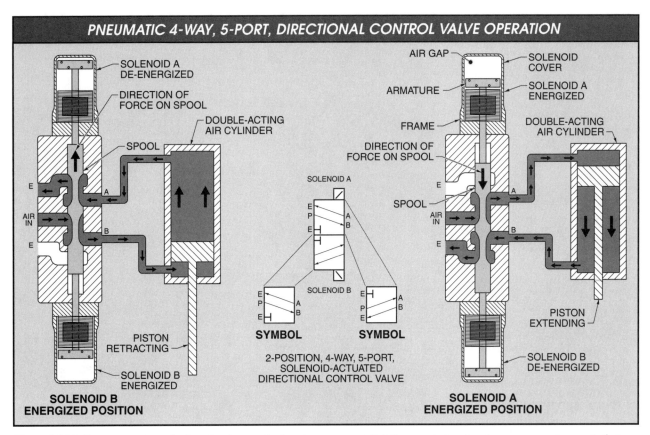

Figure 14-26. Four-way pneumatic directional control valves have one port (inlet) connected to the air compressor, two ports that are used as outlet ports to actuators (A and B), and one or two ports that are used as exhaust ports (E or EX).

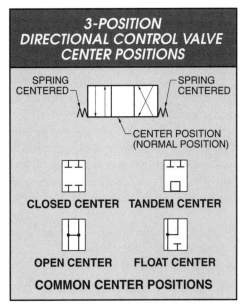

Figure 14-27. In a 3-position directional control valve, the center position (spring-centered valve) is the normal position (open center, closed center, float center, or tandem center) and can be used to stop the flow of fluid through the valve.

Solenoid-Actuated Directional Control Valves. Solenoids are often used to actuate fluid power directional control valves. When a directional control valve has two solenoids, both solenoids are shown on the electrical print. If the directional control valve is single-solenoid-operated spring offset, only one solenoid is shown on the electrical print.

For example, when a double-solenoid-actuated directional control valve is shown on a print, both solenoids will be shown. **See Figure 14-28.** The "ram extend solenoid" is energized anytime solenoid 1 is powered. The "ram retract solenoid" is energized anytime solenoid 2 is powered. When using a double-solenoid-actuated valve, only one solenoid can be actuated at a time. To avoid energizing both solenoids of a directional valve at the same time, the electrical circuit has a three-position selector switch (1LS), which directs electrical power to solenoid 1 only or solenoid 2 only, never to both.

Chapter 14—Fluid Power Systems **437**

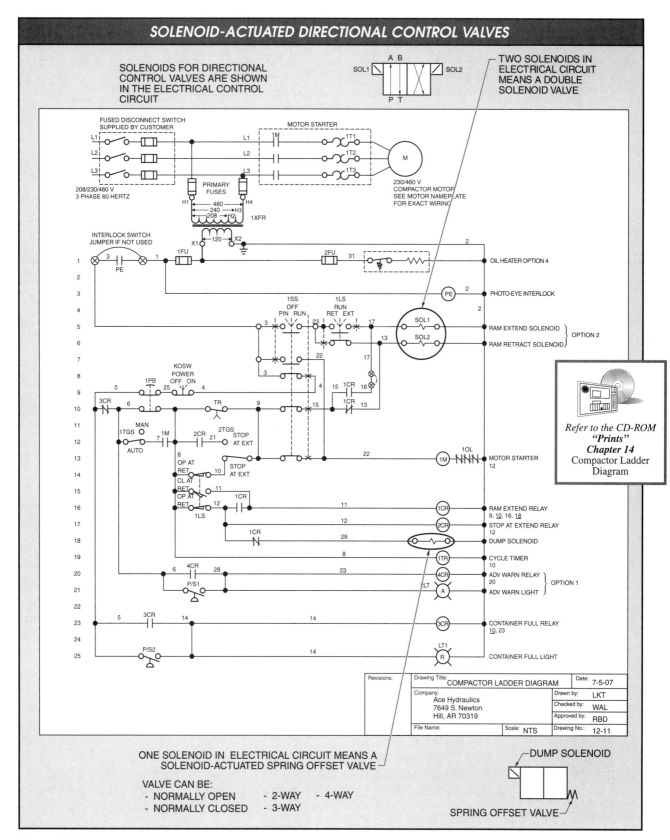

Figure 14-28. Directional control valves can be single-solenoid-actuated or double-solenoid-actuated. Depending on the application, one or two solenoids will be indicated on the electrical print.

The dump valve is controlled by the "dump" solenoid. The dump valve is a solenoid-actuated spring-offset valve. The valve can be a 2-way, 3-way, or 4-way valve. When the valve is a 2-way valve, the valve can be either normally open or normally closed. Because the electrical circuit does not indicate the number of ways or whether the valve is normally open or normally closed, a fluid power print must be used with the electrical print when troubleshooting.

Continental Hydraulics

Almost all hydraulic and pneumatic directional valves are classified as spool valves. The spool slides back and forth in the bore of the valve to open or block the valve ports.

Flow Control Valves and Check Valves

A *flow control valve* is a fluid power valve used to control the volume of fluid traveling through a system. By controlling the flow of fluid, the speed of a cylinder or motor can be controlled. Liquid flow rate is expressed in gallons per minute (gpm). Air flow is expressed in cubic feet per minute (cfm). To control the amount of fluid flowing through a flow control valve, the valve includes a fine threaded stem. The more a flow valve is opened, the greater the flow through the valve. The more a flow valve is closed, the lower the flow through the valve. **See Figure 14-29.**

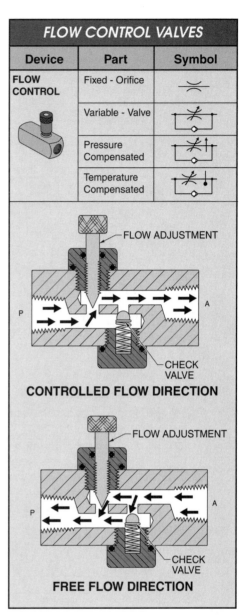

Figure 14-29. Flow control valves placed on the output of an actuator improve the speed control of the actuator.

Most flow control valves include a built-in check valve. A *check valve* is a valve that allows fluid to flow in one direction only. When a flow control valve includes a check valve, the flow control valve controls fluid flow in one direction and the check valve allows full flow in the opposite direction through the valve. **See Figure 14-30.** Fluid power check valves perform the same type of work that a diode performs in an electrical circuit.

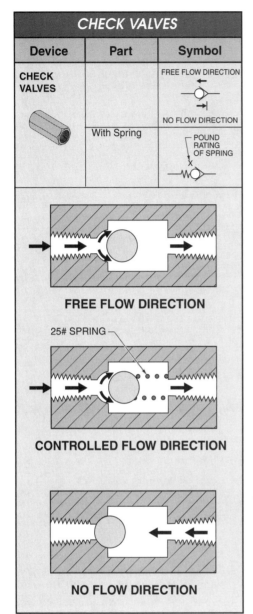

Figure 14-30. A check valve is a valve that allows fluid to flow in only one direction.

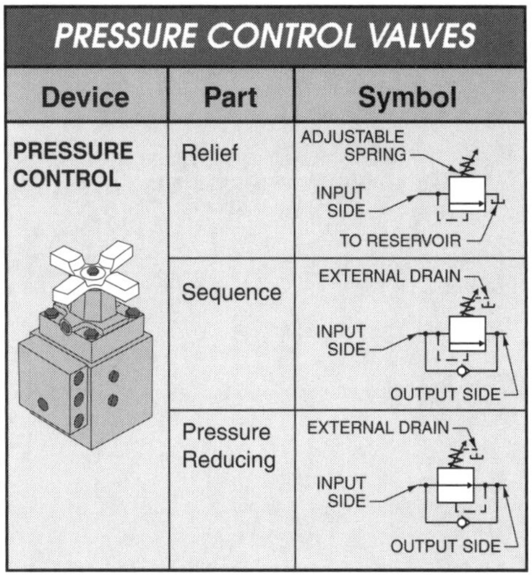

Pressure Control Valves

Fluid pressure must be controlled in all hydraulic and pneumatic applications. Pressure control valves are used to set maximum system pressure (pressure-relief valve), automatically redirect pressure when a preset pressure value is reached (sequence valve), or reduce pressure in one leg of a system (pressure-reducing valve). **See Figure 14-31.**

Figure 14-31. The three most popular pressure control valves are the NC relief valve, the NC externally drained sequence valve, and the NO externally drained pressure-reducing valve.

> In fluid power symbols, an arrow through a device means that the device is adjustable. Arrows are found on variable flow pump symbols and motor symbols, flow valves (resistances that are not adjustable are orifices), and pressure valves (most pressure valves are adjustable, some are not, such as preset safety valves).

Pressure-Relief Valves. A *pressure-relief valve* is a pressure control valve that limits the maximum pressure in a fluid power system. A pressure-relief valve protects the pump, prime mover (electric motor), and piping from overloading. A pressure-relief valve opens when the pressure has reached a preset value and serves as a safety valve to prevent damage that may be caused by excessively high pressures developing in a system.

A pressure-relief valve is a 2-way, normally closed valve that does not allow fluid through it when the pressure in the system is below the setting of the valve. The valve opens and relieves the pressure in the system to the reservoir when the pressure in the system increases to the setting of the valve. The relief valve closes when the pressure in the system drops below the setting of the relief valve. **See Figure 14-32.**

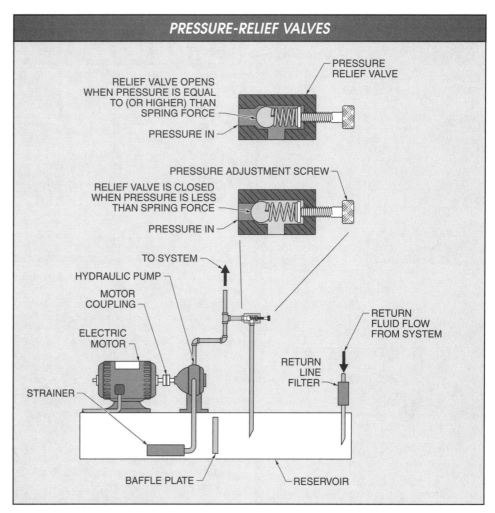

Figure 14-32. Pressure relief valves limit the maximum pressure in a fluid power system.

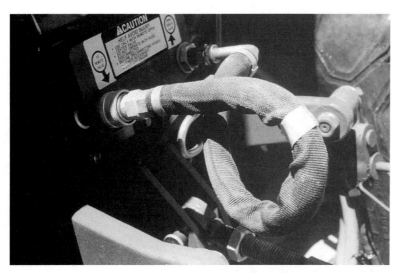

The Gates Rubber Company

Fluid power hoses are used whenever movement of piping is required. Hoses must be installed with a bending radius of at least six times the inside diameter of the hose.

Sequence Valves. A *sequence valve* is a pressure control valve that directs fluid from one part of a circuit to another part of the circuit to sequence the movement of cylinders, motors, and/or valves. Sequence valves open only after the pressure has reached a preset value. Sequence valves are designed to open at a set pressure and close when pressure drops below the set value. **See Figure 14-33.**

For example, in a plastic injection machine, flow from a hydraulic pump travels through the tandem center of the directional control valve and return line filter back to the reservoir. When solenoid A on the directional control valve is energized, fluid flows to the valve-and-die cylinder at 550 psi (the resistance of the valve-and-die cylinder loads) to extend the cylinder, and also flows to the normally closed sequence valve. When the valve-and-die cylinder is fully extended, pressure increases in the system until the pressure reaches 800 psi. When pressure in the system reaches 800 psi, the sequence valve opens, allowing fluid to flow into the injection cylinder and extending it. Energizing solenoid B on the directional valve retracts both cylinders.

> Pressure valves fall into three categories: direct-acting, pilot-operated, and compound. Direct-acting valves are designed so that the spring in the valve is in contact with the valve spool to close the valve. Pilot-operated valves have a small direct-acting relief valve that controls the spool. Compound valves consist of two pressure valves built into one housing.

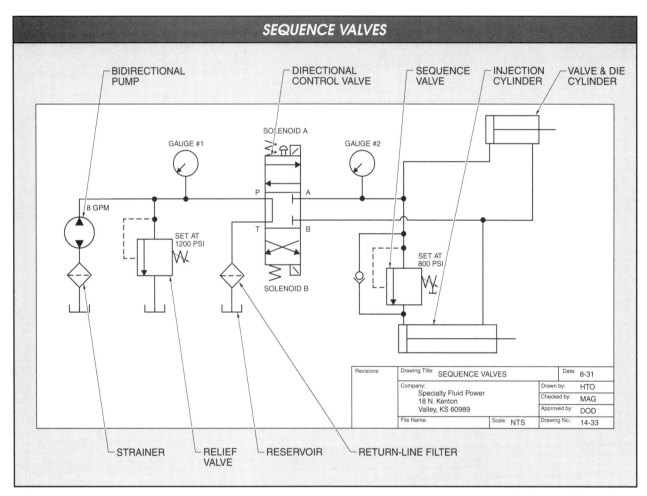

Figure 14-33. Sequence valves direct fluid from one part of a circuit to another part of a circuit to sequence the movement of cylinders, motors, and/or valves.

Pressure-Reducing Valves. A *pressure-reducing valve* is a pressure control valve that reduces the pressure in one leg of a circuit. Pressure on the output side of the valve can be adjusted from 0% to 100% of input pressure. Any number of pressure-reducing valves are used to control the different pressures required in a fluid power system. **See Figure 14-34.** As a pressure-reducing valve closes to limit pressure, the closing valve also reduces flow rate.

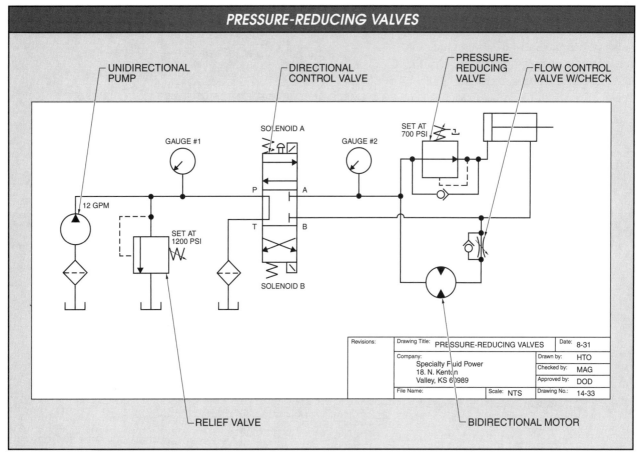

Figure 14-34. Pressure-reducing valves reduce and/or limit the pressure in one leg of a circuit.

For example, in a hexane roto-cell, flow from a hydraulic pump travels through the tandem-center directional control valve and return line filter back to the reservoir. When solenoid A on the directional control valve is energized, fluid flows through the normally open pressure-reducing valve into the cylinder, which extends the cylinder. Pump flow also flows through the bidirectional motor, causing motor rotation, and flows through the flow control valve, which limits the rotational speed of the motor.

As the resistance on the cylinder increases, the pressure will go up. When the pressure approaches 700 psi, the pressure-reducing valve starts to close, limiting the pressure to the cylinder to 700 psi. Pressure to the motor can increase to 1200 psi before the relief valve opens, dumping pump flow into the reservoir. Energizing solenoid B on the directional valve retracts the cylinder and reverses the direction of motor rotation.

> Hydraulic systems have one reservoir. No matter how many reservoir symbols appear on a hydraulic print, only one reservoir exists. Machinery with more than one reservoir has more than one hydraulic system.

Miscellaneous Fluid Power Devices and Components

In addition to the main devices and components in fluid power circuits, additional components are required for specialized functions. Miscellaneous fluid power devices and components include pressure gauges, air mufflers, manual shutoff valves, accumulators, and various types of piping for connections. **See Figure 14-35.**

A *pressure gauge* is a device used to indicate the pressure at a given point in a fluid power system. Pressure gauges monitor and display system pressure. An *air muffler* is a device that reduces the noise that is made when air is exhausted from a pneumatic system to the atmosphere. A manual shutoff valve is a type of valve that must be manually opened or closed to allow or stop fluid flow. An *accumulator* is a device that stores fluid under pressure. Accumulators allow the pump size required for a circuit to be reduced, provide additional volume in part of a circuit when required, and act as pressure shock absorbers. Accumulator pressure is maintained by a spring, a nitrogen charge, or a physical weight.

Piping is used to connect various devices and components in a fluid power circuit. Connecting lines (piping) include main pressure lines (solid lines), pilot lines (dashed lines), exhaust or drain lines (dotted lines), and flexible lines (curved and dashed lines). The hoses and tubing that are used must be rated at a higher pressure, temperature, and flow rate than the temperature, pressure, and flow rate that is present in the system. Hoses must be made of an outer material that is compatible with the environment in which the hose is being used and an inner material that is compatible with the fluid.

Because CAD software is in such wide use, fluid lines, such as pilot, pressure, and drain lines, can be color coded on fluid power prints to distinguish between the different types of piping. Whether or not colors are used on a fluid power print, solid, dashed, and dotted lines must be used to represent the different types of piping.

MISCELLANEOUS FLUID POWER SYMBOLS				
Device	Type	Symbol	
		Hydraulic	Pneumatic
GAUGES	Pressure		
	Temperature		
MUFFLER	Air		
SHUTOFF VALVE	Manual		
ACCUMULATORS	Spring		
	Weighted		
	Gas Charged		
PIPING	Pressure Line		
	Pilot Line		
	Drain Line		
	Hose		

Figure 14-35. In addition to the main components in fluid power circuits, additional devices and components are required for specialized functions.

> Pressure gauge symbols do not indicate the type of gauge used. Pressure gauges can be Bourdon tubes, which are the most accurate, or Schrader gauges, which are not as accurate, but are tougher against pressure surges.

Refer to the CD-ROM "Quick Quizzes" Chapter 14

Review Questions and Activities

Fluid Power Systems

Name _____ Date _____

True-False

T F   1. The symbol for a fluid power valve is very specific as far as how a device or component operates.

T F   2. The flow rate in a hydraulic system is easily regulated because liquid is considered noncompressible.

T F   3. Hydraulics and most pneumatic systems use non-positive-displacement pumps.

T F   4. The pressurized fluid in a hydraulic or pneumatic system must be converted back to mechanical power after the fluid is transmitted to the point of use.

T F   5. The way is the part of the directional control valve that changes the position of the spool in the valve.

T F   6. A normally open valve does not allow fluid to flow when it is in the spring-actuated position.

T F   7. Directional control valves that are 3-way are typically used to control single-acting cylinders and nonreversible fluid power motors.

T F   8. Most flow control valves do not include a built-in check valve.

T F   9. Pressure control valves are used to set maximum system pressure.

T F   10. Sequence valves open only after the pressure has reached a preset value.

Completion

_____ 1. Fluid power uses ___ or gas to produce work.

_____ 2. A(n) ___ is a device that removes solid contaminants from a fluid power system.

_____ 3. A(n) ___ valve is a fluid power valve that connects, disconnects, or directs fluid flow from one part of a circuit to another.

_____ 4. A 4-way directional control valve has four ports on hydraulic valves and four or ___ ports on pneumatic valves.

_____ 5. A(n) ___ is a device that stores fluid under pressure.

Multiple Choice

_____ 1. ___ is the transmission and control of energy by means of a pressurized fluid.
   A. Hydraulics
   B. Fluid power
   C. Pneumatics
   D. Pressure control

_____ 2. ___ systems are open systems because air is exhausted to the atmosphere after it performs its work.
   A. Flow
   B. Directional control
   C. Pneumatic
   D. none of the above

_____ 3. A ___ is a pump that delivers a finite quantity of fluid for each revolution of the shaft.
   A. fluid displacement pump
   B. variable displacement pump
   C. non-positive-displacement pump
   D. positive displacement pump

_____ 4. A ___ displacement motor is a fluid power motor that provides rotary motion that is unidirectional or bidirectional and a constant torque and speed output.
   A. bidirectional fluid power motor
   B. fixed
   C. variable
   D. double-acting

_____ 5. A ___ valve allows flow, stops flow, or allows oil back to the tank in hydraulic applications or exhausts air in pneumatic applications.
   A. way
   B. 2-way
   C. 3-way
   D. 4-way

Name _____ Date _____

Activity—Adding to a Hydraulic Control Circuit

14-1

Scenario:
Activity 14-1 is an installation activity requiring that a hydraulic milling machine in a machine shop have a 2-position, 2-way, solenoid-actuated, directional control valve added to the table's hydraulic circuit. A switching station will also be added. The machinists want independent control of table operations.

Task:
Verify understanding of hydraulic circuit operation and the function of the electrical controls.

Required Understanding to Perform Work:
1. Which pushbutton(s) must be pressed to start the table automatically cycling? _____
2. Which pushbutton(s) must be pressed to stop the table from cycling? _____
3. Which flow control valve controls the speed of the table during cylinder extension? _____
4. Which flow control valve controls the speed of the table during cylinder retraction? _____
5. If the spring were to break in directional valve #3, what would be the effect on table operation?
_____
_____
_____

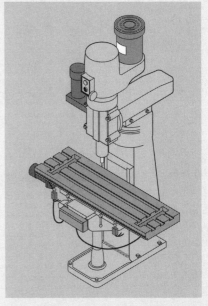

HYDRAULIC MILLING MACHINE

Reference Prints:

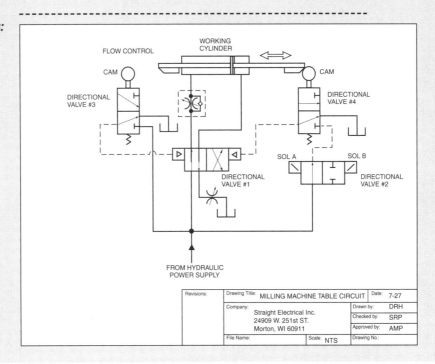

447

Activity—Adding to a Hydraulic Control Circuit

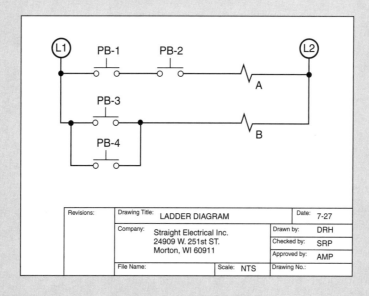

Name _____ Date _____

Activity—Troubleshooting Press Control Circuits

14-2

Scenario:
Activity 14-2 is a troubleshooting activity requiring that a hydraulic stamping press with a pneumatic table clamp be repaired. The press stopped working in the middle of a production run.

Task:
After about 150 cycles, the platen cylinder would not retract. Verify an understanding of the operation of hydraulic and pneumatic circuits and the function of the electrical controls.

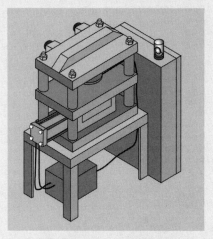

HYDRAULIC PRESS WITH AUTOMATIC TABLE VISE

Required Understanding to Perform Work:
1. Which switch or valve is used to advance the clamping cylinder? _____
2. Which switch or valve is used to advance the working cylinder? _____
3. Which switch or valve is used to retract the clamping cylinder? _____
4. Which switch or valve is used to retract the working cylinder? _____
5. What is the process for troubleshooting the problem with the hydraulic press?

_____
_____
_____
_____
_____

Reference Prints:

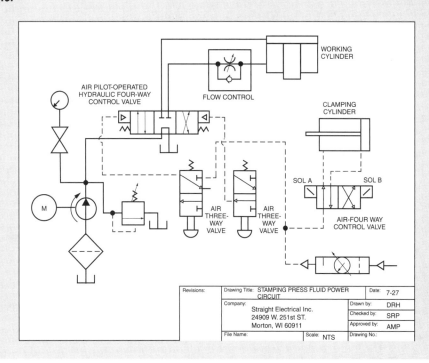

449

Activity—Troubleshooting Press Control Circuits

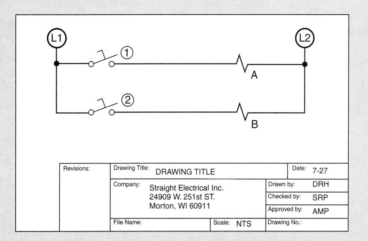

Trade Competency Test
Fluid Power Systems

Name _____ Date _____

REFERENCE PRINT #1 (Fluid Power Clamp and Eject Circuit)

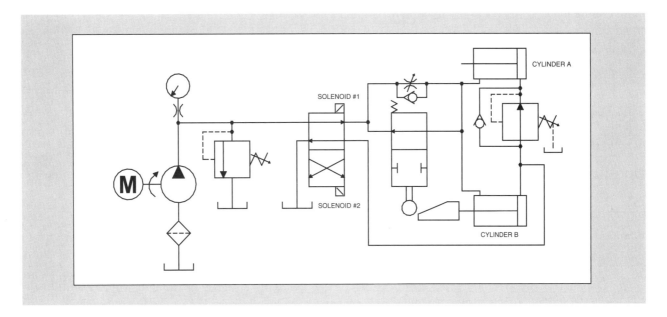

Questions – Reference Print #1

_____ 1. Is this circuit hydraulic, pneumatic, or combination hydraulic/pneumatic?

_____ 2. Which solenoid (SOL 1 or SOL 2) must be energized to extend the cylinders?

_____ 3. Do both the cylinders (Cylinder A and Cylinder B) extend with the same pressure?

_____ 4. How many ways (1, 2, 3, 4, 5, etc.) is the mechanically operated, spring-returned directional valve?

_____ 5. Is the mechanically operated directional valve a normally open or normally closed valve?

_____ 6. Does the flow control valve with check valve that is connected in parallel with the mechanically operated directional valve control the speed of the cylinders during extension or retraction?

_____ 7. What type of device is on the suction side of the pump (between the reservoir and pump)?

REFERENCE PRINT #2 (Accumulator Table Gate Circuit)

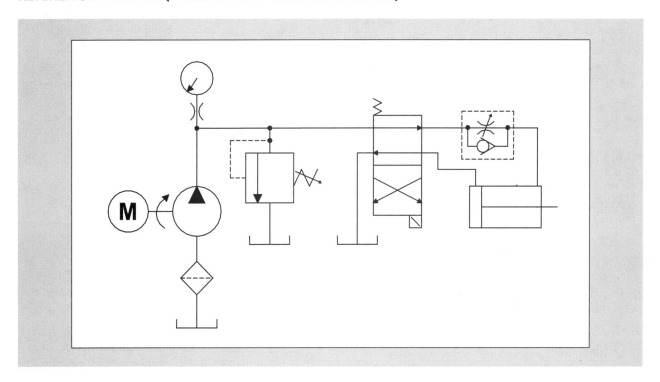

Questions – Reference Print #2

_____ 1. How many positions does the directional control valve in the circuit have?

_____ 2. How many ways (1, 2, 3, 4, 5, etc.) does the directional control valve have?

_____ 3. How many ways (1, 2, 3, 4, 5, etc.) does the valve next to the pump have?

_____ 4. If the cylinder is extending and a fuse blows in the control circuit for the directional valve solenoid (removing power), will the cylinder keep extending, stay where it is, or retract?

_____ 5. Does the flow control valve with check valve control the speed of the cylinder during extension, retraction, or both?

 ADVANCED CD-ROM PRINT QUESTION (Hydraulic Schematic)

1. The hydraulic circuit's cylinder and oscillator are controlled by solenoid-actuated directional control valves. Using standard symbols, draw the electrical control circuit in ladder (line) diagram format. Draw the electrical control circuit so that PB-1 and PB-2 must both be pressed to extend the cylinder and that PB-3 or PB-4 retract the cylinder. Use a 3-position selector switch (Left/OFF/Right) to control the oscillator. In the Left or Right position, one of the solenoids is energized and no solenoid is energized in the OFF position.

Process and Instrumentation Systems

Printreading for Installing and Troubleshooting Electrical Systems

Process industries include beverage, petrochemical, power generation, water treatment, food, and pharmaceutical facilities. Process industries manufacture a product by changing the state or form of a material. For example, a petrochemical refinery changes crude oil into various oil-based products such as gasoline, diesel fuel, and lubricating oils. Process manufacturing requires instrumentation, control elements, and process equipment to monitor and control the processes.

A variety of prints types are used in process manufacturing. One of the most commonly used prints in process manufacturing is the piping and instrumentation diagram (P&ID) print. P&ID prints are also known as process and instrumentation diagrams.

PIPING AND INSTRUMENTATION DIAGRAMS

A *piping and instrumentation diagram (P&ID)* is a print that depicts equipment that is essential to a process. Examples might be tanks, vessels, and piping. P&ID prints also include the equipment that monitors and controls the process, such as valves and actuators, instrumentation, and switching equipment. P&ID prints use lines to depict the piping that connects process equipment together, the wiring that connects instrumentation to the process equipment, and the wiring that connects the instrumentation to the control elements. **See Figure 15-1.**

P&ID prints are not drawn to scale, do not represent the exact physical location of process equipment, instruments, or control elements and do not show the chemical processes that are taking place in a facility. However, P&ID prints do indicate relative location. For example, a valve located under a vessel would be depicted on a P&ID print underneath the vessel. The complexity of the process determines the number of P&ID prints required to depict the process.

Specific abbreviations, lines, and symbols are used on P&ID prints. The ISA (Instrumentation, Systems, and Automation Society) has developed standards for P&ID prints. The ANSI/ISA-5.1-1984 (R1992) standard, *Instrumentation Symbols and Identification,* is widely used by engineers and designers throughout process industry for P&ID prints. Some process engineers use the ISA standard but add additional information to P&ID prints to meet specific facility requirements. Adding information that is not part of the standard is permissible under the standard.

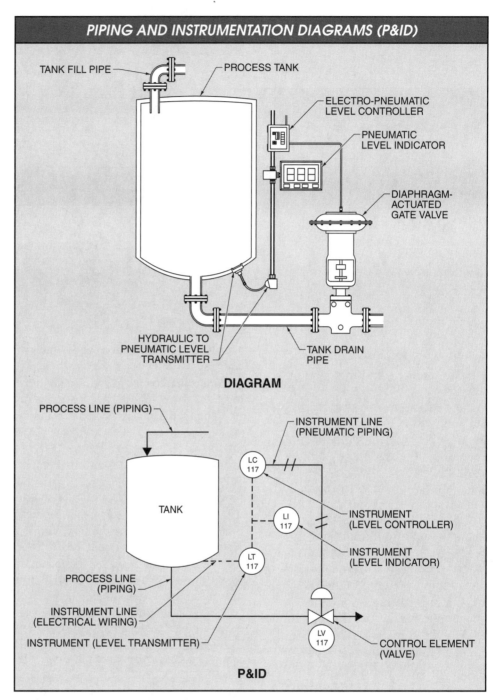

Figure 15-1. P&ID prints depict process equipment, process lines, instrument lines, instruments, and control elements.

LINES

Lines are used to depict the process piping and instrument lines on P&ID prints. Although various types of lines are used for process piping and instrument lines, certain conventions apply to both. Lines on a P&ID print do not represent actual distances or actual locations. The lines however, do represent relative location. For example, a process pipe entering the top of a vessel would be depicted on a P&ID print by a line entering the top of the vessel.

In order to avoid cluttering drawing space and to enhance P&ID print clarity, print lines are drawn at 90° to each other. Print lines with 90° turns do not represent bends in the process piping or instrument lines. Lines that touch at 90° or cross over each other are meant to depict process piping or instrument wiring that is connected. A loop in a line, a U-shaped curve, or an S-shaped curve is used to denote process piping or instrument wiring that is not connected. **See Figure 15-2.**

appropriate process line. **See Figure 15-3.** Abbreviations have been developed to denote the material flowing through a process pipe. **See Appendix.**

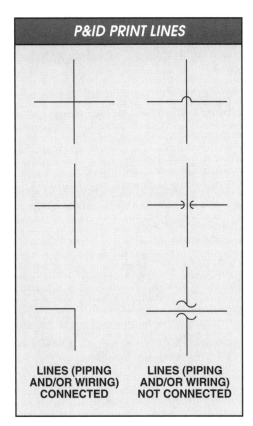

Figure 15-2. Different methods are used on P&ID prints to depict lines that are connected and lines that are not connected.

P&ID Print Process Lines

P&ID print process lines depict the piping that carries the gas or liquid being controlled and monitored as part of a manufacturing process. Process piping is depicted by bold lines on a P&ID print. Typically, the diameter of the process pipe, contents of the process pipe, and the process pipe number are drawn on a P&ID print adjacent to the

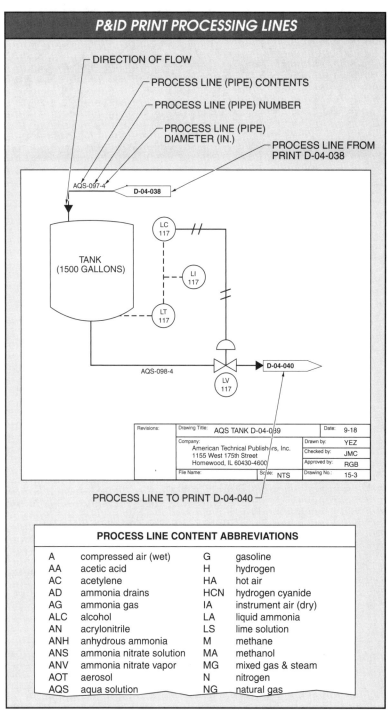

Figure 15-3. Descriptions (coded numbers) of a facility's process lines are typically shown on a P&ID print.

Cleaver-Brooks

The rising costs of energy have increased the use of pipe insulation. There are 13 different types of pipe insulation, including insulation for hot, cold, and Class 1 fire-rated applications.

Figure 15-4. *Process lines may be heat traced and/or insulated.*

Each process line on a P&ID print has a unique identifying number. A process line maintains the same number from point to point, such as from a tank to a heat exchanger. When two or more process lines exit a tank, vessel, heat exchanger, or other piece of process equipment, each line is assigned a new number. New numbers are assigned even when the process lines are the same size and carry the same material. The numbering sequence for process lines is typically determined by plant standards.

The direction of flow in a process is indicated on a P&ID print with a solid black arrow. More than one P&ID print may be required to depict a complex process. When more than one P&ID print is required, rectangles with pointed ends are used to depict the direction of flow, and the P&ID print that is the source or destination of the product flow is shown in a rectangle.

Certain industrial and commercial applications require that process piping be maintained at a constant temperature. The temperature of process piping can be maintained by using insulation, or the piping can be heat traced. *Heat tracing* is a method used to heat process piping. Heat tracing techniques used in piping include steam tracing, electrical tracing, and oil jacketing. **See Figure 15-4.**

Steam tracing is a pipe heating method where a steam jacket surrounds the process piping. *Electrical tracing* is a pipe heating method where an electrical heating element is fastened along the length of the process pipe. *Oil jacketing* is a pipe heating method where an oil jacket surrounds the process piping. Typically, heat traced process piping is also insulated. Separate symbols and abbreviations are used for process piping that is heat traced and/or insulated.

Instrument Lines

Instrument lines (piping or wiring) can connect an instrument to a process, connect instruments to one another, or connect an instrument to a control element such as a valve. Instrument lines carry the signals that monitor and control the process. Depending upon the application, various types of signals, such as electric, pneumatic, and hydraulic, are used for transmitting instrument information.

Instrument piping or wiring is depicted with fine lines on a P&ID print. Symbols have been developed to depict the various types of instrument lines used.

See **Figure 15-5.** Abbreviations have also been developed to denote the source of power for instrument lines.

Figure 15-5. Each type of instrument line has a unique symbol and abbreviation.

The length of an instrument line between instruments or between an instrument and a control element on a P&ID print does not correspond to the actual distance. The line does provide information about the functional relationship between instruments and control elements. While only a single line is used to represent the connection between instruments and/or control elements on a P&ID print, multiple wires or cables are actually used.

P&ID PRINT GRID SYSTEMS

Because P&ID drawings and prints can be quite large and complex, finding a specific pipe, valve, or piece of equipment may be difficult. The problem becomes even more difficult when a pipe run is drawn on two or three drawings. To help technicians find a pipe, valve, or piece of equipment, most large P&ID prints utilize a grid system.

A P&ID print grid consists of numbers and letters that border the grid vertically and horizontally. Each number and letter combination (grid coordinate) identifies the specific block on the print. When a pipe run is shown on more than one print, the second print is referenced on the first print using the grid coordinates to indicate the position of the pipe run on the second print.

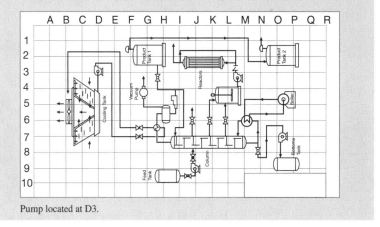

Pump located at D3.

INSTRUMENTS

Instruments are used to monitor and control a process. Instruments transmit signals to control elements, provide data to operators and other instruments, and monitor process variables.

Instruments monitor and control many different types of process variables. The most common variables found in process manufacturing are flow, level, pressure, and temperature. **See Figure 15-6.** Other variables include position, speed, vibration, and weight. Symbols and abbreviations are used to represent and identify instruments on P&ID prints.

INSTRUMENT SYMBOLS		
Device	Part	Symbol
FLOW 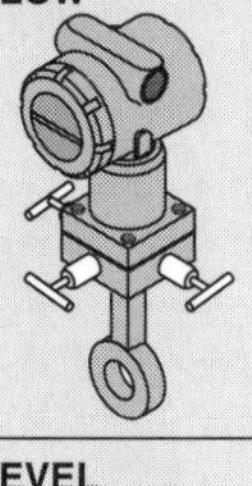	Indicator Abbr = FI	FI
	Transmitter Abbr = FT	FT
	Recorder Abbr = FR	FR
	Controller Abbr = FC	FC
LEVEL 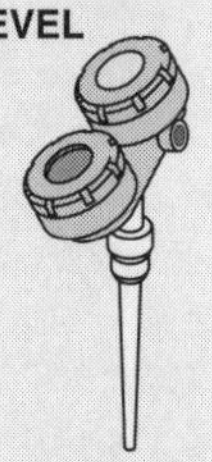	Indicator Abbr = LI	LI
	Transmitter Abbr = LT	LT
	Recorder Abbr = LR	LR
	Controller Abbr = LC	LC
PRESSURE 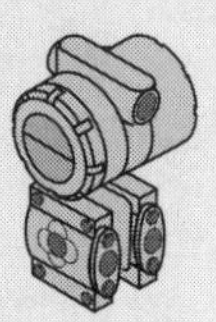	Indicator Abbr = PI	PI
	Transmitter Abbr = PT	PT
	Recorder Abbr = PR	PR
	Controller Abbr = PC	PC
TEMPERATURE 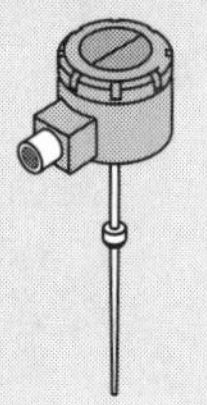	Indicator Abbr = TI	TI
	Transmitter Abbr = TT	TT
	Recorder Abbr = TR	TR
	Controller Abbr = TC	TC

Figure 15-6. Specific types of instruments are used to measure flow, level, pressure, and temperature.

Instrument Type and Location Symbols

The location of the instrument symbol on the print is not representative of the actual location of the instrument. For example, a temperature transmitter shown adjacent to a heat exchanger on a P&ID print may actually be located some distance from the heat exchanger. Instrument lines show the connection between the instrument and the process equipment. The instrument line touches the instrument symbol, and may or may not touch the process line, tank, or vessel.

P&ID prints must indicate when the diameter (size) of a process pipe is increased or decreased. A change in the diameter of a process pipe impacts how instrumentation responds.

The type of symbol used to represent an instrument depends on the actual location of the instrument and the type of instrument control being used. There are four different location classifications and four different instrument-control classifications for a total of 16 unique symbols. **See Figure 15-7.**

The four instrument location classifications are primary location, field mounted location, auxiliary location, and behind the panel location. A *primary location* is an instrument location, typically in a control room, where the instrument is accessible to operators who monitor and control the process. A *field mounted location* is an instrument location in the manufacturing plant in the vicinity of the process the instrument monitors or controls. An *auxiliary location* is an instrument location near a control room where instruments are accessible to the operators. A *behind the panel (normally inaccessible) location* is an instrument mounting location behind a control panel or control cabinet.

The four instrument-control classifications are discrete instrument, shared display instrument, computer function, and programmable logic control. A *discrete instrument* is a stand-alone piece of hardware with a single specific function. The function could be flow indication, level transmittion, pressure recording, or temperature control. A *shared display–shared–control instrument* is a stand-alone piece of hardware that provides the functions of both display and control. A *computer function* is a function that is part of a computer-controlled system that executes logic or calculations and transmits the result. An *instrument controller* is an instrument that is part programmable logic controller and part sensor, has multiple inputs and outputs, and contains a program that can be modified.

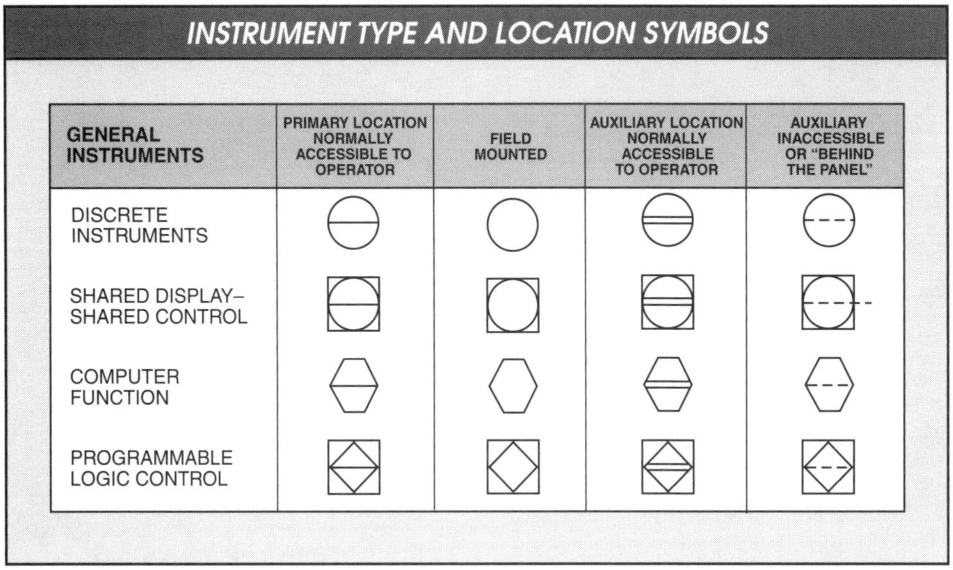

INSTRUMENT TYPE AND LOCATION SYMBOLS

GENERAL INSTRUMENTS	PRIMARY LOCATION NORMALLY ACCESSIBLE TO OPERATOR	FIELD MOUNTED	AUXILIARY LOCATION NORMALLY ACCESSIBLE TO OPERATOR	AUXILIARY INACCESSIBLE OR "BEHIND THE PANEL"
DISCRETE INSTRUMENTS	⊖	○	⊖	⊖
SHARED DISPLAY– SHARED CONTROL	⊟	▢	⊟	⊟
COMPUTER FUNCTION	⬡	⬡	⬡	⬡
PROGRAMMABLE LOGIC CONTROL	◇	◇	◇	◇

Figure 15-7. Instrument symbols are classified by location and instrument control type.

ANSI/ASI-5.5-1985, Graphic Symbols for Process Displays is a standard that is used with ANSI/ISA-5.1-1984 (R1992), Instrumentation Identification, and ISA-5.3-1983, Graphic Symbols for Distributed Control/Shared Display Instrumentation Logic and Computer Systems. ASI-5.5-1985 is compatible with the chemical, petroleum, power generation, HVAC, and refining industries.

The standard is used whenever references to process equipment on prints are required. Following ANSI/ASI-5.5-1985, Graphic Symbols for Process Displays, allows for better communication between all personnel.

Instrument and Control Element Identification

Typically a P&ID print has multiple instruments and control elements as part of the drawing. Letter abbreviations are used to identify the function of the instruments and control elements. Numbers are used to identify specific control loops. A *control loop* is a combination of instruments or a combination of instruments and control elements connected together to monitor and control a specific process.

Each instrument and control element has a unique identifier. The *control loop identification system* is a combination of the letter abbreviation and the control loop number. Loop information for control elements is contained in a circle drawn near the control element symbol. The letter abbreviation appears in the upper half of the circle, and the loop number appears in the bottom half of the circle. **See Figure 15-8.** ANSI/ISA-5.1-1984 (R1992) is the standard used for instrument abbreviation and control loop numbering.

Instrument tag identification consists of two or more letters. The letters are always upper case. The first letter of the tag identification stands for the variable being measured, such as flow, level, pressure, or temperature. The second letter of the tag identification stands for the function of the instrument, such as alarm, controller, or transmitter. A third letter may be used to modify the first or second letter. The modifying letter is placed immediately after the letter it modifies. The same letter can have multiple meanings depending on the letter's position in the tag identification. For example, the letter "T" denotes "temperature" when it is the first letter and "transmit" when it is the second letter. **See Figure 15-9. See Appendix.**

Refer to the CD-ROM "Prints" Chapter 15 Instrument Control Loop

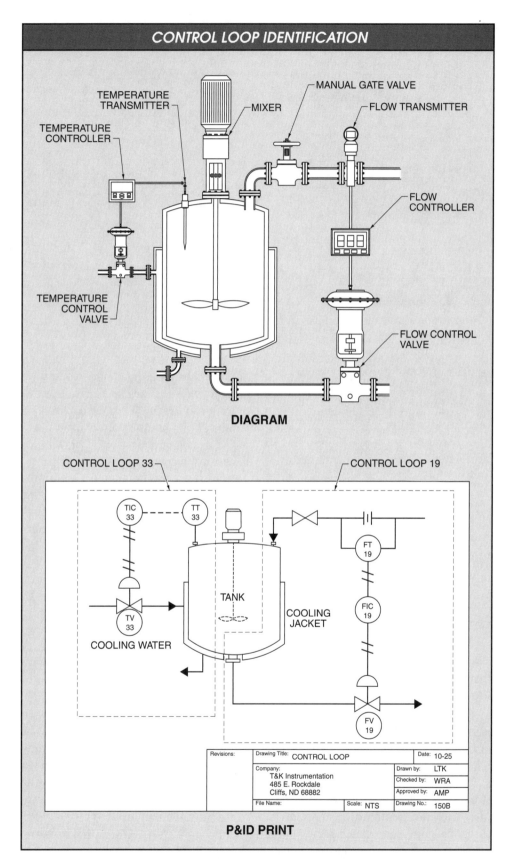

Figure 15-8. Control loops are identified using letter abbreviations and numbers.

INSTRUMENT TAG IDENTIFICATION

	FIRST LETTER		SUCCEEDING LETTERS		
	MEASURED OR INITIATING VARIABLE	MODIFIER	READOUT OR PASSIVE FUNCTION	OUPUT FUNCTION	MODIFIER
A	Analysis		Alarm		
B	Burner, Combustion		User's choice	User's choice	User's choice
C	User's choice			Control	
D	User's choice	Differential			
E	Voltage		Sensor (Primary element)		
F	Flow rate	Ratio (Fraction)			
G	User's choice		Glass, Viewing device		
H	Hand				High
I	Current (Electrical)		Indicate		
J	Power	Scan			
K	Time, Time Schedule	Time rate of change		Control station	
L	Level		Light		Low
M	User's choice	Momentary			Middle, Intermediate
N	User's choice		User's choice	User's choice	User's choice
O	User's choice		Orifice, Restriction		
P	Pressure, Vacuum		Point (Test) connection		
Q	Quantity	Integrate, Totalize			
R	Radiation		Record		
S	Speed, Frequency	Safety		Switch	
T	Temperature			Transmit	
U	Multivariable		Multifunction	Multifunction	Multifunction
V	Vibration, Mechanical analysis			Valve, Damper, Louver	
W	Weight, Force		Well		
X	Unclassified	X Axis	Unclassified	Unclassified	Unclassified
Y	Event, State or Presence	Y Axis		Relay, Compute, Convert	
Z	Position, Dimension	Z Axis		Driver, Actuator, Unclassified Final Control Element	

INSTRUMENT ABBREVIATIONS

FLOW
- FAL flow alarm low
- FAH flow alarm high
- FE flow element
- FI flow indicator
- FIC flow indicating controller
- FIT flow indicating transmitter
- FS flow switch
- FT flow transmitter

- PSE rupture disc, vacuum breaker or emergency vent valve
- PT pressure transmitter
- PV pressure control valve

TEMPERATURE
- TAL temperature alarm low
- TC temperature controller
- TE temperature element

Figure 15-9. Instrument tags are based on the ANSI/ISA-5.1 standard, Instrumentation Symbols and Identification.

Typically more than one control loop is found on a P&ID drawing. All of the instruments and control elements that are part of that specific control loop have the same number. There are two possible numbering sequences for control loops: parallel or serial. **See Figure 15-10.**

Parallel control loop numbering starts a new number sequence for each instrument abbreviation first letter, such as LI-100, PC-100, TE-100. Serial numbering has only one number sequence regardless of the instrument abbreviation's first letter, such as LI-100, PC-101, TE-102. Control loop numbers for parallel or serial numbering sequences may start at any number 1, 100, 201, or 1001. Some manufacturing facilities use loop numbering systems that provide additional information, such as facility location.

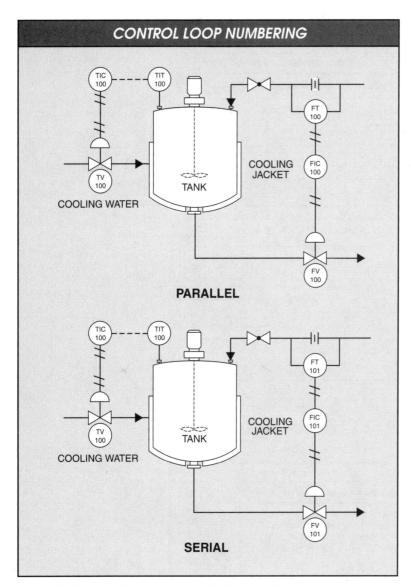

Figure 15-10. Control loops are numbered using a parallel numbering sequence or a serial numbering sequence.

CONTROL ELEMENT SYMBOLS

A control element is a component such as a valve or damper that receives a signal from an instrument to control a process variable. Control elements include valves and louvers. Valves are the most common control element. Valves directly control flow, and by controlling flow, they indirectly control level, pressure, and temperature. **See Figure 15-11.** There are two basic types of valves: manual valves and control valves.

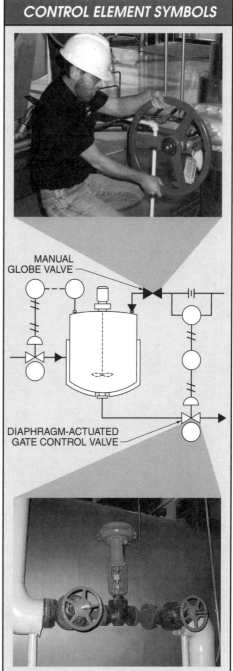

Figure 15-11. Valves are the most common control element used in facilities and are therefore the most common symbols found on P&ID prints.

Manual Valves

A *manual valve* is a valve that is opened or closed by hand. Manual valves are not considered to be part of control loops. However, manual valves do appear on P&ID prints as safety valves for systems during shutdowns.

The symbols for manual valves are the basis for control valve symbols. The symbols represent the various types of valves: gate, ball, plug, and globe. **See Figure 15-12.** Manual valves are typically used to bypass or isolate equipment for maintenance or repair. For example, maintenance personnel can use manual valves to isolate a pump for repair. Manual valves are also used as drains for vessels and tanks.

Manual valves and control valves are drawn in line with the process piping being controlled. Valve symbols are drawn horizontally or vertically depending on the orientation of the process piping and are identified by instrument abbreviations. A control valve uses the same control loop number as the instrument the valve is connected to. Manual valves are not identified by instrument abbreviations and do not have control loop numbers.

MANUAL VALVE SYMBOLS		
Device	Part	Symbol
MANUAL	Gate	
	Globe	
	Plug	
	Ball	
	Check	
	Blowdown	
	Angle Blowdown	
	Angle	
	Diaphragm	
	Hand Control	
	Stop Check	
	Butterfly	
	3-Way	
	4-Way	
	Valve w/Bleed	
	Needle Abbr = NV	NV
	Insulated Body Abbr = IhV	IhV
	Tight Shutoff Abbr = TSO	TSO

Figure 15-12. Specific symbols are used to represent the different types of manual valves.

Most industrial processes and commercial systems have control valves to control the process or system and manual valves for maintenance and safety.

Control Valves

A *control valve* is a valve that is part of a control loop and is opened or closed by an actuator. A control valve actuator receives a signal from an instrument to physically open or close the valve.

A *control valve actuator* is a component that allows a valve to be opened or closed automatically. Control valve actuators are classified by the type of control signal being used (electric, hydraulic, or pneumatic) and the actuating element in the actuator (diaphragm, piston, motor, or solenoid). The symbol for a control valve consists of a manual valve symbol and an actuator symbol. **See Figure 15-13.**

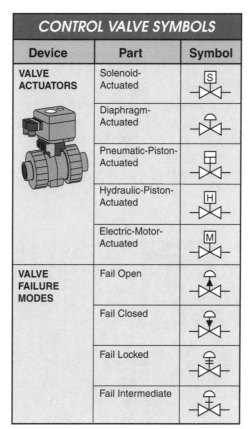

Figure 15-13. A variety of different actuator types are used with control valves, and each type has a specific symbol.

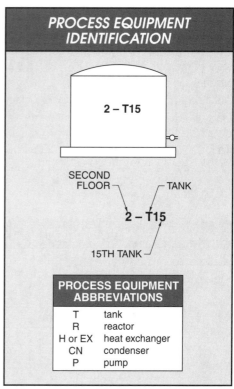

Figure 15-14. Companies modify the numbering system used to identify equipment in different buildings and on different floors to meet their specific needs.

PROCESS EQUIPMENT SYMBOLS

In addition to instrument symbols and control element symbols, process equipment symbols are found on P&ID prints. Vessels, tanks, pumps, compressors, and heat exchanger symbols are common process equipment symbols. Process equipment symbols are not part of the ANSI/ISA-5.1 standard.

Process equipment symbols resemble the actual items they represent. Process equipment symbols are not drawn to scale; however, attempts are made to depict the relative size of the process equipment. For example, two different-sized tanks would not appear as the same size on a P&ID print. Most process equipment is identified using a plant-specific numbering system. **See Figure 15-14.** The numbering system may include location information in addition to identifying the process equipment.

> Not all manual valves are suitable for use as control valves. Many applications require that the position of the valve be predetermined (fail open or fail closed) in the event of a power failure.

Process Vessels and Tanks

Vessels and tanks come in a wide variety of shapes and sizes. Vessels and tanks contain gases, liquids, or solids. Often a vessel has an agitator, which consists of a motor and a mixing blade. **See Figure 15-15.** An agitator ensures that the contents of a vessel are properly mixed. Reactors are a special type of vessel. Reactors contain gases, liquids, or solids that are undergoing a chemical reaction. Reactors are designed to withstand the pressure and temperature of the chemical reaction. Depending on the manufacturing process, vessels can be insulated or jacketed.

PROCESS VESSEL AND TANK SYMBOLS

Device	Part	Symbol
VEHICLES	Tank Truck	
	Tank Truck	
	Railroad Tank Car	
REACTORS	Abbr = R	
	Jacketed	
VESSELS	Horizontal Pressure	
	Vertical Pressure	
	Compressed Gas Cylinder	
	Spherical	
TANKS	Abbr = T	OR
	Agitator with Motor Abbr = T	
	Storage	
	Open Top	
	Floating Roof	
	Silo	

Figure 15-15. A variety of tanks and vessels are used in batch manufacturing or continuous-process manufacturing.

PROCESS FLOW DIAGRAMS

Process flow diagrams (PFDs) usually include all the equipment and piping that a process takes through a facility. Some symbols are the same as on P&ID prints, while other symbols are used only by a particular industry or facility. Process flow diagrams are designed to be simple to understand. Flow for the most part through the system is indicated as flow from left to right (beginning of process to the end of the process) and top to bottom.

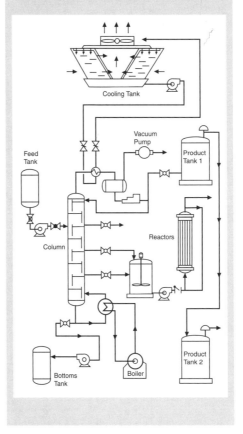

Manufacturing processes require that liquids and gases be moved from point to point. Pumps are used to move liquids. Compressors are used to move gases. A separate set of symbols is used to depict the various types of pumps and compressors. **See Figure 15-16.** Pumps and compressors are installed in the process piping. Arrows are typically used to indicate the direction of flow for pumps and compressors.

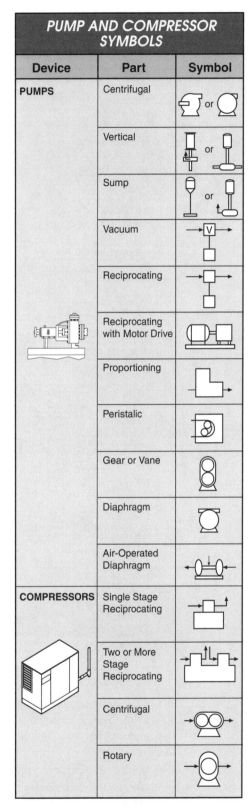

Heat exchangers are designed to add heat or remove heat from a process material. *Coolers* are heat exchangers that remove heat from, or cool, process materials and fluids. Coolers are also known as condensers. *Heaters* are heat exchangers that add heat to process materials and fluids. Heaters are also known as vaporizers. Heat exchangers come in a variety of shapes and sizes. A separate set of symbols is used for heat exchangers. **See Figure 15-17.** Heat exchangers are installed in the process piping. Arrows are used to indicate the direction of flow through the heat exchangers.

Figure 15-16. Compressors and pumps move liquid and gas process materials in batch manufacturing or continuous-process manufacturing from place to place.

Figure 15-17. Heat exchangers add or remove heat from a process material.

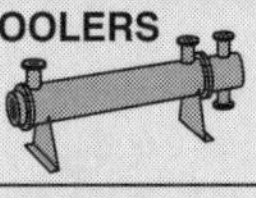

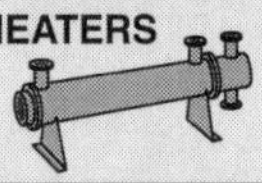

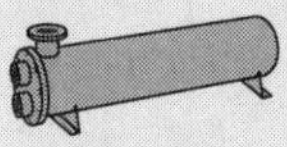

The ISA standard subcommittee SP5.5 does not meet any longer. The standard ISA-5.5, Graphic Symbols for Process Displays, is not current or valid for today's process display technology.

Review Questions and Activities

Process and Instrumentation Systems

Name _____ Date _____

True-False

T F 1. Lines on P&ID prints represent distances.

T F 2. Process piping is depicted by broken lines on a P&ID print.

T F 3. Each process line on a P&ID print has a unique identifying number.

T F 4. Oil jacketing is a pipe heating method where an oil jacket surrounds the process piping.

T F 5. Instrument lines can connect an instrument to a process, connect instruments to one another, or connect an instrument to a control element such as a valve.

T F 6. The most common variables found in heat tracing are flow, level, pressure, and temperature.

T F 7. The type of symbol used to represent an instrument depends on the instrument's actual location and the type of instrument control being used.

T F 8. Letter abbreviations are used to identify the function of instruments and control elements on a P&ID print.

T F 9. Valve symbols are drawn horizontally or vertically depending on the orientation of the process piping.

T F 10. Heaters are also known as condensers.

Completion

_____ 1. A(n) ___ location is an instrument location that is in the vicinity of the process the instrument monitors or controls.

_____ 2. A(n) ___ is a stand-alone piece of hardware with a single specific function.

_____ 3. The control loop identification system is a combination of a letter abbreviation and a ___.

_____ 4. A(n) ___ valve is a valve that is part of a control loop and is opened or closed by an actuator.

_____ 5. ___ symbols resemble the actual items they represent.

Multiple Choice

_____ 1. To avoid cluttering and to enhance P&ID print clarity, print lines are drawn at ___° to each other.
    A. 30
    B. 60
    C. 90
    D. 120

_____ 2. ___ techniques used in piping include steam tracing, electrical tracing, and oil jacketing.
    A. Heat tracing
    B. Heat power
    C. Temperature tracing
    D. Fluid power

_____ 3. A(n) ___ location is an instrument location near a control room where instruments are accessible to the operators.
    A. behind the panel
    B. auxiliary
    C. discrete
    D. primary

_____ 4. The ___ letter of tag identification stands for the variable being measured, such as flow, level, pressure, or temperature.
    A. first
    B. second
    C. third
    D. none of the above

_____ 5. ___ numbering has only one number sequence regardless of the instrument abbreviation's first letter.
    A. Open-loop
    B. Control-loop
    C. Parallel
    D. Serial

Name _____ Date _____

Activity—Troubleshooting a Process Instrumentation System 15-1

Scenario:
Activity 15-1 is a troubleshooting activity where an industrial process system has a problem. To isolate and identify the problem, information must be gathered from the process P&ID drawings.

Task:
Identify symbols that are found on the system P&ID drawings and on other drawings for the process.

INDUSTRIAL PROCESS

Required Process Instrumentation Troubleshooting Information:

Process Line Symbols

1. _____ —◯(S)◯— a. Steam-traced process line
2. _____ —S—S— b. Electrical tracing
3. _____ —J—J— c. Insulation
4. _____ ▭ d. Insulation, steam traced
5. _____ —E—E— e. Oil jacketed

Process Line Abbreviations

6. _____ WT a. Aerosol
7. _____ V b. Liquid ammonia
8. _____ SL c. Treated water
9. _____ LA d. Slurry
10. _____ AOT e. Vent
11. _____ PGC f. Cooled process gas

Instrument Line Symbols

12. _____ —⊙—⊙— a. Pneumatic signal
13. _____ —⊥—⊥— b. Electrical signal
14. _____ —○—○— c. Hydraulic signal
15. _____ —///—///— d. Mechanical link
16. _____ —//—//— e. Software or data link

Instrument Line Abbreviations

17. _____ SS a. Air supply
18. _____ GS b. Electrical supply
19. _____ AS c. Water supply
20. _____ IA d. Steam supply
21. _____ ES e. Instrument air
22. _____ WS f. Gas supply

Instrument and Control Element Symbols

23. _____ ⏃ a. Programmable logic control flow transmitter, field mounted
24. _____ (LIC 272) b. Pneumatically-actuated control valve
25. _____ (FT 112) c. Discrete pressure indicator, any location
26. _____ (PI 129) d. Shared display and control level-indicating controller, primary location

Activity—Troubleshooting a Process Instrumentation System

27. _____   a. Shared display and control temperature-indicating controller, field mounted

28. _____   b. Discrete temperature element, field mounted

29. _____   c. Solenoid-actuated control valve

30. _____   d. Pneumatically actuated control valve, fail closed

Reference Prints:

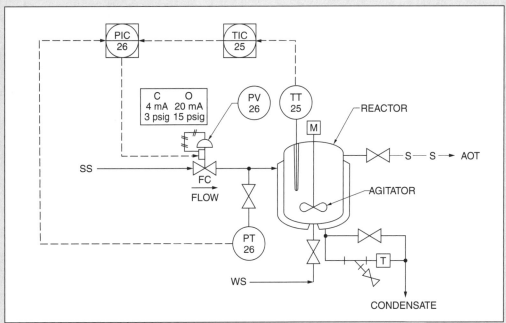

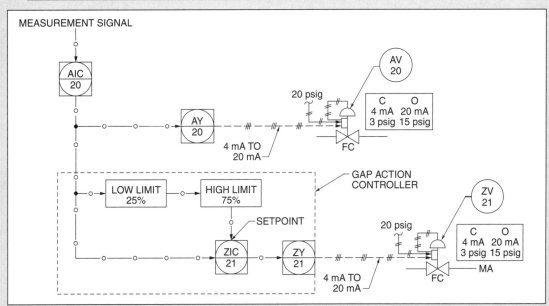

Name _____ Date _____

Activity—Installing a Process Instrumentation System   15-2

Scenario:
Activity 15-2 is an installation activity requiring that an industrial plating process that includes tanks, piping, valves, heat exchanger, instrumentation, and pumps be installed in a facility.

PLATING PREPARATION PROCESS

Task:
It is necessary to understand the plating process P&ID print in order to select the correct material and correct size piping, the proper type of valve for the position and proper actuator for the application, the proper instrumentation for the process and environment, and to connect and wire the process equipment into the proper control loops.

Required Plating Process Installation Information:
1. What is the size of the intake process line for pump M-3-P-4? _____
2. What is the size of the discharge process line from pump M-3-P-4? _____
3. Which instrument loop will pump M-3-P-4 be connected to? _____
4. Which alarm instruments will be used to monitor pump M-3-P-4 (abbreviations and names)? _____

5. Where will the alarm instruments that are monitoring pump M-3-P-4 be located? _____

6. What flow rate is pump M-3-P-4 creating? _____

7. What type of process material is pump M-3-P-5 transporting? _____
8. Which instrument loop is monitoring pump M-3-P-5? _____
9. Which instruments will be used to monitor pump M-3-P-5 (abbreviations and names)? _____

10. Which instruments are found in loop 362 (abbreviations and names)? _____

11. Exactly where can more information be found about the plant air supply? _____

ACTIVITY—Installing a Process Instrumentation System

Reference Print

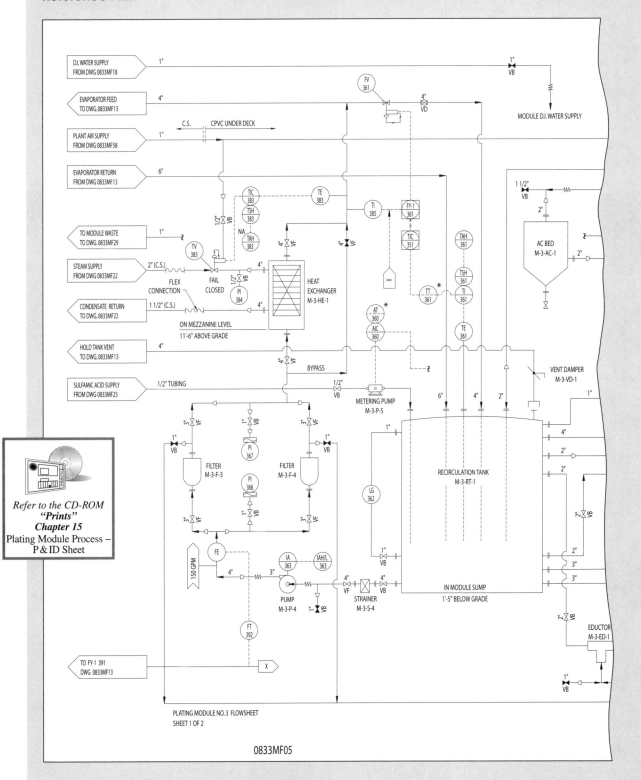

Trade Competency Test
Process and Instrumentation Systems

Name _____ Date _____

REFERENCE PRINT #1 (AQS Tank)

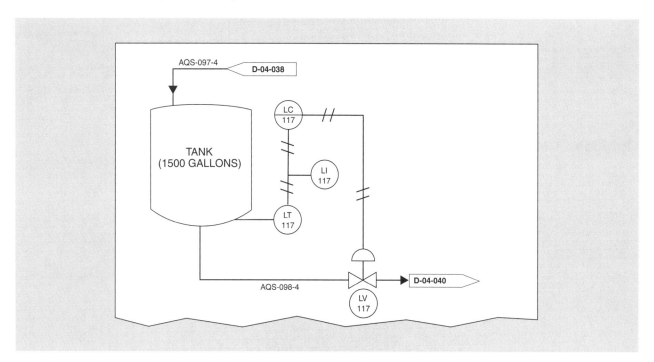

Questions – Reference Print #1

_____ 1. The capacity of the AQS tank is ___ gal.

T   F   2. LI 117 is field mounted.

T   F   3. There are two control loops in this drawing.

_____ 4. The continuation of the process line that exits the tank is found on print ___.

T   F   5. The print is drawn to scale.

T   F   6. The process lines are jacketed.

_____ 7. The tank receives product from sheet ___.

_____ 8. LV 117 is a ___ actuated valve.

REFERENCE PRINT #2 (Control Loop)

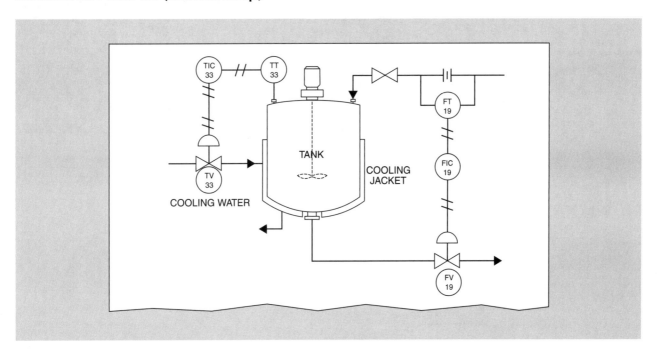

Questions – Reference Print #2

_____ 1. There is/are ___ control loop(s) shown on the print.

_____ 2. Loop 33 is a ___ circuit.

T   F   3. The agitator motor is part of control loop 33.

T   F   4. The drawing provides the sizes of the process piping.

T   F   5. TIC 33 is mounted where it is normally accessible to the operator.

T   F   6. The contents (process material) of the tank are shown on the print.

T   F   7. The tank is an open to the atmosphere tank.

_____ 8. There is a ___ valve before the process line enters the top of the tank.

Advanced CD-ROM Print Questions (Control Loop)

_____ 1. The temperature of the cooling jacket is regulated by control loop ___.

_____ 2. TIC 33 is a ___ controller.

_____ 3. FT 19 is a ___ transmitter.

_____ 4. The instrument lines in control loop 33 are ___.

Appendix

Piping and Processing Symbols	476
Pipe Fitting and Valve Symbols	478
Fluid Power Symbols	479
Industrial Electrical Symbols	481
VDV Symbols	485
Locking Wiring Devices	486
Non-Locking Wiring Devices	487
Electrical Symbols	488
Residential Electrical Symbols	490
HVAC Symbols	491
Refrigeration Symbols	492
Plumbing Symbols	493
Fire Alarm/Safety/Security Symbols	494
Architectural Symbols	495
Plot Plan Symbols	496
Print Symbols	497
Alphabet of Lines	498
Floor Plan Abbreviations	499
Section and Detail Abbreviations	500
Elevation Abbreviations	501
Process Line Content Abbreviations	503
Common Instrument Abbreviations	504
Electrical/Electronic Abbreviations/Acronyms	505
Equipment Log Sheet	509
Request for Information (RFI) Document	510
Punch List	511
Authorization Form	512
English System Measurements	513
Metric System Measurements	514
Sketching Isometric Drawings	515
Sketching Isometric Circles	516
Sketching Perspective Drawings	517
Sketching Multiview Drawings	518

PIPING AND PROCESSING SYMBOLS...

PUMPS	PROCESS VESSELS AND TANKS	HEAT EXCHANGERS
CENTRIFUGAL	TANK TRUCK VEHICLE	CONDENSER COOLER
VERTICAL	TANK TRUCK VEHICLE	VAPORIZER HEATER
SUMP	RAIL ROAD TANK CAR	EXCHANGER
VACUUM	REACTOR	EXCHANGER
RECIPROCATING	JACKETED REACTOR	FIXED TUBE EXCHANGER
RECIPROCATING WITH MOTOR DRIVE	HORIZONTAL PRESSURE VESSEL	U-TUBE EXCHANGER
PROPORTIONING	VERTICAL PRESSURE VESSEL	BOX EXCHANGER
PERISTALIC	COMPRESSED GAS CYLINDER	DOUBLE PIPE EXCHANGER
GEAR OR VANE	SPHERICAL VESSEL	FIN TUBE EXCHANGER
DIAPHRAGM		AIR COOLED EXCHANGER
AIR OPERATED DIAPHRAGM	TANK	**INSTRUMENTS**
COMPRESSORS	AGITATOR WITH MOTOR	FLOW INDICATOR — FI
		FLOW TRANSMITTER — FT
		FLOW RECORDER — FR
		FLOW CONTROLLER — FC
		LEVEL INDICATOR — LI
		LEVEL TRANSMITTER — LT
		LEVEL RECORDER — LR
		LEVEL CONTROLLER — LC
SINGLE STAGE RECIPROCATING	STORAGE TANK	PRESSURE INDICATOR — PI
		PRESSURE TRANSMITTER — PT
TWO OR MORE STAGE RECIPROCATING	OPEN TOP TANK	PRESSURE RECORDER — PR
		PRESSURE CONTROLLER — PC
CENTRIFUGAL	FLOATING ROOF TANK	TEMPERATURE INDICATOR — TI
		TEMPERATURE TRANSMITTER — TT
ROTARY	SILO	TEMPERATURE RECORDER — TR
		TEMPERATURE CONTROLLER — TC

Appendix

...PIPING AND PROCESSING SYMBOLS

MANUAL VALVES

Valve	
GATE	
GLOBE	
PLUG	
BALL	
CHECK	
BLOWDOWN	
ANGLE BLOWDOWN	
ANGLE	
DIAPHRAGM	
HAND CONTROL	
STOP CHECK	
BUTTERFLY	
3-WAY	
4-WAY	
VALVE WITH BLEED	
NEEDLE	NV
INSULATED	IhV
TIGHT SHUTOFF	TSO

CONTROL VALVES

- SOLENOID ACTUATED
- DIAPHRAGM ACTUATED
- PNEUMATIC PISTON ACTUATED
- HYDRAULIC PISTON ACTUATED
- ELECTRIC MOTOR ACTUATED
- FAIL OPEN MODE
- FAIL CLOSED MODE
- FAIL LOCKED MODE
- FAIL INTERMEDIATE MODE

HEAT TRACED

STEAM	—S—S—
ELECTRIC	—E—E—
OIL JACKET	—J—J—
ELECTRIC	
STEAM TRACED	(S)
ELECTRIC TRACED	(E)
OIL TRACED	(J)

P & ID LINES

- CONECTION TO PROCESS OR INSTRUMENT SUPPLY
- UNDEFINED SIGNAL
- PNEUMATIC SIGNAL
- HYDRAULIC SIGNAL
- ELECTRICAL SIGNAL .... OR
- CAPILLARY TUBE (FILLED THERMAL SYSTEM)
- ELECTRO-MAGNETIC OR SONIC SUGNAL
- MECHANICAL LNK
- INTERNAL SYSTEM LINK (SOFTWARE OR DATA LINK)

P & ID PRINT LINES

LINES (PIPING AND/OR WIRING) CONNECTED

LINES (PIPING AND/OR WIRING) NOT CONNECTED

INSTRUMENT TYPE AND LOCATION

GENERAL INSTRUMENTS	PRIMARY LOCATION NORMALLY ACCESSIBLE TO OPERATOR	FIELD MOUNTED	AUXILIARY LOCATION NORMALLY ACCESSIBLE TO OPERATOR	AUXILIARY INACCESSABLE OR "BEHIND THE PANEL"
DISCRETE INSTRUMENTS	⊖	○	⊖	(⌀)
SHARED DISPLAY, SHARED CONTROL				
COMPUTER FUNCTION				
PROGRAMMABLE LOGIC CONTROL				

PIPE FITTING AND VALVE SYMBOLS

	FLANGED	SCREWED	BELL & SPIGOT		FLANGED	SCREWED	BELL & SPIGOT		FLANGED	SCREWED	BELL & SPIGOT
BUSHING		—⊐—	—⊂	REDUCING FLANGE	⊣⊐—			AUTOMATIC BYPASS VALVE			
CAP		—⊐	—⊃	BULL PLUG	⊣⊳		⊂	AUTOMATIC REDUCING VALVE			
REDUCING CROSS				PIPE PLUG		—⊲	⊂	STRAIGHT CHECK VALVE			
STRAIGHT-SIZE CROSS				CONCENTRIC REDUCER	⊳⊲	⊳⊲	⊳⊃	COCK			
CROSSOVER				ECCENTRIC REDUCER				DIAPHRAGM VALVE			
45° ELBOW				SLEEVE	⫪---⫪	+---+	⊃---⊂	FLOAT VALVE			
90° ELBOW				STRAIGHT-SIZE TEE				GATE VALVE			
ELBOW — TURNED DOWN				TEE — OUTLET UP				MOTOR-OPERATED GATE VALVE			
ELBOW — TURNED UP				TEE — OUTLET DOWN				GLOBE VALVE			
BASE ELBOW				DOUBLE-SWEEP TEE				MOTOR-OPERATED GLOBE VALVE			
DOUBLE-BRANCH ELBOW				REDUCING TEE				ANGLE HOSE VALVE			
LONG-RADIUS ELBOW				SINGLE-SWEEP TEE				GATE HOSE VALVE			
REDUCING ELBOW				SIDE OUTLET TEE — OUTLET DOWN				GLOBE HOSE VALVE			
SIDE OUTLET ELBOW — OUTLET DOWN				SIDE OUTLET TEE — OUTLET UP				LOCKSHIELD VALVE			
SIDE OUTLET ELBOW — OUTLET UP				UNION	—⫪—	—+—		QUICK-OPENING VALVE			
STREET ELBOW				ANGLE CHECK VALVE				SAFETY VALVE			
CONNECTING PIPE JOINT	—⫪—	—+—	—⊂	ANGLE GATE VALVE — ELEVATION							
EXPANSION JOINT				ANGLE GATE VALVE — PLAN				GOVERNOR-OPERATED AUTOMATIC VALVE			
LATERAL				ANGLE GLOBE VALVE — ELEVATION							
ORIFICE FLANGE	—⫪—			ANGLE GLOBE VALVE — PLAN							

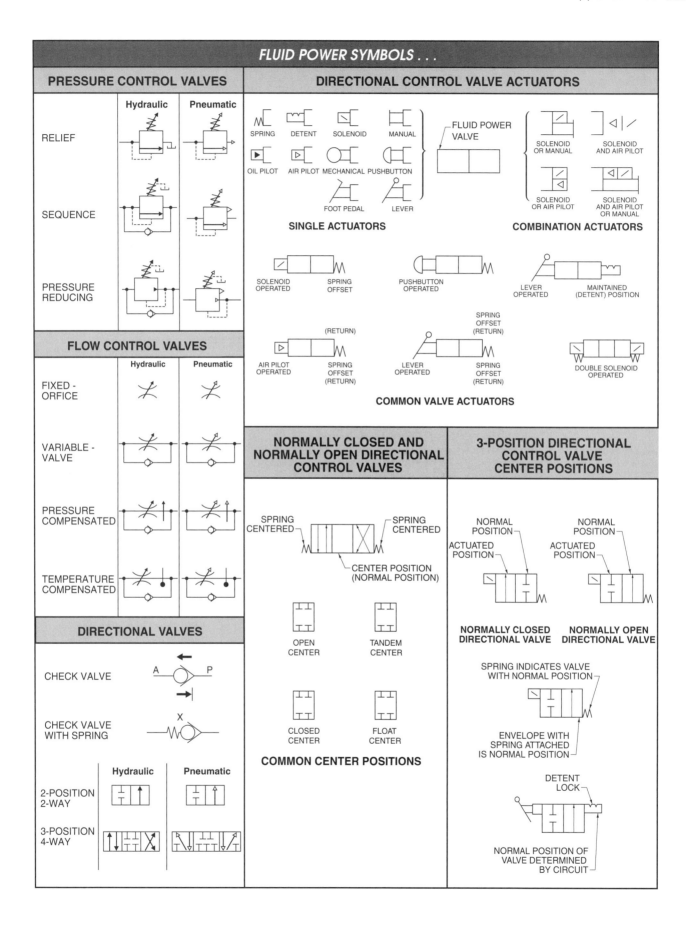

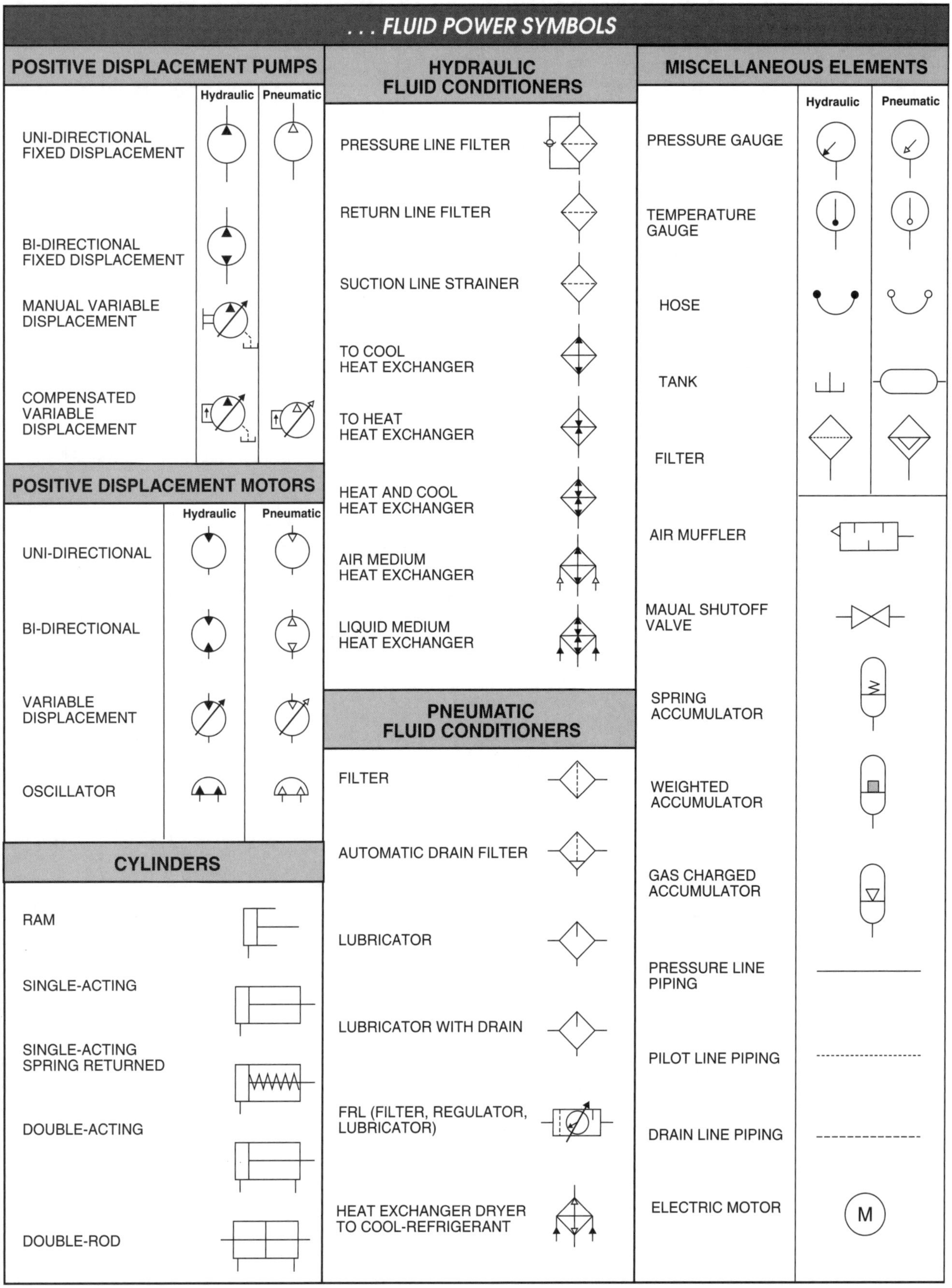

...FLUID POWER SYMBOLS

POSITIVE DISPLACEMENT PUMPS

	Hydraulic	Pneumatic
UNI-DIRECTIONAL FIXED DISPLACEMENT		
BI-DIRECTIONAL FIXED DISPLACEMENT		
MANUAL VARIABLE DISPLACEMENT		
COMPENSATED VARIABLE DISPLACEMENT		

POSITIVE DISPLACEMENT MOTORS

	Hydraulic	Pneumatic
UNI-DIRECTIONAL		
BI-DIRECTIONAL		
VARIABLE DISPLACEMENT		
OSCILLATOR		

CYLINDERS

- RAM
- SINGLE-ACTING
- SINGLE-ACTING SPRING RETURNED
- DOUBLE-ACTING
- DOUBLE-ROD

HYDRAULIC FLUID CONDITIONERS

- PRESSURE LINE FILTER
- RETURN LINE FILTER
- SUCTION LINE STRAINER
- TO COOL HEAT EXCHANGER
- TO HEAT HEAT EXCHANGER
- HEAT AND COOL HEAT EXCHANGER
- AIR MEDIUM HEAT EXCHANGER
- LIQUID MEDIUM HEAT EXCHANGER

PNEUMATIC FLUID CONDITIONERS

- FILTER
- AUTOMATIC DRAIN FILTER
- LUBRICATOR
- LUBRICATOR WITH DRAIN
- FRL (FILTER, REGULATOR, LUBRICATOR)
- HEAT EXCHANGER DRYER TO COOL-REFRIGERANT

MISCELLANEOUS ELEMENTS

	Hydraulic	Pneumatic
PRESSURE GAUGE		
TEMPERATURE GAUGE		
HOSE		
TANK		
FILTER		

- AIR MUFFLER
- MAUAL SHUTOFF VALVE
- SPRING ACCUMULATOR
- WEIGHTED ACCUMULATOR
- GAS CHARGED ACCUMULATOR
- PRESSURE LINE PIPING
- PILOT LINE PIPING
- DRAIN LINE PIPING
- ELECTRIC MOTOR — (M)

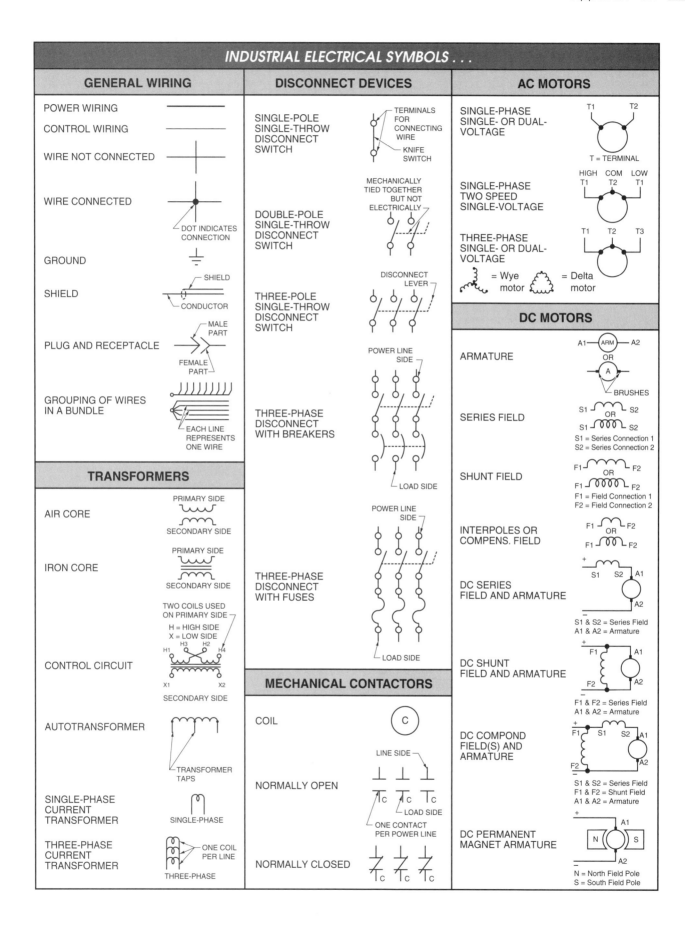

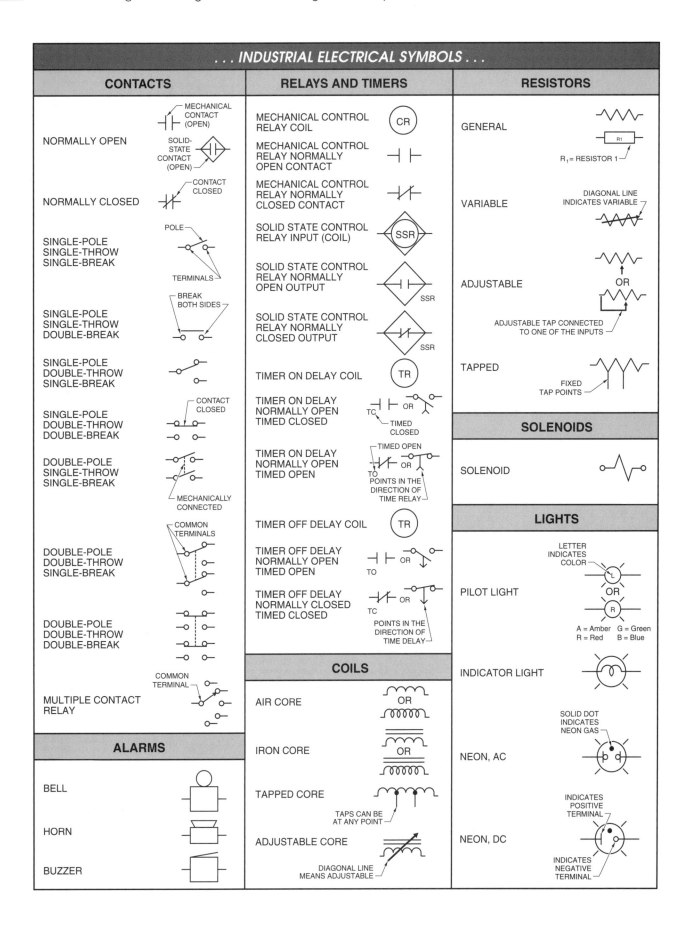

...INDUSTRIAL ELECTRICAL SYMBOLS...

MOTOR STARTERS

- MECHANICAL MOTOR STARTER COIL SINGLE VOLTAGE
- MECHANICAL MOTOR STARTER COIL DUAL VOLTAGE
  - HIGH VOLTAGE (Coils connected in Series)
  - LOW VOLTAGE (Coils connected in Parallel)
- MECHANICAL MOTOR STARTER THERMAL OVERLOAD
  - OL — HEATING ELEMENT
- MECHANICAL MOTOR STARTER OVERLOAD CONTACT
  - CONTACTS CLOSE ON OVERLOAD
  - CONTACTS OPEN ON OVERLOAD
- MECHANICAL MOTOR STARTER NORMALLY OPEN CONTACTS
  - STARTER COIL, L1 L2 L3, T1 T2 T3
- MECHANICAL MOTOR STARTER NORMALLY CLOSED CONTACTS
  - L1 L2 L3, T1 T2 T3
- SOLID STATE MOTOR STARTER NORMALLY OPEN CONTACTS
- SOLID STATE MOTOR STARTER NORMALLY CLOSED CONTACTS
- SOLID STATE MOTOR STARTER COMBINATION DIAGRAM
  - LINE SIDE
  - OVERLOAD CONTACT
  - MOTOR CONNECTIONS

MANUALLY OPERATED CONTROL SWITCHES

- SINGLE-CIRCUIT NORMALLY OPEN PUSHBUTTON
  - MANUAL OPERATION
  - TERMINALS
- SINGLE-CIRCUIT NORMALLY CLOSED PUSHBUTTON
- DOUBLE-CIRCUIT NORMALLY OPEN AND CLOSED PUSHBUTTON
- MUSHROOM HEAD PUSHBUTTON
  - MUSHROOM OPERATOR
- MAINTAINED
  - MECHANICAL LINK
- ILLUMINATED
  - R = RED
  - G = GREEN
  - A = AMBER
  - B = BLUE
  - LAMP INSIDE OPERATOR
- TWO POSITION SELECTOR SWITCH
  - JOG / RUN, A1, A2

	J	R
A1	X	
A2		X

- THREE POSITION SELECTOR SWITCH
  - JOG / STOP / RUN, A1, A2

	J	S	R
A1	X		
A2			X

  J = Jog
  R = Run
  X = Contacts closed
  Blank = Contacts open

- NORMALLY OPEN FOOT SWITCH
  - FOOT OPERATOR
- NORMALLY CLOSED FOOT SWITCH

MECHANICALLY OPERATED CONTROL SWITCHES

- NORMALLY OPEN MECHANICAL LIMIT SWITCH — MECHANICAL OPERATOR
- NORMALLY OPEN HELD CLOSED MECHANICAL LIMIT SWITCH — MECHANICAL OPERATOR IN CLOSED POSITION
- NORMALLY CLOSED MECHANICAL LIMIT SWITCH
- NORMALLY CLOSED HELD OPEN MECHANICAL LIMIT SWITCH — OPERATOR IN OPEN POSITION
- NORMALLY OPEN SOLID STATE LIMIT SWITCH — SOLID-STATE — OR
- NORMALLY CLOSED SOLID STATE LIMIT SWITCH — OR

AUTOMATICALLY OPERATED CONTROL SWITCHES

- NORMALLY OPEN TEMPERATURE SWITCH — TEMPERATURE OPERATOR
- NORMALLY CLOSED TEMPERATURE SWITCH
- NORMALLY OPEN PRESSURE SWITCH — PRESSURE OPERATOR
- NORMALLY CLOSED PRESSURE SWITCH
- NORMALLY OPEN FLOW SWITCH — FLOW OPERATOR
- NORMALLY CLOSED FLOW SWITCH
- NORMALLY OPEN LEVEL SWITCH — LEVEL OPERATOR
- NORMALLY CLOSED LEVEL SWITCH
- NORMALLY OPEN PHOTOELECTRIC SWITCH
  - THREE LEAD / TWO LEAD
  - **NPN**
- NORMALLY CLOSED PHOTOELECTRIC SWITCH
  - **PNP**

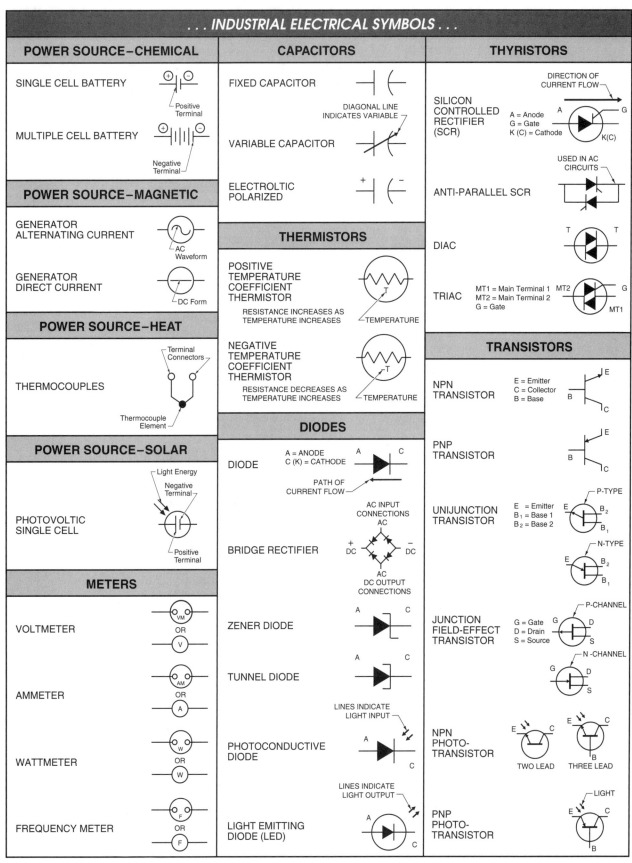

...INDUSTRIAL ELECTRICAL SYMBOLS

DIGITAL LOGIC GATES

- AND GATE
- OR GATE
- NOT (INVERTER) GATE
- NOR GATE
- NAND GATE
- EXCLUSIVE GATE

CONTROL CIRCUIT LOGIC

Device	Hardwired	Statement	Boolean Expression	Read As	Symbol	Function/Notes
AND GATE		Allows electricity to flow only if Switches 1 and 2 are pressed	$Y = A \bullet B$ ($\bullet$ = AND)	Y = A AND B		To provide logic level 1 only if all inputs are at logic level 1 HIGH = 1 LOW = 0
OR GATE		Allows electricity to flow only if Switch 1 or 2 is pressed	$Y = A + B$ (+ = OR)	Y = A OR B		To provide logic level 1 if one or more inputs are at logic level 1 HIGH = 1 LOW = 0
NOT (INVERTER) GATE		Allows electricity to flow only if Switch 1 is not pressed	$Y = \overline{A}$ (− = NOT)	Y = NOT A		To provide an ouput that is the opposite of the input HIGH = 1 LOW = 0
NOR GATE		Allows electricity to flow if neither Switch 1 nor Switch 2 is pressed	$Y = \overline{A+B}$	Y = NOT (A OR B)		To provide logic level 0 if one or more inputs are at logic level 1 HIGH = 1 LOW = 0
NAND GATE		Allows electricity to flow only if Switch 1 and 2 are not pressed	$Y = \overline{A \bullet B}$	Y = NOT (A AND B)		To provide logic level 0 only if all inputs are at logic level 1 HIGH = 1 LOW = 0

VDV SYMBOLS

DATA AND TELEPHONE OUTLET

- WALL MOUNTED DATA OUTLET
- FLUSH FLOOR MOUNTED DATA OUTLET — F
- SURFACE FLOOR MOUNTED DATA OUTLET — S
- WALL MOUNTED TELEPHONE OUTLET
- FLUSH FLOOR MOUNTED TELEPHONE OUTLET — F
- SURFACE FLOOR MOUNTED TELEPHONE OUTLET — S
- WALL MOUNTED TELEPHONE/DATA OUTLET
- FLUSH FLOOR MOUNTED TELEPHONE/DATA OUTLET — F
- SURFACE FLOOR MOUNTED TELEPHONE/DATA OUTLET — S

MISCELLANEOUS SIGNALING SYSTEMS

- COMPUTER DATA OUTLET
- COMPUTER DATA SYSTEM DEVICE
- PAGING SYSTEM DEVICE
- ELECTRIC DOOR OPENER — D
- TELEVISION OUTLET — TV
- SOUND SYSTEM
- TELEPHONE
- PRIVATE TELEPHONE SYSTEM DEVICE — PVT

MISCELLANEOUS VDV EQUIPMENT

- CABLE ANTENNA OUTLET (CATV) — TV
- MASTER ANTENNA OUTLET (MATV) — TV M
- EQUIPMENT CABINET
- WALL MOUNTED EQUIPMENT RACK
- FREE STANDING EQUIPMENT RACK
- TERMINAL CABINET — TCC
- PLYWOOD BACKBOARD
- TELEPHONE UNDERCARPET FLAT CONDUCTOR — UCT
- DATA UNDERCARPET FLAT CONDUCTOR — UCD
- POWER POLE — P2
- TELECOM MAST — T
- TELECOM POWER TOWER — TP2

LOCKING WIRING DEVICES

2-POLE, 3-WIRE

WIRING DIAGRAM	NEMA ANSI	RECEPTACLE CONFIGURATION	RATING
	ML2 C73.44		15 A 125 V
	L5-15 C73.42		15 A 125 V
	L5-20 C73.72		20 A 125 V
	L6-15 C73.74		15 A 250 V
	L6-20 C73.75		20 A 250 V
	L6-30 C73.76		30 A 250 V
	L7-15 C73.43		15 A 277 V
	L7-20 C73.77		20 A 277 V
	L8-20 C73.79		20 A 480 V
	L9-20 C73.81		20 A 600 V

3-POLE, 4-WIRE

WIRING DIAGRAM	NEMA ANSI	RECEPTACLE CONFIGURATION	RATING
	L14-20 C73.83		20 A 125/250 V
	L14-30 C73.84		30 A 125/250 V
	L15-20 C73.85		20 A 3φ 250 V
	L15-30 C73.86		30 A 3φ 250 V
	L16-20 C73.87		20 A 3φ 480 V
	L16-30 C73.88		30 A 3φ 480 V
	L17-30 C73.89		30 A 3φ 600 V

3-POLE, 3-WIRE

WIRING DIAGRAM	NEMA ANSI	RECEPTACLE CONFIGURATION	RATING
	ML3 C73.30		15 A 125/250 V
	L10-20 C73.96		20 A 125/250 V
	L10-30 C73.97		30 A 125/250 V
	L11-15 C73.98		15 A 3φ 250 V
	L11-20 C73.99		20 A 3φ 250 V
	L12-20 C73.101		20 A 3φ 480 V
	L12-30 C73.102		30 A 3φ 480 V
	L13-30 C73.103		30 A 3φ 600 V

4-POLE, 4-WIRE

WIRING DIAGRAM	NEMA ANSI	RECEPTACLE CONFIGURATION	RATING
	L18-20 C73.104		20 A 3φY 120/208 V
	L18-30 C73.105		30 A 3φY 120/208 V
	L19-20 C73.106		20 A 3φY 277/480 V
	L20-20 C73.108		20 A 3φY 347/600 V

4-POLE, 5-WIRE

WIRING DIAGRAM	NEMA ANSI	RECEPTACLE CONFIGURATION	RATING
	L21-20 C73.90		20 A 3φY 120/208 V
	L22-20 C73.92		20 A 3φY 277/480 V
	L23-20 C73.94		20 A 3φY 347/600 V

NON-LOCKING WIRING DEVICES

2-POLE, 3-WIRE

WIRING DIAGRAM	NEMA ANSI	RECEPTACLE CONFIGURATION	RATING
	5-15 C73.11		15 A 125 V
	5-20 C73.12		20 A 125 V
	5-30 C73.45		30 A 125 V
	5-50 C73.46		50 A 125 V
	6-15 C73.20		15 A 250 V
	6-20 C73.51		20 A 250 V
	6-30 C73.52		30 A 250 V
	6-50 C73.53		50 A 250 V
	7-15 C73.28		15 A 277 V
	7-20 C73.63		20 A 277 V
	7-30 C73.64		30 A 277 V
	7-50 C73.65		50 A 277 V

4-POLE, 4-WIRE

WIRING DIAGRAM	NEMA ANSI	RECEPTACLE CONFIGURATION	RATING
	18-15 C73.15		15 A 3φ Y 120/208 V
	18-20 C73.26		20 A 3φ Y 120/208 V
	18-30 C73.47		30 A 3φ Y 120/208 V
	18-50 C73.48		50 A 3φ Y 120/208 V
	18-60 C73.27		60 A 3φ Y 120/208 V

3-POLE, 3-WIRE

WIRING DIAGRAM	NEMA ANSI	RECEPTACLE CONFIGURATION	RATING
	10-20 C73.23		20 A 125/250 V
	10-30 C73.24		30 A 125/250 V
	10-50 C73.25		50 A 125/250 V
	11-15 C73.54		15 A 3φ 250 V
	11-20 C73.55		20 A 3φ 250 V
	11-30 C73.56		30 A 3φ 250 V
	11-50 C73.57		50 A 3φ 250 V

3-POLE, 4-WIRE

WIRING DIAGRAM	NEMA ANSI	RECEPTACLE CONFIGURATION	RATING
	14-15 C73.49		15 A 125/250 V
	14-20 C73.50		20 A 125/250 V
	14-30 C73.16		30 A 125/250 V
	14-50 C73.17		50 A 125/250 V
	14-60 C73.18		60 A 125/250 V
	15-15 C73.58		15 A 3φ 250 V
	15-20 C73.59		20 A 3φ 250 V
	15-30 C73.60		30 A 3φ 250 V
	15-50 C73.61		50 A 3φ 250 V
	15-60 C73.62		60 A 3φ 250 V

ELECTRICAL SYMBOLS...

LIGHTING OUTLETS

	CEILING	WALL
OUTLET BOX AND INCANDESCENT LIGHTING FIXTURE	○ or ○L	○ / ○L
RECESSED CEILING FIXTURE (OUTLINE INDICATES SHAPE)	▫	
INCANDESCENT TRACK LIGHTING	▭ooo▭	▭ooo▭
BLANKED OUTLET	Ⓑ	Ⓑ
DROP CORD	Ⓓ	
EXIT LIGHT AND OUTLET BOX. SHADED AREAS DENOTE FACES.	⊗	⊗
OUTDOOR POLE-MOUNTED FIXTURES	○—○	
JUNCTION BOX	Ⓙ	Ⓙ
LAMPHOLDER WITH PULL SWITCH	Ⓛ$_{PS}$	Ⓛ$_{PS}$
MULTIPLE FLOODLIGHT ASSEMBLY	ΨΨΨ	
EMERGENCY BATTERY PACK WITH CHARGER		
INDIVIDUAL FLUORESCENT FIXTURE	▭	▭
OUTLET BOX AND FLUORESCENT LIGHTING TRACK FIXTURE	⊢⊣⊢⊣	
CONTINUOUS FLUORESCENT FIXTURE	▭▭	
SURFACE-MOUNTED FLUORESCENT FIXTURE	▭	▭

PANELBOARDS

FLUSH-MOUNTED PANELBOARD AND CABINET	▬
SURFACE-MOUNTED PANELBOARD AND CABINET	▬

CONVENIENCE OUTLETS

SINGLE RECEPTACLE OUTLET	⊖
DUPLEX RECEPTACLE OUTLET	⊜
TRIPLEX RECEPTACLE OUTLET	⊕ or ⊜$_3$
ISOLATED GROUND OUTLET	⊖$_{IG}$
ISOLATED GROUND WEATHER PROOF OUTLET	⊖$_{IG\ WP}$
GROUND FAULT CIRCUIT INTERRUPTER OUTLET	⊖$_{GFCI}$
GFCI WITH SWITCH	⊖$^S_{GFCI}$
SPLIT-WIRED DUPLEX RECEPTACLE OUTLET	⊜
SPLIT-WIRED TRIPLEX RECEPTACLE OUTLET	⊕
SINGLE SPECIAL-PURPOSE RECEPTACLE OUTLET	⊖
DUPLEX SPECIAL-PURPOSE RECEPTACLE OUTLET	⊜
RANGE OUTLET	⊜$_R$
SPECIAL-PURPOSE CONNECTION	⊙ DW
CLOSED-CIRCUIT TELEVISION CAMERA	▭◁
CLOCK HANGER RECEPTACLE	Ⓒ or 🕐
JUNCTION BOX CEILING	Ⓙ
DROP CORD	Ⓓ
FAN HANGER RECEPTACLE	Ⓕ
FLOOR SINGLE RECEPTACLE OUTLET	⊙ or ▣
FLOOR DUPLEX RECEPTACLE OUTLET	▣
FLOOR SPECIAL-PURPOSE OUTLET	△
UNDERFLOOR DUCT AND JUNCTION BOX FOR TRIPLE, DOUBLE, OR SINGLE DUCT SYSTEM AS INDICATED BY NUMBER OF PARALLEL LINES	✥

BUSDUCT AND WIREWAYS

SERVICE, FEEDER, OR PLUG-IN BUSWAY	B	B	B
CABLE THROUGH LADDER OR CHANNEL	C	C	C
WIREWAY	W	W	W

TEE	ELBOW	CROSS
W W W / W	W W / W	W W W W / W W

SWITCH OUTLETS

SINGLE-POLE SWITCH	S or S$_1$
DOUBLE-POLE SWITCH	S$_2$
THREE-WAY SWITCH	S$_3$
FOUR-WAY SWITCH	S$_4$
AUTOMATIC DOOR SWITCH	S$_D$
KEY-OPERATED SWITCH	S$_{KO}$
CIRCUIT BREAKER	S$_{CB}$
WEATHERPROOF CIRCUIT BREAKER	S$_{WCB}$
DIMMER	S$_{DM}$
REMOTE CONTROL SWITCH	S$_{RC}$
SWITCH WITH PILOT LIGHT	S$_P$
WEATHERPROOF SWITCH	S$_{WP}$
FUSED SWITCH	S$_F$
WEATHERPROOF FUSED SWITCH	S$_{WF}$
TIME SWITCH	S$_T$
CEILING PULL SWITCH	Ⓢ
SWITCH AND SINGLE RECEPTACLE	⊖$_S$
SWITCH AND DOUBLE RECEPTACLE	⊜$_S$
A STANDARD SYMBOL WITH AN ADDED LOWERCASE SUBSCRIPT LETTER IS USED TO DESIGNATE A VARIATION IN STANDARD EQUIPMENT	○$_{a,b}$ ⊖$_{a,b}$ S$_{a,b}$

...ELECTRICAL SYMBOLS

COMMERCIAL AND INDUSTRIAL SYSTEMS

- PAGING SYSTEM DEVICE
- FIRE ALARM SYSTEM DEVICE
- COMPUTER DATA SYSTEM DEVICE
- PRIVATE TELEPHONE SYSTEM DEVICE
- SOUND SYSTEM
- FIRE ALARM CONTROL PANEL — FACP

SIGNALING SYSTEM OUTLETS FOR RESIDENTIAL SYSTEMS

- PUSHBUTTON
- BUZZER
- BELL
- BELL AND BUZZER COMBINATION
- TELEPHONE
- PRIVATE TELEPHONE SYSTEM DEVICE — PVT
- COMPUTER DATA OUTLET
- BELL RINGING TRANSFORMER — BT
- ELECTRIC DOOR OPENER — D
- SMOKE/CARBON MONOXIDE DETECTOR — SD OR SD
- CHIME — CH
- TELEVISION OUTLET — TV
- THERMOSTAT — T

UNDERGROUND ELECTRICAL DISTRIBUTION OR ELECTRICAL LIGHTING SYSTEMS

- MANHOLE — M
- HANDHOLE — H
- TRANSFORMER-MANHOLE OR VAULT — TM
- TRANSFORMER PAD — TP
- UNDERGROUND DIRECT BURIAL CABLE
- UNDERGROUND DUCT LINE
- STREET LIGHT STANDARD FED FROM UNDERGROUND CIRCUIT

ABOVE-GROUND ELECTRICAL DISTRIBUTION OR LIGHTING SYSTEMS

- POLE
- STREET LIGHT AND BRACKET
- PRIMARY CIRCUIT
- SECONDARY CIRCUIT
- DOWN GUY
- HEAD GUY
- SIDEWALK GUY
- SERVICE WEATHERHEAD

PANEL CIRCUITS AND MISCELLANEOUS

- LIGHTING PANEL
- POWER PANEL OR SWITCH BOARD
- WIRING – CONCEALED IN CEILING OR WALL
- WIRING – CONCEALED IN FLOOR
- WIRING EXPOSED
- HOME RUN TO PANEL BOARD
  Indicate number of circuits by number of arrows. Any circuit without such designation indicates a two-wire circuit. For a greater number of wires indicate as follows: ―///― (3 wires) ―////― (4 wires), etc.
- FEEDERS
  Use heavy lines and designate by number corresponding to listing in feeder schedule
- WIRING TURNED UP
- WIRING TURNED DOWN
- GENERATOR — G
- MOTOR — M
- INSTRUMENT (SPECIFY) — I
- TRANSFORMER — T
- CONTROLLER
- EXTERNALLY-OPERATED DISCONNECT SWITCH
- PULL BOX

Appendix 489

RESIDENTIAL ELECTRICAL SYMBOLS

GENERAL OUTLETS

Symbol	Description
○ or Ⓛ	Lighting (wall)
or ▫Ⓛ▫	Lighting (ceiling)
⌐○¬	Ceiling lighting outlet for recessed fixture (Outline shows shape of fixture.)
▭○▭	Continuous wireway for fluorescent lighting on ceiling, in coves, cornices, etc. (Extend rectangle to show length of installation.)
Ⓛ	Lighting outlet with lamp holder
Ⓛ$_{PS}$	Lighting outlet with lamp holder and pull switch
Ⓕ	Fan outlet
Ⓙ	Junction box
Ⓓ	Drop-cord equipped outlet
—Ⓒ	Clock outlet

To indicate wall installation of above outlets, place circle near wall and connect with line as shown for clock outlet.

AUXILIARY SYSTEMS

Symbol	Description
▪	Pushbutton
◳	Buzzer
◖	Bell
◖▷	Combination bell-buzzer
CH	Chime
◇	Annunciator
D	Electric door opener
M	Maid's signal plug
□	Interconnection box
T	Thermostat
▶	Outside telephone
▷	Telephone
R	Radio outlet
TV	Television outlet

CONVENIENCE OUTLETS

Symbol	Description
⊜	Duplex convenience receptacle
⊜$_3$	Triplex convenience outlet (Substitue other numbers for other variations in number of plug positions.)
⊜	Duplex convenience outlet — split wired
⊜$_{GR}$	Duplex convenince outlet for grounding-type plugs
⊜$_{WP}$	Weatherproof convenience outlet
⊜ X″	Multioutlet assembly (Extend arrows to limits of installation. Use appropriate symbol to indicate type of outlet. Also indicate spacing of outlets as X inches.)
⊜$_S$	Combination switch and convenience outlet
⊜$_R$	Combination radio and convenience outlet
⊙	Floor outlet
⊜$_R$	Range outlet
⬤$_{DW}$	Special-purpose outlet. (Use subscript letters to indicate function. DW–dishwasher, CD–clothes dryer, etc.)

MISCELLANEOUS

Symbol	Description
◣	Heating panel
▨	Service panel
■	Distribution panel
------	Switch leg indication. (Connects outlets with control points.)
Ⓜ	Motor
⏚	Ground connection
—‖—	2-conductor cable
—‖‖—	3-conductor cable
—‖‖‖—	4-conductor cable
◀—	Cable returning to service panel
—⌒—	Fuse
—o o—	Circuit breaker
⊜$_{a,b}$ ⊜$_{a,b}$ ⬤$_{a,b}$ □$_{a,b}$	Special outlets. (Any standard symbol given above may be used with the addition of subscript letters to designate some special variation of standard equipment for a particular architectural plan. When so used, the variation should be explained in the key to symbols and, if necessary, in the specifications.)
⊜ GFCI	Ground-fault circuit interrupter
⊜ WP	Weatherproof

SWITCH OUTLETS

Symbol	Description	Symbol	Description
S—S$_1$	Single-pole switch	S$_D$	Dimmer switch
S$_2$	Double-pole switch	S$_P$	Switch and pilot light
S$_3$	Three-way switch	S$_{WP}$	Weatherproof switch
S$_4$	Four-way switch		

Appendix 491

HVAC SYMBOLS...

EQUIPMENT SYMBOLS	EQUIPMENT SYMBOLS	DUCTWORK
EXPOSED RADIATOR	AQUASTAT SENSOR — A	SHEET METAL DUCT — OR
RECESSED RADIATOR	FIRESTAT SENSOR — F	DUCT (1ST FIGURE, WIDTH; 2ND FIGURE, DEPTH) — 12 X 20
FLUSH ENCLOSED RADIATOR	HUMIDISTAT SENSOR — H	DIRECTION OF FLOW
PROJECTING ENCLOSED RADIATOR	THERMOSTAT — T	FLEXIBLE CONNECTION
UNIT HEATER (PROPELLER) – PLAN	LINE VOLTAGE THERMOSTAT — T L	DUCTWORK WITH ACOUSTICAL LINING
UNIT HEATER (CENTRIFUGAL) – PLAN	LOW VOLTAGE THERMOSTAT — T LV	FIRE DAMPER WITH ACCESS DOOR — FD / AD
UNIT VENTILATOR – PLAN	TIME SWITCH — TS	MANUAL VOLUME DAMPER — VD
STEAM	PIPE TRACE HEATER	AUTOMATIC VOLUME DAMPER
DUPLEX STRAINER	PNEUMATIC DAMPER	EXHAUST, RETURN OR OUTSIDE AIR DUCT – SECTION — 20 X 12
PRESSURE-REDUCING VALVE	ELECTRIC DAMPER	SUPPLY DUCT – SECTION — 20 X 12
AIR LINE VALVE	FAN	CEILING DIFFUSER SUPPLY OUTLET — 20" DIA CD / 1000 CFM
STRAINER	PUMP	CEILING DIFFUSER SUPPLY OUTLET — 20 X 12 CD / 700 CFM
THERMOMETER	ELECTRIC CONTROLLING PNEUMATIC SWITCH	LINEAR DIFFUSER — 96 X 6-LD / 400 CFM
PRESSURE GAUGE AND COCK	PNEUMATIC CONTROLLING ELECTRIC SWITCH	FLOOR REGISTER — 20 X 12 FR / 700 CFM
RELIEF VALVE	TWISTED SHIELDED PAIR CABLE	TURNING VANES
AUTOMATIC 3-WAY VALVE	HEATING COIL — H / C	FAN AND MOTOR WITH BELT GUARD
AUTOMATIC 2-WAY VALVE	COOLING COIL — C / C	
SOLENOID VALVE — S	FILTER	LOUVER OPENING — 20 X 12-L / 700 CFM
MAGNETIC MOTOR STARTER AND DISCONNECT		
ADJUSTABLE SPEED DRIVE — ASD		
AUTOMATIC TEMPERATURE CONTROL PANEL — ATC		

...HVAC SYMBOLS

HEATING PIPING		HEATING PIPING		AIR CONDITIONING PIPING	
HIGH-PRESSURE STEAM	—HPS—	FEEDWATER PUMP DISCHARGE	—PPD—	REFRIGERANT LIQUID	—RL—
MEDIUM-PRESSURE STEAM	—MPS—	MAKEUP WATER	—MU—	REFRIGERANT DISCHARGE	—RD—
LOW-PRESSURE STEAM	—LPS—	AIR RELIEF LINE	—V—	REFRIGERANT SUCTION	—RS—
HIGH-PRESSURE RETURN	—HPR—	FUEL OIL SUCTION	—FOS—	CONDENSER WATER SUPPLY	—CWS—
MEDIUM-PRESSURE RETURN	—MPR—	FUEL OIL RETURN	—FOR—	CONDENSER WATER RETURN	—CWR—
LOW-PRESSURE RETURN	—LPR—	FUEL OIL VENT	—FOV—	CHILLED WATER SUPPLY	—CHWS—
BOILER BLOW OFF	—BD—	COMPRESSED AIR	—A—	CHILLED WATER RETURN	—CHWR—
CONDENSATE OR VACUUM PUMP DISCHARGE	—VPD—	HOT WATER HEATING SUPPLY	—HW—	MAKEUP WATER	—MU—
		HOT WATER HEATING RETURN	—HWR—	HUMIDIFICATION LINE	—H—
				DRAIN	—D—

REFRIGERATION SYMBOLS

GAUGE		PRESSURE SWITCH	DRYER
SIGHT GLASS		HAND EXPANSION VALVE	FILTER AND STRAINER
HIGH SIDE FLOAT VALVE		AUTOMATIC EXPANSION VALVE	COMBINATION STRAINER AND DRYER
LOW SIDE FLOAT VALVE		THERMOSTATIC EXPANSION VALVE	EVAPORATIVE CONDENSOR
IMMERSION COOLING UNIT		CONSTANT PRESSURE VALVE, SUCTION	HEAT EXCHANGER
COOLING TOWER		THERMAL BULB	AIR-COOLED CONDENSING UNIT
NATURAL CONVECTION, FINNED TYPE EVAPORATOR		SCALE TRAP	WATER-COOLED CONDENSING UNIT
FORCED CONVECTION EVAPORATOR		SELF-CONTAINED THERMOSTAT	

PLUMBING SYMBOLS

FIXTURES	FIXTURES (Continued)	PIPING (Continued)	
STANDARD BATHTUB	LAUNDRY TRAY	CHILLED DRINKING WATER SUPPLY	—— DWS ——
OVAL BATHTUB	BUILT-IN SINK	CHILLED DRINKING WATER RETURN	—— DWR ——
WHIRLPOOL BATH	DOUBLE OR TRIPLE BUILT-IN SINK	HOT WATER SUPPLY	— — — —
SHOWER STALL	COMMERCIAL KITCHEN SINK	HOT WATER RETURN	— — — —
SHOWER HEAD	SERVICE SINK (SS)	SANITIZING HOT WATER SUPPLY (180°)	—+—+—+
TANK-TYPE WATER CLOSET	CLINIC SERVICE SINK	SANITIZING HOT WATER RETURN (180°)	—+—+—+
WALL-MOUNTED WATER CLOSET	FLOOR-MOUNTED SERVICE SINK	DRY STANDPIPE	—— DSP ——
FLOOR-MOUNTED WATER CLOSET	DRINKING FOUNTAIN (DF)	COMBINATION STANDPIPE	—— CSP ——
LOW-PROFILE WATER CLOSET	WATER COOLER	MAIN SUPPLIES SPRINKLER	—— S ——
BIDET	HOT WATER TANK (HWT)	BRANCH AND HEAD SPRINKLER	—o——o—
WALL-MOUNTED URINAL	WATER HEATER (WH)	GAS — LOW PRESSURE	— G — G —
FLOOR-MOUNTED URINAL	METER (M)	GAS — MEDIUM PRESSURE	—— MG ——
TROUGH-TYPE URINAL	HOSE BIBB (HB)	GAS — HIGH PRESSURE	—— HG ——
WALL-MOUNTED LAVATORY	GAS OUTLET (G)	COMPRESSED AIR	—— A ——
PEDESTAL LAVATORY	GREASE SEPARATOR (G)	OXYGEN	—— O ——
BUILT-IN LAVATORY	GARAGE DRAIN	NITROGEN	—— N ——
WHEELCHAIR LAVATORY	FLOOR DRAIN WITH BACKWATER VALVE	HYDROGEN	—— H ——
CORNER LAVATORY	**PIPING**	HELIUM	—— HE ——
FLOOR DRAIN	SOIL, WASTE, OR LEADER — ABOVE GRADE	ARGON	—— AR ——
FLOOR SINK	SOIL, WASTE, OR LEADER — BELOW GRADE	LIQUID PETROLEUM GAS	—— LPG ——
	VENT	INDUSTRIAL WASTE	—— INW ——
	COMBINATION WASTE AND VENT —— SV ——	CAST IRON	—— CI ——
	STORM DRAIN —— S ——	CULVERT PIPE	—— CP ——
	COLD WATER	CLAY TILE	—— CT ——
		DUCTILE IRON	—— DI ——
		REINFORCED CONCRETE	—— RCP ——
		DRAIN — OPEN TILE OR AGRICULTURAL TILE	= = = =

FIRE ALARM/SAFETY/SECURITY SYMBOLS

FIRE ALARM		MODIFIED FIRE ALARM		SECURITY SYSTEM	
ANNUNCIATOR PANEL	ANN	NODE	NODE X (NODE IDENTIFIER)	CCTV FIXED CAMERA	
STROBE/HORN	F (OR) XXcd — S	TERMINAL CABINET	FATC X	CCTV FIXED CAMERA IN DOME HOUSING	
BELL	F CEILING MOUNT OR F WALL MOUNT	HEAT DETECTOR	HD F (OR) HD R WALL MOUNT	CCTV WEATHERPROOF EXTERIOR CAMERA	C WP
MANUAL PULL STATION	F	LCD ANNUNCIATOR	ANN X	CCTV COAXIAL CABLE OUTLET	CCTV
CONTROL PANEL	FACP	FAN CONTROL PANEL	FAN X	DOOR CHIME	B
PHOTOELECTRIC SMOKE DETECTOR	SD	CABINET WITH UIO MODULE	MFC-A CT1 SINGLE INPUT MODULE OR MFC-A CT2 DUAL INPUT MODULE	SIREN	
TAMPER SWITCH	TS OR (TAMPER SWITCH)	ADDRESSABLE SMOKE DETECTOR	P	FLAT PANEL MONITOR	OR PV PUBLIC VIEW
FLOW SWITCH	FS OR	PRIVATE TELEPHONE SYSTEM DEVICE	FSD-SA SUPPLY AIR OR FSD-EA EXHAUST AIR	KEYPAD	KPD OR K
		POST INDICATOR VALVE	PIV XX VALVE NUMBER	CONTROL PANEL	SCP
SMOKE DETECTOR	S	FIREMAN'S PHONE JACK	J	CARD READER	CR OR CR WP

ARCHITECTURAL SYMBOLS . . .

MATERIAL	ELEVATION	PLAN	SECTION
Earth			
Brick	WITH NOTE INDICATING TYPE OF BRICK (COMMON, FACE, ETC.)	COMMON OR FACE / FIREBRICK	SAME AS PLAN VIEWS
Concrete		LIGHTWEIGHT / STRUCTURAL	SAME AS PLAN VIEWS
Concrete Masonry Unit			OR
Stone	CUT STONE / RUBBLE	CUT STONE / RUBBLE / CAST STONE (CONCRETE)	CUT STONE / CAST STONE (CONCRETE) / RUBBLE OR CUT STONE
Wood	SIDING / PANEL	WOOD STUD / REMODELING	ROUGH MEMBER / TRIM MEMBER / PLYWOOD
Plaster		METAL LATH AND PLASTER / SOLID PLASTER	LATH AND PLASTER
Roofing	SHINGLES	SAME AS ELEVATION	
Glass	OR / GLASS BLOCK	GLASS / GLASS BLOCK	SMALL SCALE / LARGE SCALE

Appendix 495

...ARCHITECTURAL SYMBOLS

MATERIAL	ELEVATION	PLAN	SECTION
Facing Tile	CERAMIC TILE	FLOOR TILE	CERAMIC TILE LARGE SCALE / CERAMIC TILE SMALL SCALE
Structural Clay Tile	(brick pattern)	(cross-hatch pattern)	SAME AS PLAN VIEW
Insulation		LOOSE FILL OR BATTS / RIGID / SPRAY FOAM	SAME AS PLAN VIEWS
Sheet Metal Flashing	(vertical lines)	OCCASIONALLY INDICATED BY NOTE	
Metals Other Than Flashing	INDICATED BY NOTE OR DRAWN TO SCALE	SAME AS ELEVATION	SMALL SCALE / STEEL / CAST IRON / ALUMINUM / BRONZE OR BRASS
Structural Steel	INDICATED BY NOTE OR DRAWN TO SCALE	—— — —— OR ———— — — ————	REBAR / SMALL SCALE / LARGE SCALE / L-ANGLES, S-BEAMS, ETC.

PLOT PLAN SYMBOLS

↑N	NORTH	●	FIRE HYDRANT	═══	WALK	—E— OR —●—	ELECTRIC SERVICE
⊕	POINT OF BEGINNING (POB)	⊠	MAILBOX	———	IMPROVED ROAD	—G— OR —●—	NATURAL GAS LINE
▲	UTILITY METER OR VALVE	○	MANHOLE	- - - -	UNIMPROVED ROAD	—W— OR -----	WATER LINE
●→	POWER POLE AND GUY	⊕	TREE	BL	BUILDING LINE	—T— OR —●—	TELEPHONE LINE
☼	LIGHT STANDARD	○	BUSH	PL	PROPERTY LINE	— — —	NATURAL GRADE
⊂○	TRAFFIC SIGNAL	⌇⌇⌇	HEDGE ROW	———	PROPERTY LINE	———	FINISH GRADE
—○	STREET SIGN	—●—●—	FENCE	- - - -	TOWNSHIP LINE	+ XX.00′	EXISTING ELEVATION

PRINT SYMBOLS

PLOT PLANS		PRIVATE PROPERTY		PUBLIC PROPERTY	
WALK	═══	NORTH	N ↑	LIGHT STANDARD	⊖
BUILDING LINE	BL			TRAFFIC SIGNAL	⊙
PROPERTY LINE	PL OR —··—	POINT OF BEGINNING	◓	STREET SIGN	—○
TOWNSHIP LINE	— — — —	UTILITY METER OR VALVE	▲	FIRE HYDRANT	●—
ELECTRIC SERVICE	—E— OR —·—·—	MAILBOX	⊠	MANHOLE	M
NATURAL GAS SERVICE	—G— OR —··—··—	TREE	✿		
WATER LINE	—W— OR ········	BUSH	◯		
TELEPHONE LINE	—T— OR —·—·—	HEDGE ROW	⌬		
NATURAL GRADE	— — —				
FINISH GRADE	———				
EXISTING ELEVATION	+XX.00'	FENCE	—•—•—		

GRID LINES

NUMBERED GRID LINES — ① ②
LETTERED GRID LINES — Ⓓ Ⓒ
32'-11"

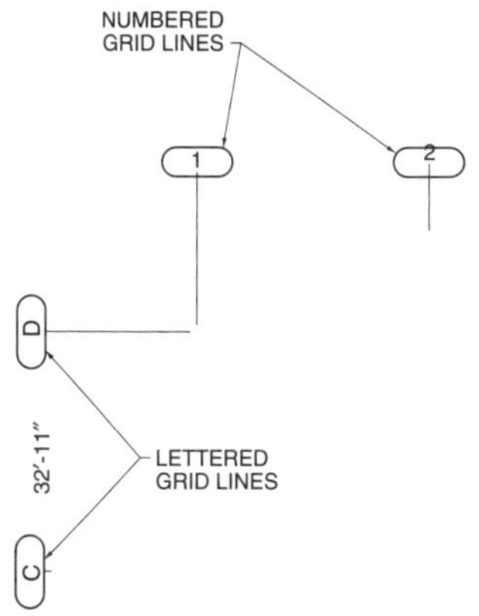

PRINT DIVISIONS

A = ARCHITECTURAL
S = STRUCTURAL
M = MECHANICAL
E = ELECTRICAL
C = CIVIL

ALPHABET OF LINES

NAME AND USE	CONVENTIONAL REPRESENTATION	EXAMPLE
Object Line — Define shape. Outline and detail objects.	THICK	OBJECT LINE
Hidden Line — Show hidden features.	$\frac{1}{8}''$ (3 mm); $\frac{1}{32}''$ (0.75 mm); THIN	HIDDEN LINE
Centerline — Locate centerpoints of arcs and circles.	$\frac{1}{16}''$ (1.5 mm); $\frac{1}{8}''$ (3 mm); $\frac{3}{4}''$ (18 mm) TO $1\frac{1}{2}''$ (36 mm); THIN	CENTERLINE; CENTERPOINT
Dimension Line — Show size or location. **Extension Line** — Define size or location.	DIMENSION LINE; DIMENSION; 2'-6"; EXTENSION LINE; THIN	DIMENSION LINE; $1\frac{3}{4}$; EXTENSION LINE
Leader — Call out specific features.	OPEN ARROWHEAD; CLOSED ARROWHEAD; X; 3X; THIN	$1\frac{1}{2}$ DRILL; LEADER
Cutting Plane — Show internal features.	$\frac{1}{8}''$ (3 mm); $\frac{1}{16}''$ (1.5 mm); A—A; $\frac{3}{4}''$ (18 mm) TO $1\frac{1}{2}''$ (36 mm); THICK	LETTER IDENTIFIES SECTION VIEW; CUTTING PLANE LINE
Section Line — Identify internal features.	$\frac{1}{16}''$ (1.5 mm); THIN	SECTION LINES
Break Line — Show long breaks. **Break Line** — Show short breaks.	$\frac{3}{4}''$ (18 mm) TO $1\frac{1}{2}''$ (36 mm); THIN; FREEHAND; THICK	LONG BREAK LINE; SHORT BREAK LINE

FLOOR PLAN ABBREVIATIONS...

ABBREVIATION	TERM	ABBREVIATION	TERM
ACS	access	HVAC	heating, ventilating, and air conditioning
ACSP or AP	access panel	HW	hot water
ADD or ADH	adhesive	INTR or INT	interior
AT.	asphalt tile	K or KIT	kitchen
B	bathroom	KD	knocked down
BC	between centers	LAU	laundry
BCL	broom closet	LAV	lavatory
BLK	block	LCL or LCLO	linen closet
BLKG	blocking	LDG	landing
BPL or BDG PL	bearing plate	LIB	library
BR	bedroom	LINO or LINOL	linoleum
BT	bathtub	LKT	lookout
BYP	bypass	LNG	lining
CA	cold air	LR	living room
CFM	cubic feet per minute	MC	medicine cabinet
CLG	cooling or ceiling	OP	operator
CLO	closet	OVHD DR	overhead door
CNCL	concealed	P	porch
CT	ceramic tile	PAN.	pantry
CW	cold water	PASS.	passage
DBL ACT. or DA	double-acting	PFN	prefinished
DD	dutch door	PL	plate
DK	deck	PLAT	platform
DMR	dimmer	PORC	porcelain
DNG R or DR	dining room	PREFAB or PFB	prefabricated
DW	dishwasher	PTN	partition
EMT	electric metallic tubing	PW	projected window
ENTR	entrance	QT	quarry tile
EQPT	equipment	R	riser
EXT	extinguisher	R or RM	room
FC	furred ceiling	R or RNG	range
FL	flush	RAD	radiator
FLMT	flush mount	RBT or R TILE	rubber tile
FLR or FLG	flooring	REC	recessed
FLUR or FLUOR	fluorescent	REF or REFR	refrigerator
FUR.	furring	REG	register
GB	gypsum board	RGH	rough
GFCI	ground-fault circuit interrupter	RO	rough opening
GRB	garbage	S	scuttle
GYP	gypsum	SC	sillcock
HA	hot air	SC	solid core
HC	hollow core	SCH or SCHED	schedule
HD	head	SD or SL DR	sliding door
HDR	header	SH	shower
HTD	heated	SH & RD	shelf and rod

...FLOOR PLAN ABBREVIATIONS

ABBREVIATION	TERM	ABBREVIATION	TERM
SHELV	shelving	U RM	utility room
SK or S	sink	V	vent
SP	standpipe	V TILE or VT	vinyl tile
ST	stairs	VENT	ventilation
STOR	storage	VS	vapor seal
STWY	stairway	VT	vaportight
SUSP	suspend	WA	warm air
TBG	tubing	WC	water closet
THERMO or T	thermostat	WV	wall vent
TR or T	tread		

SECTION AND DETAIL ABBREVIATIONS

ABBREVIATION	TERM	ABBREVIATION	TERM
ACT or AT	acoustical tile	FNSH	finish
ACST	acoustic	FBCK	firebrick
CAB.	cabinet	FP	fireplace
CSG	casing	FRWK	framework
CM	center-matched	FR	frame
CER	ceramic	HBD	hardboard
CHM	chimney	JB or JMB	jamb
CO	cleanout	JT	joint
COMB.	combination	LAM	laminate
CTR	counter	MIR	mirror
CO	cut out	MLD or MLDG	molding
DMPR	damper	MRB or MR	marble
DET OR DTL	detail	NOS	nosing
DWL	dowel	RTN	return
D & M	dressed and matched	SEC or SECT.	section
E to E	end-to-end	SURF.	surface
FAB	fabricate	X SECT	cross section

ELEVATION ABBREVIATIONS...

ABBREVIATION	TERM	ABBREVIATION	TERM
AC	air conditioner	GALV	galvanized
ALUM	aluminum	GALVI	galvanized iron
ANT	antenna	GALVS	galvanized steel
AWN	awning	GL	glass
BC	bookcase	GLB	glass block
BD	board	GLZ	glaze
BEV	beveled	HMD	hollow metal door
BK SH	book shelves	KPL	kickplate
BRK	brick	LMS or LS	limestone
BRS	brass	LNTL	lintel
BRZ	bronze	LT	light
BUR	built-up roofing	LTH	lath
CAN.	canopy	LV OR LVR	louver
CI	cast iron	MET F	metal flashing
CL GL	clear glass	MET J	metal jalousie
CLG	ceiling	MGS	metal gravel stop
CLKG	caulking	OB	obscure
CMPSN	composition	OBSC GL or OGL	obscure glass
CMU	concrete masonry unit	OPG or OPNG	opening
COR	cornice	OVHG	overhang
CORR	corrugated	P	pitch
CPLR	center of pillar	PK	peak
CRE	corrosion-resistant	PL GL	plate glass
CST or CS	cast stone	PLAS	plaster
CSMT	casement	PLR	pillar
CTD	coated	PLSTC	plastic
CUTS.	cut stone	PNT	paint
DH	double-hung	PTD	painted
DHW	double-hung window	QTR	quarry tile roof
DR	door	RD	roof drain
DRM	dormer	RDG	ridge
DS	downspout	RF	roof
DSG	double-strength glass	RFG	roofing
ENAM	enamel	RR	roll roofing
EXHV	exhaust vent	RS	rubble stone
EXPSR	exposure	RSTPF	rustproof
EXT	exterior	SB	splash block
EXT GR	exterior grade	SCR OR SCRN	screen
FB	face brick	SDG	siding
FL	flashing	SDL	saddle
FL	floor line	SDLT	sidelight
FLD	field	SF	soffit
FNSH FL	finished floor	SH OR SHT	sheet
FR	front	SHGL	shingle
FX WDW	fixed window	SHK	shake

... ELEVATION ABBREVIATIONS

ABBREVIATION	TERM	ABBREVIATION	TERM
SHTNG	sheathing	VAL	valley
SHTR	shutter	VCT	vitrified clat tile
SL	slate	VP	vent pipe
SLT	skylight	VS	vent stack
SM	sheet metal	WD	wood door
SS	steel sash	WH	weep hole
SSG	single-strength glass	WI	wrought iron
STK	stack	WIND	window
STN	stained	WJ	wood jalousie
STN	stone	WS	weatherstripping
TC	terra-cotta	WSCT OR WAIN	wainscot
TH	threshold	WSR	wood-shingle roof
THRM	thermal	WT	wood threshold
TMPD	tempered	WTHPRF	weatherproof
TSR	tile-shingled roof		

PROCESS LINE CONTENT ABBREVIATIONS

ABBR/ACRONYM	MEANING	ABBR/ACRONYM	MEANING
A	compressed air (wet)	FSL	Ferrous sulfate
AA	acetic acid	G	gasoline
AC	acetylene	H	gydrogen
AD	ammonia drains	HA	got air
AG	ammonia gas	HCN	gydrogen cyanide
ALC	acohol	IA	instrument air (dry)
AN	acrylonitrile	LA	liquid ammonia
ANH	anhydrous ammonia	LS	lime solution
ANS	ammonia nitrate solution	M	methane
ANV	ammonia nitrate vapor	MA	methonol
AOT	aerosol	MG	mixed gas and steam
AQS	aqua solution	N	nitrogen
AS	acid slurry	NG	natural gas
ATV	atmospheric vent	NTA	ammonium nitrate
BD	boiler blow down	O	oxygen
BG	burner gas	OF	off gas
BI	blend	PGC	cooled process gas
BR	brine	PRG	process gas
C	condensate	PV	pressure vent
CA	conditioning agent	S	steam
CAT	catalyst	SL	slurry
CF	chemical feed	SLG	sludge
CLG	chlorine gas	STG	stack gas
CLS	chlorine solution	SV	solvent
CNG	converter/converted gas	SH	sodium hydroxide
CS	chemical sewer	SUR	surfactant
D	drains	TOL	toluene
EA	exhaust air	UA	utility air
EQ	equalizer line	V	vent
ES	exhaust steam	VAC	vacuum
F	filtrate (cold)	VC	vent cold
FG	fuel gas	VCL	vinyl chloride
FLRG	flare gas	VG	vent gas
FO	fuel oil	WC	cooling water supply
FR	freon	WCR	cooling water return
FRG	freon gas	WT	treated water
FRL	freon liquid	WTOL	wet toluene
FS	freon solvent	WWH	wash water

COMMON INSTRUMENT ABBREVIATIONS

ABBR/ACRONYM	MEANING	ABBR/ACRONYM	MEANING
FLOW		**...PRESSURE OR VACUUM**	
FAL	flow alarm low	PSV	pressure safety relief valve
FAH	flow alarm high	PSE	rupture disc, vacuum breaker or emergency vent valve
FE	flow element		
FI	flow indicator	PT	pressure transmitter
FIC	flow indicating controller	PV	pressure control valve
FIT	flow indicating transmitter	**TEMPERATURE**	
FS	flow switch	TAL	temperature alarm low
FT	flow transmitter	TC	temperature controller
FV	fLow control valve	TE	temperature element, thermocouple, resistance bulb, or thermopile
LEVEL		TI	temperature indicator
LAL	level alarm low	TIC	temperature indicating controller
LAH	level alarm high	TIS	temperature indicating switch
LC	level controller	TIT	temperature indicating transmitter
LCV	level control valve self-operated	TR	temperature recorder
LE	level element	TS	temperature switch
LG	level glass	TT	temperature transmitter
LI	level indicator	TV	temperature control valve
LIT	level indicating transmitter	TW	temperature well
LR	level recorder	TY	temperature interlock solenoid valve, relay or converter
LS	level switch		
LT	level transmitter	**MISCELLANEOUS**	
LV	levl control valve	BC	burner controller
LY	level interlock solenoid valve, relay, or converter	HC	manual (hand) controller
PRESSURE OR VACUUM ...		HV	manual (hand) control valve
PAL	pressure alarm low	II	current indicator
PAH	pressure alarm high	ZI	position indicator
PC	pressure controller	A	analysis
PCV	pressure control valve, self-operated	I	current
PDI	pressure differential indicator	J	power
PE	pressure element	K	time, program, or counting
PI	pressure indicator or manometer	Q	quantity
PIT	pressure indicating transmitter	WI	weight indicator

ELECTRICAL/ELECTRONIC ABBREVIATIONS/ACRONYMS . . .

ABBREVIATION	MEANING	ABBREVIATION	MEANING
A	amps	CH	chapter
AC	air conditioner	CKT	circuit
AC/DC	alternating current/direct current	CM	circular mills
A/D	analog to digital	CMP	code-making panel
AEGCP	assured equipment grounding program	COMPT	compartment
AF	audio frequency	COND	conduit
AFC	automatic frequencycontrol	CONT	continuous; control
AFF	abopve finished floor	CP	control panel
Ag	silver	CPS	cycles per second
A/H	air handler	CPT	control power transformer
AHJ	authority having jurisdiction	CPU	central processing unit
AIR	ampere interrupting rating	CR	control relay
Al	aluminum	CRM	control relay master
ALM	alarm	CT	current transformer
AM	ammeter; amplitude moduation	CU	copper
AM/FM	amplitude modulation; frequency modulation	CW	clockwise
ANN	annunciator	D	diameter; diode; down
AMP	amperes, amperage	D/A	digital to analog
ANSI	American National Standards Institute	DB	dynamic braking contactor; relay
ARM	armature	DC	direct current
AS	ammeter switch	DIA	diameter
AT	ampere trip	DIO	diode
ATCB	adjustable-trip circuit breaker	DISC	disconnect switch
ATS	automatic transfer switch	DISTR	distribution
Au	gold	DMM	digital multimeter
AU	automatic	DP	double-pole
AUTO	automatic	DPDT	double-pole, double throw
AUX	auxiliary	DPST	double-pole, sigle-throw
AVC	automatic volume control	DS	drum switch
AWG	american wire gauge	DT	double-throw
BAT	battery (electric)	DVM	digital voltmeter
BCD	binary coded decimal	DWG	drawing
BJ	bonding jumper	E	voltage
BJT	bipolar junction transistor	EBJ	equipment bonding jumper
BK	black	E_{ff}	efficiency
BKR	breaker	EGC	equipment grounding conductor
BL	blue	ELEV	elevation
BR	break relay; brown	EMERG	emergency
C	celsius; capacitance; capacitor	EMF	electromotive force
CAB	cabinet	EMT	electrical metallic tubing
CAP	capacitor	ENCL	enclosure
CB	circuit breaker	EP	explosion proof
CCW	counterclockwise	EQPT	equipment
CE	common-emitter configuration	ER	conductance relay
CEM	counter electromotive force	ETM	elapsed time relay

...ELECTRICAL/ELECTRONIC ABBREVIATIONS/ACRONYMS...

ABBREVIATION	MEANING	ABBREVIATION	MEANING
EXH	exhaust	I	current
EXIST	existing	IC	integrated circuit
F	fahrenheit; fast; field; forward; fuse	IDCI	immersion detection circuit interrupter
FDR	feeder	IG	isolated ground
FET	field-effect transistor	IMC	intermediate metal conduit
FF	flip-flop	INCAND	incandescent
FLA	full-load amps	IND	indication, indicating
FLC	full-load current	I/O	input/output
FLEX	flexible	INST	instantaneous
FLS	flow switch	INSTR	instrument
FLT	full-load torque	Isc	short circuit current
FLUOR	fluorescent	INT	intermediate; interrupt
FM	frequency modulation	INTLK	interlock
FR	frame	IOL	instantaneous overload
FREQ	frequency	IR	infrared
FS	float switch	ITB	inverse time breaker
FTS	foot switch	ITCB	instantaneous trip circuit breaker
FU	fuse	JB	junction box
FUT	future	JFET	junction field-effect transistor
FVR	full voltage reversing	k	kilo (1000)
FVNR	full voltage, non-reversing	kcmil	1000 circular mils
FWD	forward	kVA	kilovolt amps
G	gate; giga; green; conductance	kW	kilowatt
GALV	galvanized	kWh	kilowatt-hour
GEN	generator	L	line; load; coil; impedance
GEC	grounding electrode conductor	LB-FT	pounds per foot
GES	grounding electrode system	LB-IN	pounds per inch
GFCI	ground fault circuit interrupter	LC	inductance-capacitance
GFPE	ground fault protection of equipment	LCB	local control board
GR	green	LCD	liquid crystal display
GRD	ground	LCP	local control panel
GY	gray	LCR	inductance-capacitance-resistance
H	Henry; high side of transformer; magnetic flux	LED	light emitting diode
HACR	heating; air-conditioning; refrigeration	LOC	local
HDA	hand-off-automatic	LO	lockout
HF	high frequency	LOS	lockout stop
HG	mercury	LP	lighting panel
HID	high intensity discharge	LRA	locked-rotor ampacity
HP	horsepower	LRC	locked rotor current
HPS	hours	LS	limit switch
HRS	hertz	LT	lamp
HTR	heater	LTG	lighting
HT TR	heat traced	LT	light
HVAC	heating, ventilating, A/C	M	motor; motor starter; motor starter contacts
Hz	hertz	MA	milliamps

...ELECTRICAL/ELECTRONIC ABBREVIATIONS/ACRONYMS...

ABBREVIATION	MEANING	ABBREVIATION	MEANING
MAG	magnetic	OZ/IN	ounces per inch
MAN	manual	P	peak; positive; pole; power; power consumed
MAX	maximum	PB	pushbutton
MB	magnetic brake	PC	personal computer
MBCCGF	motor branch-circuit, short-circuit, ground fault	PCB	printed circuit board
MBJ	main bonding jumper	PCM	process control module
MCB	main control board	PCP	process control panel
MCC	main control board	PF	power factor
MCP	main circuit protector	PH; φ	phase
MCS	motor circuit switch	PLS	plugging switch
MD	motorized dasmper	PNL	panel
MEM	memory	PNLBD	panelboard
MED	medium	PNP	positive-negative-positive
MH	mounting hrights	POS	positive
MIN	minimum	POT	potentiometer
MLO	main lugs only	P-P	peak-to-peak
MN	manual	PQM	power quality meter
MOS	metal-oxide semiconductor	PRI	primary switch
MOSFET	metal-oxide semiconductor field effect transistor	PS	pressure switch
MS	motor starter	PSI	pounds per square inch
MMS	manual motor starter	PT	potential transformer
MTD	mounted	PUT	pull-up torque
MTR	motor	PW	part winding
MUX	multiplexing panel	Q	transistor
N	north; negative; neutral	R	red; radius; resistance; resistor; reverse
NA	non-automatic	RAM	random-access memory
NATCB	nonadjustable-trip circuit breaker	RC	resistance-capacitance
NC	normally closed	RCL	resistance-inductance-capacitance
NEC®	National Electrical Code	REC	rectifier; receptacle
NEMA	National Electrical Manufacturers Association	RES	resistor
NESC	National Electrical Safety Code	RECPTS	receptacles
NEUT	neutral	REV	reverse
NEG	negative	RGS	required galvanized steel
NFPA	National Fire Protection Association	RF	radio frequency
NIC	not in contract	RH	rheostat
NO	normally open	rms	root mean square
NP	nameplate	ROM	read-only memory
NPN	negative-positive-negative	rpm	revolutions per minute
NTS	not to scale	RPS	revolutions per second
NTDF	non-time delay fuse	RTU	remote terminal unit
O	open; orange	RVAT	reduced voltage autotransformer
OC	on center	RVNR	reduced voltage non-reversing
OCPD	overcurrent protection device	RVSS	reduced voltage solid state
OHM	ohmmmeter	S	series; slow; south; switch
OL	overload relay	SCH	schedule

...ELECTRICAL/ELECTRONIC ABBREVIATIONS/ACRONYMS

ABBREVIATION	MEANING	ABBREVIATION	MEANING
SCR	silicon controlled rectifier	TR	timing relay
SEC	secondary	TS	temperature switch
SECT	section	TTL	transistor-trnasistor logic
SEL SW	selector switch	TYP	typical
SEQ	sequence	U	up
SF	service factor	UCL	unclamp
SH	shield	UF	underground feeder
SHT	sheet	UG	underground
SHLD	shielded	UL	Underwriter's Laboratory
1 PH; 1φ	single-phase	UON	unless otherwise noted
SIG	signal	UHF	ultrahigh frequency
S1, S2	start contactor coils	UJT	unijunction transistor
SOC	socket	US	unit substation
SOL	solenoid	UV	ultraviolet; undervoltage
SP	single-pole	V	violet; volt
SPCB	single-pole circuit breaker	VA	volt amps
SPDT	single pole double throw	VAC	volts alternating current
SPECS	specifications	VAR	var meter
SPST	single pole single throw	VD	voltage drop
SS	selector switch	VDC	volts direct current
SSW	safety switch	VFD	variable frequency drive
STA	station	VHF	very high frequency
STD	standards	VLF	very low frequency
STL	steel	VOM	volts-ohm-milliammeter
STR	starter	VP	vapor proof
SW	switch	VS	voltmeter switch; variable speed
SWD	switched disconnect	W	watts; white
SYM	symmetrical	WP	weatherproof
SYS	system	w/	with
T	tera; terminal; torque; transformer	WHD	watthourground meter
TACH	tachometer	WHM	watthour meter
TB	terminal board	WP	weatherproof
TERM	terminal	X	low side of transformer
3 PH; 3φ	three-phase	XD	transducer
TD	time delay	XFWR	transformer
TDON	time delay-on	XFER	transfer
TDOFF	time delay-off	XWTR	transducer
TDF	time delay fuse	XPDR	transponder
TEMP	temperature	Y	yellow
THS	thermostat switch	Z	impedance
TP	thermally protected		

EQUIPMENT LOG SHEET

EQUIPMENT INFORMATION:			
NAME:		LOCATION	
MANUFACTURE:	BUILDING:	FLOOR:	
MODEL:	SYSTEM:		
SERIAL NUMBER:	MACHINE:		
OIL:	GREASE:	BEARINGS	
TYPE:	TYPE:	SHAFT #1:	REAR:
AMOUNT:	AMOUNT:	SHAFT #2:	REAR:
DRIVE MOTOR		SEALS	
HP:	SHAFT #1:	REAR:	
FRAME:	SHAFT #2:	REAR:	

SPECIAL COMPONENTS		

MAINTENANCE LOG		
DATE	PROBLEM FOUND	SOLUTION

XYZ ELECTRIC
REQUEST FOR INFORMATION

TO: _____ R.F.I. #: _____ DATED: _____
     _____ FROM: _____
     _____ DATE INFORMATION
     _____ REQUIRED: _____

SUBJECT: _____

CATEGORY:
- ❏ INFORMATION NOT SHOWN ON CONTRACT DOCUMENTS
- ❏ INTERPRETATION OF CONTRACT REQUIREMENTS
- ❏ CONFLICT IN CONTRACT REQUIREMENTS
- ❏ COORDINATION ISSUES

CONTRACT DRAWING REFERENCE _____
SHOP DRAWING REFERENCE _____
SPECIFICATION REFRENCE _____
- ❏ POSSIBLE COST IMPACT
- ❏ POSSIBLE TIME IMPACT

_____
_____
_____
_____
_____
_____
_____
_____
_____
_____

REPLY:
_____
_____
_____
_____
_____
_____
_____

ANSWERED BY: _____
DATE: _____

CC:
- ❏ GENERAL CONTRACTOR   ❏ SUPERINTENDENT   ❏ PROJECT MNGR
- ❏ ARCHITECT   ❏ ENGINEER   ❏ PURCHASING

XYZ ARCHITECTS
PUNCH LIST

PROJECT: _____
XYZ ASSOCIATE: _____
DATE: _____

ITEM #1
LOCATION: _____
DESCRIPTION: _____
RESPONSBILE PARTY: _____ REQUIRED COMPLETION DATE: _____

COST IMPACT:
❏ NO IMPACT, INCLUDED IN CONTRACT _____
❏ IMPACT, NOT INCLUDED IN CONTRACT. $ _____

ITEM #2
LOCATION: _____
DESCRIPTION: _____
RESPONSBILE PARTY: _____ REQUIRED COMPLETION DATE: _____

COST IMPACT:
❏ NO IMPACT, INCLUDED IN CONTRACT _____
❏ IMPACT, NOT INCLUDED IN CONTRACT. $ _____

ITEM #3
LOCATION: _____
DESCRIPTION: _____
RESPONSBILE PARTY: _____ REQUIRED COMPLETION DATE: _____

COST IMPACT:
❏ NO IMPACT, INCLUDED IN CONTRACT _____
❏ IMPACT, NOT INCLUDED IN CONTRACT. $ _____

ITEM #4
LOCATION: _____
DESCRIPTION: _____
RESPONSBILE PARTY: _____ REQUIRED COMPLETION DATE: _____

COST IMPACT:
❏ NO IMPACT, INCLUDED IN CONTRACT _____
❏ IMPACT, NOT INCLUDED IN CONTRACT. $ _____

REQUIRED COMPLETION DATE: _____

AUTHORIZATION FORM

CITY OF _____

AUTHORIZED

Address _____

Type of Work _____

Date of Permit Issued _____

Date Permit Expires _____

Building Commissioner _____

INSPECTION APPROVALS

1. Sanitary Lateral _____ Date _____
2. Footing _____ Date _____
3. Plumbing Ground Rough _____ Date _____
4. Plumbing Rough _____ Date _____
5. Electrical Rough _____ Date _____
6. Building Framing _____ Date _____
7. Mechanical Systems _____ Date _____
8. Plumbing Final _____ Date _____
9. Electrical Final _____ Date _____
10. Building Final _____ Date _____

Any Questions Regarding Above Call City Hall

ENGLISH SYSTEM

		UNIT	ABBREVIATION	EQUIVALENTS
LENGTH		mile	mi	5280?, 320 rd, 1760 yd
		rod	rd	5.50 yd, 16.5?
		yard	yd	3′, 36?
		foot	ft or ?	12″, .333 yd
		inch	in. or ?	.083′, .028 yd
AREA $A = l \times w$		square mile	sq mi or mi^2	640 A, 102,400 sq rd
		acre	A	4840 sq yd, 43,560 sq ft
		square rod	sq rd or rd^2	30.25 sq yd, .00625 A
		square yard	sq yd or yd^2	1296 sq in., 9 sq ft
		square foot	sq ft or ft^2	144 sq in., .111 sq yd
		square inch	sq in. or in^2	.0069 sq ft, .00077 sq yd
VOLUME $V = l \times w \times t$		cubic yard	cu yd or yd^3	27 cu ft, 46,656 cu in.
		cubic foot	cu ft or ft^3	1728 cu in., .0370 cu yd
		cubic inch	cu in. or in^3	.00058 cu ft, .000021 cu yd
CAPACITY WATER, FUEL, ETC.	*U.S. liquid measure*	gallon	gal.	4 qt (231 cu in.)
		quart	qt	2 pt (57.75 cu in.)
		pint	pt	4 gi (28.875 cu in.)
		gill	gi	4 fl oz (7.219 cu in.)
		fluidounce	fl oz	8 fl dr (1.805 cu in.)
		fluidram	fl dr	60 min (.226 cu in.)
		minim	min	1/6 fl dr (.003760 cu in.)
VEGETABLES, GRAIN, ETC.	*U.S. dry measure*	bushel	bu	4 pk (2150.42 cu in.)
		peck	pk	8 qt (537.605 cu in.)
		quart	qt	2 pt (67.201 cu in.)
		pint	pt	1/2 qt (33.600 cu in.)
DRUGS	*British imperial liquid and dry measure*	bushel	bu	4 pk (2219.36 cu in.)
		peck	pk	2 gal. (554.84 cu in.)
		gallon	gal.	4 qt (277.420 cu in.)
		quart	qt	2 pt (69.355 cu in.)
		pint	pt	4 gi (34.678 cu in.)
		gill	gi	5 fl oz (8.669 cu in.)
		fluidounce	fl oz	8 fl dr (1.7339 cu in.)
		fluidram	fl dr	60 min (.216734 cu in.)
		minim	min	1/60 fl dr (.003612 cu in.)
MASS AND WEIGHT COAL, GRAIN, ETC.	*avoirdupois*	ton		2000 lb
		short ton	t	2000 lb
		long ton		2240 lb
		pound	lb or #	16 oz, 7000 gr
		ounce	oz	16 dr, 437.5 gr
		dram	dr	27.344 gr, .0625 oz
		grain	gr	.037 dr, .002286 oz
GOLD, SILVER, ETC.	*troy*	pound	lb	12 oz, 240 dwt, 5760 gr
		ounce	oz	20 dwt, 480 gr
		pennyweight	dwt or pwt	24 gr, .05 oz
		grain	gr	.042 dwt, .002083 oz
DRUGS	*apothecaries'*	pound	lb ap	12 oz, 5760 gr
		ounce	oz ap	8 dr ap, 480 gr
		dram	dr ap	3 s ap, 60 gr
		scruple	s ap	20 gr, .333 dr ap
		grain	gr	.05 s, .002083 oz, .0166 dr ap

METRIC SYSTEM			
	UNIT	ABBREVIATION	NUMBER OF BASE UNITS
LENGTH	kilometer	km	1000
	hectometer	hm	100
	dekameter	dam	10
	*meter	m	1
	decimeter	dm	.1
	centimeter	cm	.01
	millimeter	mm	.001
AREA	square kilometer	sq km or km²	1,000,000
	hectare	ha	10,000
	are	a	100
	square centimeter	sq cm or cm²	.0001
VOLUME	cubic centimeter	cu cm, cm³, or cc	.000001
	cubic decimeter	dm³	.001
	*cubic meter	m³	1
CAPACITY	kiloliter	kl	1000
	hectoliter	hl	100
	dekaliter	dal	10
	*liter	l	1
	cubic decimeter	dm³	1
	deciliter	dl	.10
	centiliter	cl	.01
	milliliter	ml	.001
MASS AND WEIGHT	metric ton	t	1,000,000
	kilogram	kg	1000
	hectogram	hg	100
	dekagram	dag	10
	*gram	g	1
	decigram	dg	.10
	centigram	cg	.01
	milligram	mg	.001

SKETCHING ISOMETRIC DRAWINGS

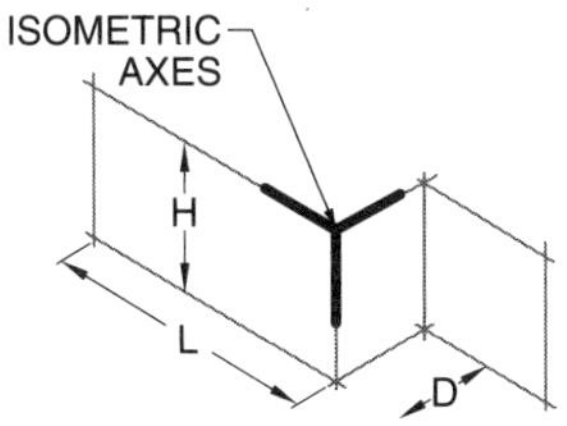

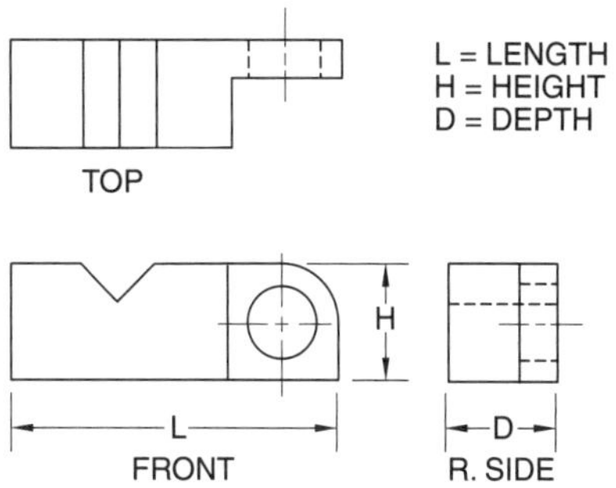

L = LENGTH
H = HEIGHT
D = DEPTH

TOP
FRONT R. SIDE

MULTIVIEW

ISOMETRIC AXES

① "Block in" front view. Use measurements from the multiview.

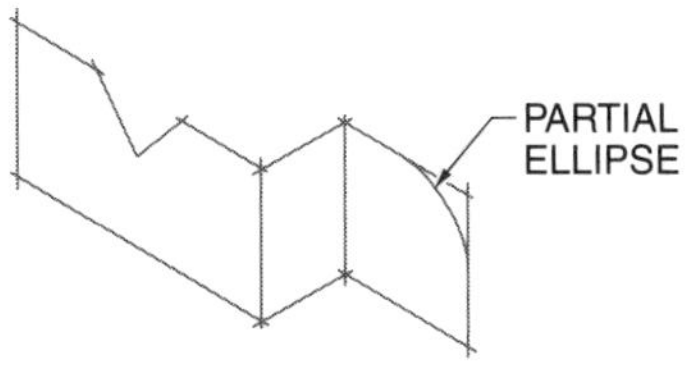

② Sketch outline shape of front view. — PARTIAL ELLIPSE

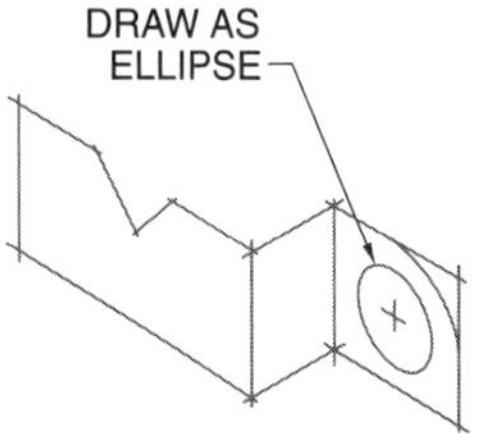

③ Locate centerpoint shown in front view. Sketch ellipse for drilled hole. — DRAW AS ELLIPSE

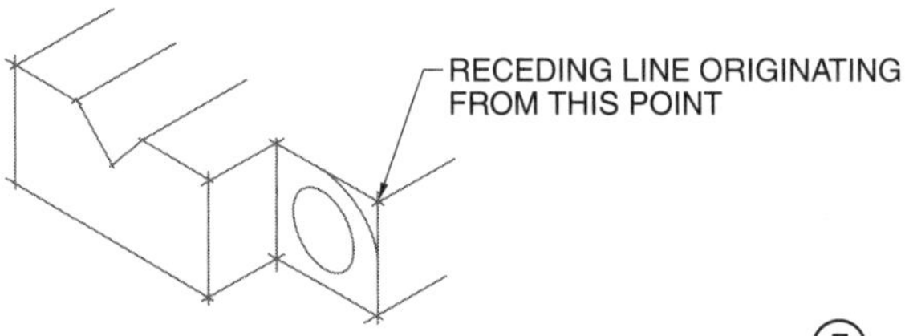

④ Draw receding lines. — RECEDING LINE ORIGINATING FROM THIS POINT

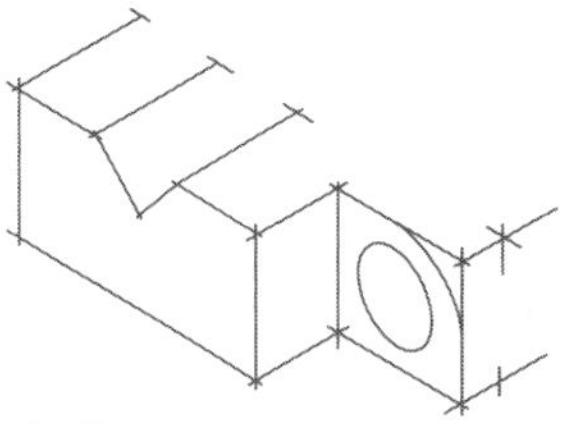

⑤ Establish depth.

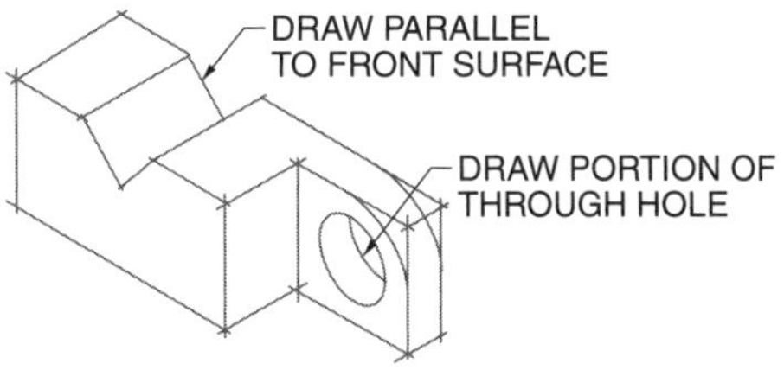

⑥ Draw lines to establish back surface. — DRAW PARALLEL TO FRONT SURFACE / DRAW PORTION OF THROUGH HOLE

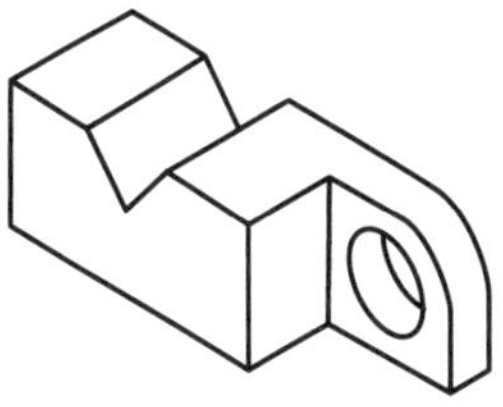

⑦ Darken all object lines.

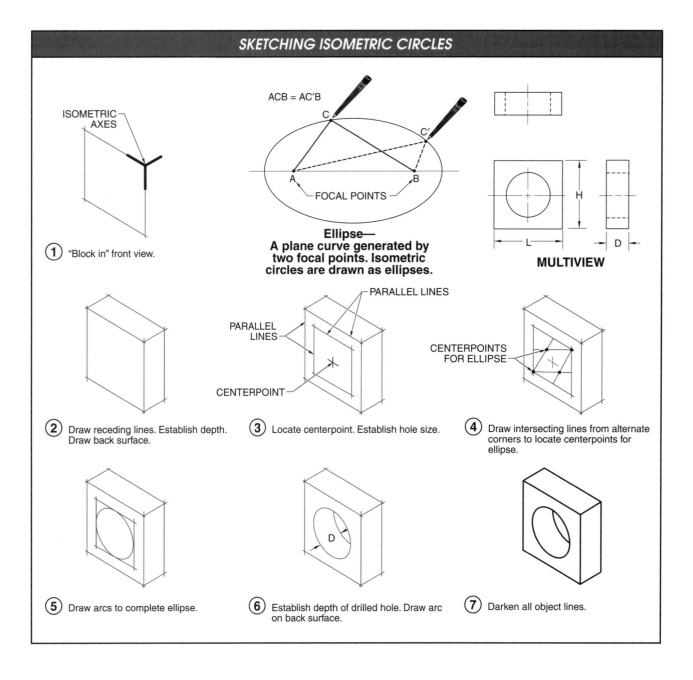

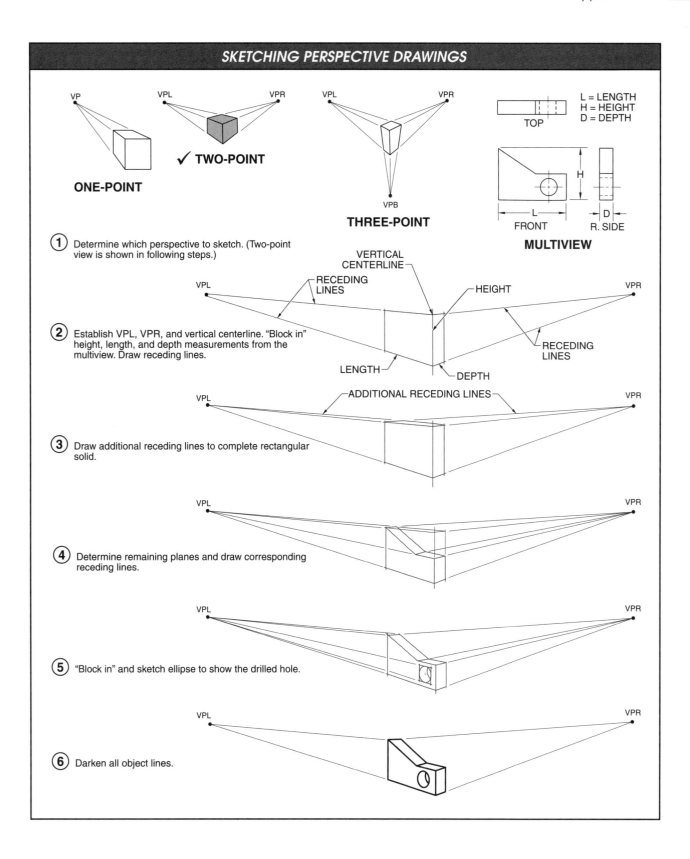

SKETCHING MULTIVIEW DRAWINGS

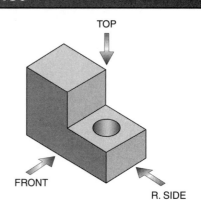

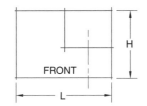

① "Block in" length and height dimensions of front view.

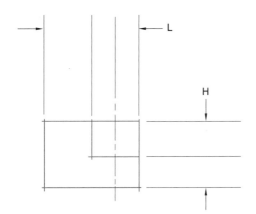

② Project length and height dimnsions of front view to top and right side views.

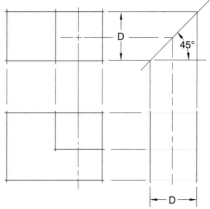

③ Establish depth dimensions of top and right side views. Note use of 45° turn.

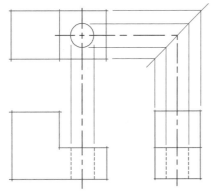

④ Sketch drilled hole in top view. Project size of hole to front and right side views and sketch hidden lines.

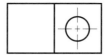

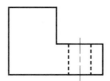

⑤ Darken all object lines. Show centerlines for drilled hole.

Glossary

A

abbreviation: A letter or group of letters that represents a term or phrase.

accumulator: A device that stores fluid under pressure.

actuator: The part of the directional control valve that changes the position of the spool in the valve.

air: The mixture of gases that are present in the atmosphere.

air muffler: A device that reduces the noise that is made when air is exhausted from a pneumatic system to the atmosphere.

alarm: An electrical component that provides an audible signal that is used as a safety feature, typically to indicate a problem or a potential problem.

American National Standards Institute (ANSI): A standards-developing organization that adopts and copublishes standards that represent the needs of its members and other standards organizations from around the world.

ampacity: The current-carrying capacity (in amperes) of an electrical component.

application drawing: A type of drawing that shows the use of a specific piece of equipment or product in an application.

assembly drawing: A type of drawing that shows as closely as possible the way individual parts or components are placed together to produce a finished piece of equipment or result.

automatically operated switch: A switch that maintains the operation of electrical circuits with little or no manual input.

autotransformer: A transformer that changes voltage levels using the same common coil for both the primary and secondary sides.

auxiliary location: An instrument location near a control room where instruments are accessible to the operators.

B

battery: A chemical power source that produces electricity through a chemical reaction between battery plates and battery acid.

behind the panel (normally inaccessible) location: An instrument mounting location behind a control panel or control cabinet.

bidirectional fluid power motor: A fluid power motor that can drive a load in both directions.

block diagram: A type of electrical drawing that shows the relationship between individual sections, or blocks, of a circuit or system.

branch circuit: The portion of a distribution system between the circuit breaker or fuse and all receptacles, lamps, and loads connected to the circuit breaker or fuse.

break: The number of separate places on the contacts where the circuit opens and closes.

break line: A line used to indicate internal features or to avoid showing continuous features of long or large objects.

bucket: A motor control center module that consists of a circuit breaker, motor starter, and circuit control components.

busduct: The metal housing of a busbar distribution system and is available in prefabricated tees, elbows, crosses, and straight sections.

busway: An electrical distribution system made up of busbars inside metal-enclosed boxes (busducts) that are available in prefabricated sections.

C

capacitance: The ability of an object to store energy in the form of an electrical charge.

capacitor: An electronic device used to store an electrical charge.

centerline: A line that locates the center of an object.

check valve: A valve that allows fluid to flow in one direction only.

coil: A winding of insulated conductors arranged to produce a magnetic field.

compound motor: A DC motor that has one stationary field connected in series with the armature and the other stationary field connected in parallel with the armature.

computer function: A function that is part of a computer-controlled system that executes logic or calculations and transmits the result.

Construction Specifications Institute (CSI): An organization that develops standardized construction specifications.

contact: The conducting part of a switch or relay that operates with another conducting part to make or break a circuit.

contact life: The number of times the contacts of a switch or relay can be actuated before malfunctioning.

contactor: An electrically operated switch that uses a low control-circuit current to energize or de-energize a high-current-load circuit.

contact position: The position of the contacts before the contacts are activated.

control loop: A combination of instruments or a combination of instruments and control elements connected together to monitor and control a specific process variable.

control loop identification system: A combination of the letter abbreviation and the control loop number.

control switch: A switch that controls the flow of current in a circuit.

control transformer: A transformer used to step-down the voltage of a power circuit to 120 VAC or 24 VAC and provide power for the control circuit loads (motor starters, lamps, and solenoids).

control valve: A valve that is part of a control loop and is opened or closed by an actuator.

control valve actuator: A component that allows a valve to be opened or closed automatically.

coolers: Heat exchangers that remove heat from, or cool, process materials and fluids.

CSA International (CSA): A product-testing organization (headquartered in Ontario, Canada) that certifies products meeting safety and performance levels set by standards.

current transformer: A transformer in which a coil is placed around a wire that carries current.

cutting-plane line: A line that indicates the path through which an object will be cut so that its internal features can be seen.

cylinder: A component that converts fluid energy into linear mechanical force.

D

dashed line method: A method of identifying the mechanically connected contacts that are on different lines of a ladder diagram by placing a dashed line between the contacts.

detail drawing: A type of drawing that provides all information needed to produce a part.

diac: A thyristor that triggers in either direction when the breakover voltage of the diac is exceeded.

dielectric material: A medium in which an electrical field is maintained with little or no outside energy supply.

digital logic gate: A circuit that performs a special logic operation such as AND, OR, NOT, NAND, or NOR.

dimension line: A line that is used with a written dimension to indicate size or location.

diode: A semiconductor device that allows current to flow in only one direction by offering very high opposition to current in one direction and very low opposition to current in the other.

directional control valve: A fluid power valve that connects, disconnects, or directs fluid flow from one part of a circuit to another.

disconnect switch (disconnect): A switch that removes electrical power from motors and machines.

discrete instrument: A stand-alone piece of hardware with a single specific function.

double-acting cylinder: A type of cylinder that provides a high force in both directions of movement and requires fluid pressure to extend and fluid pressure to retract.

double-break contact: A contact that breaks the electrical circuit in two places.

drawing: An assembly of lines, dimensions, and notes used to convey general or specific information as required by the application and use.

E

electrical detail: A precise drawing that provides in-depth information that cannot be shown on a site plan, power print, or lighting print.

electrical resistance: The opposition to electron flow of any material.

electrical service: The point of electrical connection between the local power utility and the building.

electrical tracing: A pipe heating method where an electrical heating element is fastened along the length of the process pipe.

electromechanical relay: A relay with multiple contacts that are typically used in a control circuit to switch low currents.

elevation drawing: An orthographic projection of a structure's vertical surfaces.

equipment logbook: A binder or electronic file that documents all work performed by maintenance personnel on a piece of equipment, beginning with installation.

extension line: A line that extends from the surface features of an object and is used to terminate dimension lines.

F

facility maintenance technician: A maintenance person who operates, maintains, and repairs building systems and equipment in hotels, schools, office buildings, and hospitals.

feeder: Conduit and wire, cable tray and wire, metal-clad cable, or busduct that is used to supply power to electrical loads.

field mounted location: An instrument location in the manufacturing plant in the vicinity of the process the instrument monitors or controls.

filter: A device that removes solid contaminants from a fluid power system.

filter/regulator/lubricator (FRL): A special unit that is typically located at each workstation to condition, regulate air pressure, and lubricate air from the main supply line.

fire alarm system: A fire protection system designed to emit a sound and/or signal to building occupants and fire fighting personnel on the detection of a fire by combustion monitoring devices.

fixed capacitor: A capacitor that has one value of capacitance.

fixed displacement motor: A fluid power motor that provides rotary motion that is unidirectional or bidirectional and a constant torque and speed output.

fixed displacement pump: A positive displacement pump that develops a fixed amount of flow for each revolution of the shaft.

fixed resistor: A resistor with a set value, such as 250 Ω.

floor plan: 1. A type of drawing that shows exterior walls, all room partitions, doors, windows, fireplaces, stairs, bathrooms, cabinetry, and any fixtures or appliances. **2.** A drawing that provides a plan view of each floor of a building.

flow control valve: A fluid power valve used to control the volume of fluid traveling through a system.

fluid: A liquid or a gas that can move and change shape without separating when under pressure.

fluid conditioner: A device that maintains clean fluid in a system.

fluid power: The transmission and control of energy by means of a pressurized fluid.

fluorescent lamp: A low-pressure discharge lamp in which the ionization of mercury vapor transforms ultraviolet energy into light.

foreman: A worker who manages labor and material for the portion of the project relevant to the foreman's trade.

foundation plan: A type of drawing that indicates a building's foundation, structural supports, dimensions, and building materials.

four-way directional control valve: A fluid power valve used to control the movement of a double-acting cylinder and bidirectional motor.

four-way switch: Double-pole, double-throw (DPDT) switch that is used between two three-way switches when controlling a lamp from three or more switch locations.

front view: The view when looking directly at an object from the same height as the object.

G

general foreman: A foreman who supervises a group of project foremen.

general note: A note that applies to the entire print that the note appears on.

general wiring: The wiring used to connect electrical components in a circuit.

generator: A magnetic power source that produces electricity when a rotating conductor cuts magnetic lines of force.

ground: The connection of all exposed non-current-carrying metal parts of an electrical system to earth.

ground-fault circuit interrupter (GFCI): A type of receptacle designed to help protect people by detecting ground faults and quickly disconnecting the power from a circuit that has developed a ground fault.

H

hall effect sensor: A type of sensor that detects the proximity of a magnetic field.

heaters: Heat exchangers that add heat to process materials and fluids.

heat exchanger: A device that removes heat from fluid in a hydraulic or pneumatic system.

heat tracing: A method used to heat process piping.

hidden line: A line that represents the shape of an object that cannot be seen.

high-intensity discharge (HID) lamp: A lamp that produces light from an arc tube similar to the tungsten filament of an incandescent bulb.

HVAC wiring diagram: A diagram that shows the connections between sensors, actuators, HVAC control panels, motor starters, VFDs, and mechanical equipment control panels.

hydraulic system: A system that transmits and controls energy using a liquid (typically oil).

I

incandescent lamp: A gas-filled bulb that produces light using the flow of current through a tungsten filament.

inductance: The opposition to a change of current in an AC circuit.

industrial maintenance technician: A maintenance person who operates, maintains, and repairs production systems and equipment in industrial settings.

instructional drawing: A type of drawing that is intended to indicate how to do work using the simplest and/or safest method.

interconnecting diagram: A type of electrical drawing that shows the external connections between all system devices and components.

International Organization for Standardization (ISO): An international standards-developing organization that develops standards for worldwide use.

instrument controller: An instrument that is part programmable logic controller and part sensor, has multiple inputs and outputs, and contains a program that can be modified.

isolated ground receptacle: A type of receptacle that provides a grounding path for each individual outlet that is not part of the normal grounding system used by standard receptacles.

L

ladder (line) diagram: A drawing that typically shows, using multiple rungs and graphic symbols, the logic of an electrical control circuit.

landscape plan: A type of drawing that indicates land contours along with buildings and other structural information so that landscaping designs can be created.

layout and location drawing: A drawing that helps locate devices and components for testing during troubleshooting and for system modification.

leader line: A line with a bent knee that connects a written description such as a dimension, note, or specification with a specific feature of a drawn object.

life safety system: A building or facility system that aids occupants and fire fighting personnel in the event of a fire, terrorist attack, or other catastrophe that could endanger the lives of building personnel.

lighting floor plan: A print that shows light fixtures, exit lights, switches, panels, and transformers.

limit switch: A switch that detects the physical presence of an object and is typically used as a safety device.

line: A straight mark that begins at a starting point and stops at an endpoint.

liquid: A fluid substance that can flow readily and assume the shape of its container.

load: Any electrical component in a line diagram that consumes electrical power from L1.

location drawing: A type of drawing used to show the position of switches, buttons, terminal connections, and other features found on a device or component.

lockout: The process of removing the source of power and installing a lock that prevents the power from being turned ON until the lock is removed.

lubricator: 1. A device that adds lubrication (oil) to a pneumatic system. 2. A device that is used to add small droplets of oil to compressed air.

luminaire: A complete unit consisting of a lamp or lamps, the parts that connect the lamp to the power source, and the parts that distribute the light.

M

manually operated switch: A switch that is activated by a person and is the most common type of switch used in control circuits.

manual valve: A valve that is opened or closed by hand.

motor: 1. A component that develops a rotating mechanical force (torque) on a shaft, which is used to produce work. 2. A hydraulic or pneumatic component that converts fluid energy into rotary mechanical energy.

motor control center (MCC): 1. A sheet metal enclosure that houses and protects fuses or circuit breakers, motor starters, overloads, and wiring. **2.** An electrical panel that is fed power from a power distribution panel or electrical service panel and contains several individual control units designed specifically for controlling motors.

motor starter: An electrically operated switch (contactor) for use with motors that includes 3φ motor overload protection.

mounting and installation detail drawing: A product-specific drawing that provides additional information to an electrician about product mounting and installation.

N

National Electrical Manufacturers Association (NEMA): An organization that develops technical standards and government regulations for electrical equipment.

National Fire Protection Association (NFPA): A world leader in writing codes and standards for fire prevention and public safety.

negative temperature coefficient (NTC) thermistor: A thermistor whose resistance value decreases with an increase in temperature.

non-positive-displacement (centrifugal) pump: A pump that circulates or transfers fluid using rotational speed.

normally closed (NC) valve: A valve that does not allow fluid to flow in the spring-actuated (normal) position.

normally open (NO) valve: A valve that allows fluid to flow in the spring-actuated (normal) position.

note: A sentence or two that provides drawing information that does not fit within the space of the drawing.

numerical cross-reference method: A method of identifying various mechanically connected contacts of a device that are on different lines of a complex ladder diagram by using special symbols.

O

object line: A line that indicates the visible shape of an object.

oil jacketing: A pipe heating method where an oil jacket surrounds the process piping.

one-line diagram: An electrical drawing that uses a single line and basic symbols to show the current path, voltage values, circuit disconnect, overcurrent protection devices, transformers, and panelboards for a circuit or system.

operational diagram: An electrical drawing that shows the operation of individual devices and components used in circuits.

orthographic projection: A type of drawing where all faces (front, top, bottom, and side) of an object are projected onto flat planes that generally are at 90° right angles to one another.

oscillator: A type of motor actuator that moves in fixed rotational increments each time fluid pressure is applied, then automatically reverses in direction.

overcurrent protection device (OCPD): A circuit breaker or fuse that provides overcurrent protection to a circuit.

P

panel enclosure detail: A detail drawing that provides an elevation view of the front and side of a control system panel.

panel layout detail: A detail drawing that provides an overall view of the components mounted inside a control panel.

panel mounting detail: A detail drawing that provides an enlarged view of the mounting supports of a control panel.

permanent-magnet motor: A DC motor that has armature connections but no field connections.

phototransistor: A transistor that controls the amount of current flowing through the emitter to the base junction based on the amount of light encountered.

photovoltaic cell: A solar power source that produces electricity when light strikes the surface of the cell.

pictorial drawing: Any three-dimensional drawing that resembles a picture.

pilot light: An electrical component that provides a visual indication of the presence or absence of power in a circuit.

piping and instrumentation diagram (P&ID): A print that depicts equipment that is essential to a process.

plan: A drawing of an object as it is viewed from above.

PLC programming diagram: A type of ladder diagram that is created on a computer and downloaded to a programmable logic controller (PLC).

plot plan: An aerial view of one building lot and provides specific information about the lot.

pneumatic system: A system that transmits and controls energy using a gas such as air.

pole: The number of isolated circuit contacts that are used to activate individual circuits.

position: An envelope within a valve in which the spool can be placed to direct fluid flow through the valve.

positive displacement pump: A pump that delivers a finite quantity of fluid for each revolution of the shaft.

positive temperature coefficient (PTC) thermistor: A thermistor whose resistance value increases with an increase in temperature.

power distribution panel: A panel that can be located anywhere in a building to serve as a nearby power source for branch-circuit panels, transformers, or motor control centers.

power floor plan: A print that shows all the power circuits for a specific floor of a building.

power panel: A wall-mounted distribution cabinet used in large commercial and industrial buildings.

predictive maintenance (PDM): The monitoring of wear conditions and equipment characteristics against a predetermined tolerance to predict possible malfunctions or failures.

pressure gauge: A device used to indicate the pressure at a given point in a fluid power system.

pressure-reducing valve: A pressure control valve that reduces the pressure in one leg of a circuit.

pressure regulator: A pressure valve that controls the pressure in one leg of a system.

pressure-relief valve: A pressure control valve that limits the maximum pressure in a fluid power system.

preventive maintenance (PM): A combination of unscheduled and scheduled work required to maintain equipment in peak operating condition.

primary location: An instrument location, typically in a control room, where the instrument is accessible to operators who monitor and control the process.

print convention: An agreed-upon method of displaying information on prints.

prints: Reproductions of original drawings created by an architect or engineer.

programmable logic controller (PLC): A solid-state control device that can be programmed and reprogrammed to control and monitor electrical circuits.

pump: A fluid power component that converts mechanical energy into hydraulic or pneumatic energy by pushing fluid into a system.

punch list: A formal document generated by an architect listing items that a contractor missed, partially completed, or did not complete per the prints and specifications.

R

rectification: The process of changing AC electricity into DC electricity.

reflected ceiling plan: A plan view with the viewpoint of a mirror placed on the floor to reflect ceiling-mounted objects.

regulatory agency: A federal, state, or local government organization that establishes rules and regulations related to safety, equipment installation, equipment operation, and health.

relay: An interface device that controls one electrical circuit by opening and closing contacts with another low-voltage circuit.

request for information (RFI): A formal document generated by a contractor requesting information about items that are missing or unclear on a print or specification.

resistance wire: Wire with a fixed resistance that is designed not to melt while providing high heat.

resistor: A component with a specific amount of electrical resistance.

S

schedule: A chart used to conserve space and display information in a concise and organized format.

schematic diagram: An electrical drawing that shows the electrical connections and functions of a specific circuit arrangement using graphic symbols.

sectional drawing: A type of drawing that indicates the internal features of an object.

section line: A line that identifies the materials cut by a cutting plane line in a section view.

sequence valve: A pressure control valve that directs fluid from one part of a circuit to another part of the circuit to sequence the movement of cylinders, motors, and/or valves.

series motor: A DC motor that has the stationary field connected in series with the armature and produces the highest torque of all DC motors.

service panel: A power panel used for a residential structure.

shared display–shared control instrument: A stand-alone piece of hardware that provides the functions of both display and control.

sheet note: A note that applies to a specific item in the drawing that the note appears with.

shunt motor: A DC motor that has the stationary field connected in parallel with the armature and is used where constant or adjustable speed is required.

silicon-controlled rectifier (SCR): A three-terminal semiconductor thyristor that is normally open until a signal applied to the gate terminal switches the SCR into the conducting state for one direction.

single-break contact: A contact that breaks the electrical circuit in one place.

site plan: A drawing that depicts a complete building site, the layout of the planned buildings, and all of the utility items installed below ground.

solenoid: An electrical output component that converts electrical energy into linear mechanical force.

solid-state relay: A relay that uses electronic switching devices, such as SCRs and triacs, in place of mechanical contacts to switch current flow.

specification: Additional information that is included with a set of prints.

steam tracing: A pipe heating method where a steam jacket surrounds the process piping.

step-down transformer: A transformer in which the secondary-coil output voltage is less than the primary-coil input voltage.

step-up transformer: A transformer in which the secondary-coil output voltage is greater than the primary-coil input voltage.

Standards Council of Canada (SCC): An organization that aids in the development and use of standards.

standards organizations: National and international organizations that work with governmental standards groups.

structural plan: A type of drawing that indicates the type, amount, placement, and fabrication of all materials used as structural supports of a building, bridge, or other structure.

survey plan: A type of drawing that is prepared by a licensed surveyor or civil engineer that accurately provides land contour information, dimensions, and other important feature information about a piece of property and adjacent properties.

switch: A device used to control the flow of current in an electrical circuit.

switchboard: A power panel that is freestanding (not wall mounted).

symbol: A graphic representation of a device, component, or object on a print.

T

tagout: The process of placing a danger tag on the source of electrical power, which indicates that the equipment cannot be operated until the person who placed the tag removes it.

technical society: An organization that is composed of groups of engineers and technical personnel united by a professional interest, such as creating standards.

thermal overload: A device that detects the amount of current flowing in a motor circuit by sensing the heat generated by the current flow.

thermocouple: A heat power souce that produces electricity when two different metals that are joined together are heated.

three-way directional control valve: A fluid power valve used to start and stop the flow of a liquid or gas, allow fluid to return to the reservoir in a hydraulic system, or allow air to exhaust to atmosphere in a pneumatic system.

three-way switch: Single-pole, double-throw (SPDT) switch that is used to control lamps from two switch locations.

throw: The number of closed contact positions per pole.

thyristor: A solid-state switching device that turns current ON when it receives a quick pulse of control current at its gate.

time log: A binder or electronic file that documents all work using job numbers and the time taken to perform the work by maintenance personnel.

timer: A control device that uses a preset time period as part of the control function.

title block: The area of a print that contains important information about the contents of the print.

trade association: An organization that represents the manufacturers of a specific type of product.

transformer: An electrical interface device that has no moving parts and is designed to change AC electricity from one voltage level to another voltage level.

transistor: A solid-state device that is used as a switch or as a signal amplifier.

triac: A three-terminal semiconductor thyristor that is triggered into conduction in either direction through a small amount of current to its gate terminal.

two-way directional control valve: A fluid power valve used to start and stop the flow of a fluid (liquid or gas) and has two ports located on the valve.

two-way switch: A single-pole, single-throw (SPST) switch that has an ON (closed) position and an OFF (open) position.

U

Underwriters Laboratories, Inc.® (UL): A not-for-profit product-safety testing and certification organization.

United States Military Standards (Mil Standards): Department of Defense standards used by the armed forces, but are not restricted to the armed forces.

utility plan: A type of drawing that indicates the location of and intended path of utilities such as electrical, water, sewage, gas, and communication cables.

V

variable capacitor: A capacitor that varies in capacitance value.

variable displacement motor: A fluid power motor that is unidirectional and provides a variable torque and speed output.

variable displacement pump: A positive displacement pump in which the amount of flow can be manually changed or automatically changed for each revolution of the shaft.

variable resistor: A resistor with a set range of values, such as 0 Ω to 1000 Ω.

W

wiring detail drawing: A detail drawing that provides information on terminating conductors at specific sensors or actuators.

wiring diagram: A type of electrical drawing that shows the connection of input devices and output components in a circuit.

work order: A document that describes the work a maintenance person is required to perform.

Index

A

abbreviations, 5–7, *6, 7*
   fire alarm systems, *282*
   HVAC systems, 304–306, *306*
   VDV systems, *259*
aboveground electrical distribution, 61, *61, 62*
accumulators, 443, *443*
actuators, *424,* 424–425, *428*
   defined, 428
air, 418
air mufflers, 443, *443*
alarms, 100, *100*
alternating current (AC), 81
American National Standards Institute (ANSI), 44
ampacity, 221
AND circuit logic, 339–340, *340*
ANSI, 44
application drawings, 120, *121*
architects, *181,* 181–182, *182*
architect's scale, *19,* 19–20, *20*
assembly drawings, 122–123, *124*
automatically operated switches, 87, *87*
autotransformers, 98
auxiliary locations, 458

B

behind the panel locations, 458
bidirectional fluid power motors, 424, *424*
bidirectional pumps, *420*
bill of materials, 314, *314*
block diagrams, 161, *161*
   troubleshooting, *162*
boiler system prints, *308*
branch circuits, 56
branch-circuit panels, 223–224, *224*
breaks, 85, *85*
break lines, 5, *5*
buckets, 227
building construction
   core construction, 185–186, *186*
   documentation, *190,* 190–191, *191*
   electrical systems, 186, *187, 188*
   mechanical systems, 187–189, *188, 189*
   personnel responsibilities, 181
      architects, *181,* 181–182, *182*
      building inspectors, 184, *184*
      contractors, 183, *183*
      engineers, 183, *183*
      pipefitters, 187, *188*
      sheet-metal workers, *188*
      sprinkler fitters, *188,* 189
      tradesworkers, 183–184, *184*
   process, 185
   rules and regulations, 197
   site preparation, 185, *185*
building core, 185–186, *186*
building inspectors, 184, *184*
building maintenance
   documentation, 195–197, *196*
   personnel classifications, 192, *193*
   predictive maintenance (PDM), 194, *195*
   preventive maintenance (PM), 194, *194*
   rules and regulations, 197
busduct, 56, *57*
busway, 56, *57,* 370–372, *371, 372*

C

capacitance, 91
capacitors, 91–92, *92*
centerlines, 4, *4*
centrifugal pumps, 420
check valves, 438, *439*
chemical power sources, 77, *79*
circuit logic functions, *339,* 339–344, *344*
circuit wiring, 385–386, *386, 387*
   methods, 387–391
coils, 96, *96*
combination cables, *266*
combination circuit logic, 343–344
component layout and location, *392,* 392–394, *393*
compound motors, 98, *99*
compressors, 465, *466*
   dual, *401,* 401–403
computer function, 458, *459*
condensers, *466*
conductor color-coding, *369,* 369–370
construction documentation, *190,* 190–191, *191*
Construction Specifications Institute (CSI), 22
contact life, 84
contactors, 88–89, *89*
contact positions, 84
contacts, *84,* 84–85, *85*
contractors, 183, *183*
control circuits, 325–328, *326, 344, 393*
   ladder (line) diagrams, 328–331
   logic functions, *339,* 339–344, *344*
   numbering systems, *332,* 332–338, *334, 335, 338*
   water tower application, *327*
control device connections, 331, *331*
control element symbols, 462, *462*
   control valves, 463
   manual valves, 462–463, *463*
control loops, 459
control loop identification system, 459, *460, 462*
   instrument tags, 459, *461*
control switches, 85
   automatic, 87, *87*

manual, 85, *86*
mechanical, *86*
control transformers, 325, *388*
control valves, 463, *464*
control valve actuators, 463, *464*
coolers, 466, *466*
cross-reference system, 333
   mechanical contacts, 336–338, *337, 338*
CSA International (CSA), 44
CSI MasterFormat™, 22–25, *23, 24, 25*
   fire alarm systems, 279, *280*
   HVAC systems, 303, *304*
   VDV systems, 255
current transformers, 98
customer power distribution, *354*
cutting-plane lines, 4, *4*
cylinders, 424, *424*

D

dashed line method, 338, *338*
detail drawings, 122, *123, 233,* 233–237
   fire alarm systems, 288–291, *289, 290*
   HVAC systems, 311–314
     bill of materials, 314, *314*
     panel, 311–312, *312*
     wiring, 313, *313*
   mounting and installation, 235, *236*
   security systems, 293–294, *294*
   VDV systems, 266–269, *267, 268, 269*
diacs, *93*, 94
diagrams, 237, *238*
   block, 161, *161, 162*
   function-block, 162, *163*
   interconnecting, 158, *159*
   ladder (line), *153,* 153–154, *154, 157,* 328–331, *329, 330*
   manufacturer-provided, *331*
   one-line, 151, *152, 220,* 220–227
   operational, *160,* 160–161
   piping and instrumentation (P&ID), 453, *454*
   PLC programming, *155,* 155–156
   schematic, 157–158, *158*
   single-line, 151, *152, 220,* 220–227
   VDV system, 260, *261*
   wiring, 156, *156, 157*
dielectric material, 91

digital logic gate, *95,* 95–96
dimension line, 3, *3*
diode, 92, *93*
direct current (DC), 81–82
direct hardwiring, *387,* 387–391, *394,* 394
directional control valve, *426,* 426–427
   actuators, 428, *428*
   defined, 426
   normally open/closed, 428–438, *429, 430*
   positions, *426,* 426–427
   ways, 427, *427*
disconnect switches (disconnects), *82,* 82–83
discrete instruments, 458, *459*
double-acting cylinders, 424, *424*
double-break contacts, 85
drawings, *2,* 117, *118*
   application, 120, *121*
   assembly, 122–123, *124*
   detail, 122, *123, 233,* 233–237
   electrical elevation, *234,* 234–235, *235*
   elevation, 126, *127*
   instructional, *125,* 125–126, *126*
   layout and location, *392, 392, 393*
   location, 121, *122*
   orthographic, *119,* 119, *120*
   pictorial, 117, *118*
   sectional, *128,* 128, *129*
dual compressors, *401,* 401–403

E

electrical detail, *233,* 233–234
   diagrams, 237, *238*
   elevation drawings, *234,* 234–235
   mounting and installation details, 235, *236*
   schedules, 236, *237*
electrical panels, 56, *57*
electrical power, 61, *61, 62*
electrical prints, *375*
electrical resistance, 90
electrical service, 221, *221*
electrical tracing, 456, *456*
electric motor drives, *390,* 404, *405*
electromechanical relays, 87
elevation drawings, 126, *127*
   electrical, *234,* 234–235, *235*
energy transmission system, *417*

engineers, 183, *183*
engineer's scale, *19,* 20
equipment logbooks, 196, *196*
extension lines, 3, *3*

F

facility distribution systems, *356*
facility maintenance technicians, 192
feeders, *222,* 222–223
field mounted locations, 458
filters, 421, *422*
filter/regulator/lubricator (FRL), 423, *423*
fire alarm systems, 279–291, *281*
   abbreviations, *282*
   CSI MasterFormat™ division, 279, *280*
   detail drawings, 288–291, *289, 290*
   legends, *287*
   symbols, 280–281, *282, 283*
fixed capacitors, 91
fixed displacement motors, 424, *424*
fixed displacement pumps, 420, *420*
fixed resistors, 90
fixture schedule
   lighting, *231,* 231–232, *232*
floor plans, *130, 132,* 132–133, *133*
   fire alarm systems, 285–287, *286*
   lighting and power, 215–220, *217*
   security systems, 292, *293*
   VDV systems, 261–266, *262*
flow control valves, *425,* 438, *438*
fluids, 416–417
fluid conditioner, 421–423, *422, 423*
   filter/regulator/lubricator (FRL), 423, *423*
   hydraulic systems, 422, *422*
   pneumatic systems, *422,* 422–423
fluid power, 416–417, *417, 443*
   actuators, *424,* 424–425
   check valves, 438, *439*
   conditioners, 421–423, *422, 423*
   devices, 442–443, *443*
   directional control valves, *426,* 426–428, *427, 428*
   flow control valve, 438, *438*
   hydraulic systems, 417, *417, 418*
   normally open/closed valves, 428–438, *429*
   pneumatic systems, 418–420, *419*
   pumps, 420, *420, 421*
   symbols, 415, *416, 443*

fluorescent lamps, 49, *50*
foreman, 184
foundation plans, *134,* 134–135
four-way directional control valves, 434–435, *435*
four-way switches, 50, *52*
FRL, 423, *423*
function-block diagrams, 162, *163*

G

general foreman, 184
general notes, 12, *12*
general wiring, 101, *102*
ground-fault circuit interrupter (GFCI), 54, *55*
ground(ing), *362*

H

Hall effect sensors, 120, *121*
hardwiring, *387,* 387–391, *390,* 394, *394*
  dual compressors, *401,* 401–403
  electric motor drives, 404, *405*
  motor reversing circuits, 396–397, *397, 398*
  PLC wiring, *391,* 404–406, *405*
  pump motor control circuits, 403–404
  terminal strips, *399,* 399–404, *400*
  wiring variations, 395–396
hash marks, 218, *218*
heaters, 466, *466*
heat exchanger, 421, 466, *422, 466*
heat power source, 80, *80*
heat tracing, 456, *456*
hidden lines, 3, *3*
high-intensity discharge (HID) lamps, 50, *51*
high-phase marking, 362–363
homeruns, *218*
HVAC systems, 303–316
  abbreviations, 304–306, *306*
  CSI MasterFormat™ division, 303, *304*
  detail drawings, 311–314
  prints, *305, 308,* 308–309
  sequence of operation, 315, *316*
  symbols, 306, *307*
  wiring diagrams, 309–310, *310, 311*
hydraulic system, 417, *417, 418*
  fluid conditioners, 421–423, *422*

I

incandescent lamps, 48–50, *49*
inductance, 96
industrial electrical prints, 77–102, *78*
industrial maintenance technicians, 192
installation detail drawing, 235, *236*
instructional drawings, *125,* 125–126, *126*
instrument controller, 458, *459*
instrument lines, 456–457, *457*
instruments, 457, *458*
  control loop identification, 459–461, *460, 462*
  type and location symbols, 458, *459*
instrument tags, 459, *461*
interconnecting diagrams, 158, *159*
International Organization for Standardization (ISO), 44
isolated ground receptacles, 52, *54*

L

ladder (line) diagrams, *153,* 153–154, *157,* 328
  connections, 329–330, *330*
  control device connections, 331, *331*
  load, 328–329, *329*
  rules, *403,* 404
landscape plans, 132
layout and location drawing, 392, *392, 393*
leader lines, 3, *3*
life safety systems. *See* fire alarm systems
lighting floor plan, 228–231, *229*
lighting prints, 228
lighting symbols, 47–50, *48*
  aboveground and underground, 61, *61*
limit switches, 86
line, 2–5
  instrument lines, 456–457, *457*
  piping, 443, *443*
  piping and instrumentation, 454–455, *455*
  process lines, *455,* 455–456
line diagrams. *See* ladder (line) diagrams
line reference number, *332,* 332–333
liquids, 417

load
  connections, 329–330, *330*
  control circuit power lines, 328–329, *329*
  control device connections, 329–331, *331*
location drawing, 121, *122*
lockout, 83, *83*
logic functions, *339,* 339–344, *344*
lubricator, 421, *422,* 423
luminaire, 232, *232*

M

magnetic power source, *79,* 79–80
maintenance documentation, 195–197, *196*
maintenance personnel, 192, *193*
maintenance technicians, 192
manually operated switches, 85, *86*
manual shutoff valves, 443, *443*
manual valves, 462–463, *463*
manufacturer-provided drawings, *344*
manufacturer terminal numbers, *337*
mechanically operated switches, *86*
meters, *101*
Mil Standards, 43, *43*
motors, 98, *99*
  fluid power application, *424,* 424–425
motor control center (MCC), 123, 226–228, *227*
motor control circuits
  dual compressors, *401,* 401–403, *402*
  electric motor drives, *390,* 404, *405*
  pumps, *403,* 403–404
motor drives, *390,* 404, *405*
motor speed, *425*
motor starters, 89, *89*
mounting and installation detail drawings, 235, *236*

N

NAND circuit logic, 343, *343*
National Electrical Manufacturers Association (NEMA), 45, *45*
National Fire Protection Association (NFPA), 45, *45*
negative temperature coefficient (NTC) thermistor, 91, *91*
non-positive-displacement (centrifugal) pump, 420

NOR circuit logic, 342, *342*
normally closed (NC) valve, 428–429, *429*, *430*
normally inaccessible locations, 458
normally open (NO) valve, 428–429, *429*, *430*
NOT circuit logic, 340, *341*
notes, 11–12, *12*, *13*
numbering systems
　for control circuits, *332*, 332–338, *334*, *335*
　numerical cross-reference, 333, *334*, *338*
　method, 338, *338*

O

object lines, 2, *2*
oil jacketing, 456, *456*
one-line diagrams. *See* single-line diagrams
operational diagrams, *160*, 160–161
operator stations, *373*
OR circuit logic, 340, *341*
orthographic projection, 119, *119*, *120*
oscillators, 425, *426*
overcurrent protection device (OCPD), 82–83, *83*

P

panelboard, 54, *55*
panel detail drawings, 311–312, *312*
panel enclosure detail, 312
panel layout detail, 311
panel mounting detail, 312
permanent-magnet motors, 98, *99*
phase arrangement, 362–363, *363*
phase markings, 362–363
phototransistors, 95
pictorial drawings, 117, *118*
pilot lights, 100, *100*
piping, 443, *443*
piping and instrumentation diagram (P&ID), 453, *454*
　compressors, 465, *466*
　control element identification, 459–461, *460*, *461*
　control element symbols, *462*, 462–463
　instruments, 457–461, *458*, *459*

lines, 454–455, *455*
process equipment symbols, *464*, 464–466
process lines, *455*, 455–456
pumps, 465, *466*
tanks, 464, *465*
vessels, 464, *465*
plans, 129, *130*
　floor, *130*, 132–133, *133*, *215* 215–220,
　foundation, *134*, 134–135
　site, 129–132, *130*, *131*, 213–214, *214*
　structural, 135, *135*
　utility, 136, *136*
PLC programming diagrams, *155*, 155–156
PLC wiring, *391*, 404–406, *405*
plug configuration, 359–360, *360*, *362*
pneumatic systems, 418–420, *419*
　fluid conditioners, 421–423, *422*, *423*
poles, 84, *85*
positions, *426*, 426–427
positive displacement pumps, 420
positive temperature coefficient (PTC) thermistors, 91, *91*
power circuits, 325–328, *326*, *327*, *344*, *393*
　application, *372*, 372–376
power distribution, 353, *354*, *355*
power distribution panels, 223, *223*
power distribution systems, 353–357
　busways, 370–372, *371*, *372*
　conductor color-coding, *369*, 369–370
　customer power, *354*, *356*
　grounding, 361, *362*
　high-phase marking, 362
　phase arrangement, 362–363, *363*
　plug-to-receptacle configurations, 359–360, *360*, *361*
　service entrances, *358*, 358–359, 363–369, *364*, *366*, *368*
　utility company power distribution, *354*
power floor plan, 215, *216*
　commercial, 216–218
　residential, *219*, 219–220
power panels, 56, *57*
power prints, 214
　floor plans, 215–220

　single-line diagrams, *220*, 220–227
power sources, 77, 79–80, *79*, *80*, *81*
power symbols, 54–56, *55*, *56*, *57*
power transmission, *355*
predictive maintenance (PDM), 194, *195*
pressure control valves, *439*, 439–442, *440*
pressure gauges, 443, *443*
pressure-reducing valves, 441–442, *442*
pressure regulators, 423
pressure-relief valves, 439, *440*
preventive maintenance (PM), 194, *194*
primary locations, 458
print conventions, 11
　column numbers and letters, 17, *18*
　notes, 11–12, *12*, *13*
　detail drawing symbols, 13, 16, *14*, *16*, *17*
　section view symbols, 13–15, *14*, *15*
print divisions, 9, *9*
prints, 1–2, *2*, 115–117, *116*
　commercial, 45–62, *46*
　conflict resolution, 191, *191*
　defined, 1
　fire alarm systems, 280–291
　HVAC sequence of operation, 315, *316*
　HVAC systems, *305*, *308*, 308–314
　industrial, 77–102, *78*
　lighting, 228–232, *229*, *230*
　power, 214–227, *216*, *217*, *219*
　residential, 45–62, *46*
　VDV systems, 255–269, *256*
print scales, *18*, 18–19
　architect's, *19*, 19–20, *20*
　engineer's, *19*, 20
process equipment symbols, *464*, 464–466, *465*, *466*
process lines, *455*, 455–456
　heat tracing symbols, 456, *456*
programmable logic controller (PLC), 155
project conflict resolution, 191, *191*
property symbols, 58, *59*
　private, 58, *60*
　public, 60, *60*
pump motor control circuits, *403*, 403–404
pumps, 420, *420*, 465, *466*

fixed displacement, 420, *420*
non-positive displacement, 420
positive displacement, 420
variable displacement, 420, *420*
punch lists, 190, *191*

R

receptacle configuration, 359–360, *360, 362*
receptacle symbols, 52–54, *53*
rectification, 81
rectifiers, *81*
reflected ceiling plans, 228, *230*, 287, *288*
regulatory agency, 197
relays, 87–88, *88*
request for information (RFI), 190, *190*
resistance wires, 90
resistors, *90*, 90–91
reversing motor controls
  direct hardwiring, 396–397, *397, 398*
  PLC wiring, 404–406, *405*
riser diagram
  fire alarm systems, 283–285, *284, 285*
  VDV systems, 260, *261*

S

scales, 18–20
  architect's, *19*, 19–20, *20*
  engineer's, *19*
schedules, 8, *8*, 236, *237*
  for light fixtures, *231*, 231–232
schematic diagrams, 157–158, *158*
sectional drawings, 128, *128*
  orthographic, 128
  pictorial, 129
section lines, 4, *4*
security systems, 291–294
  abbreviations, *291*
  detail drawings, 293–294, *294*
  floor plans, 292, *293*
  symbols, 291, *292*
sequence of operation, 315, *316*
sequence valves, 440–441, *441*
series motors, 98, *99*
service entrances
  120/208 V, 3ϕ, 4-wire, 363–365, *364*

120/240 V, 1ϕ, 3-wire, *358*
120/240 V, 3ϕ, 4-wire, 367–369, *368*
277/480 V, 3ϕ, 4-wire, 365, *366*
service panels, 56, *57*
shared display—shared control instrument, 458, *459*
sheet notes, 12, *13*
shunt motors, 98, *99*
shutoff valves, 443, *443*
signaling system symbols, 58, *58*
silicon-controlled rectifier (SCR), 93, *93*
single-acting cylinders, *424*
single-break contact, 85
single-line diagrams, 151, *152*, 220, 220–221
  branch-circuit panels, 223–224
  electrical service, 221, *221*
  feeder, *222*, 222–223
  motor control center (MCC), 226–227, *227*
  power distribution panel, 223, *223*
  transformer, 224–225, *225, 226*
site plans, 129–132, *130, 131*, 213–214, *214*
  symbols, 58–60, *59*
site preparation, 185, *185*
solar power source, 80, *81*
solenoids, 96, *97*
solenoid valves, 436–438, *437*
solid-state relays, 87
specifications, *21*, 21–22
  CSI MasterFormat™, 22–25, *23, 24, 25*
Standards Council of Canada (SCC), 44
standards organizations, *44*, 44–45
  international, 44, *44*
  national, 44, *44*
  private organizations, 45, *45*
  technical societies, 42–43, *43*
  trade associations, 42, *42*
  United States government departments, *43*, 43–44
steam tracing, 456, *456*
step-down transformers, 98
step-up transformers, 98
strainers, *422*
structural plans, *135*, 135–136
survey plans, 129
switchboard, 56, *57*
switches
  defined, 50

symbols for, 50, *51–52*
symbol, *42*, 45, 47, *47*
  aboveground electrical distribution, 61, *61, 62*
  alarms, 100, *100*
  capacitors, 91–92, *92*
  coils, 96, *96*
  commercial prints, 45–62, *46*
  contactors, 88–89, *89*
  contacts, *84*, 84–85, *85*
  control elements, 462, *462*
  control switches, 85–87, *86, 87*
  defined, 45
  digital logic gates, *95*, 95–96
  diodes, 92, *93*
  disconnects, *82*, 82–83
  fire alarm system, 280–281, *282, 283*
  fluid power, 416–420, *417, 418, 419*
  HVAC systems, 306, *307*
  industrial prints, 77–102, *78*
  life safety system, 280–281, *282, 283*
  lighting, 47–50, *48*
  lights, 100, *100*
  meters, 100, *101*
  motors, 98, *99*
  motor starters, 89, *89*
  outlets, *48*
  overcurrent protection devices, 82–83, *83*
  plot plans, 58, *59, 60*
  power, 54–56, *55–56, 57*
  power sources, 77, *79*, 79–80, *80, 81*
  receptacles, 52–54, *53–54*
  relays, 87–88, *88*
  residential prints, 45–62, *46*
  resistors, *90*, 90–91
  signaling systems, 58, *58, 59*
  solenoids, 96, *97*
  switches, 50, *51–52*
  telephone outlets, *257*
  thyristors, *93*, 93–94
  timers, 88, *88*
  transformers, *97*, 97–98
  transistors, *94*, 94–95
  underground electrical distribution, 61, *61, 62*
  valves, 426–442, *427, 438, 439, 441*, 462–464, *464*
  VDV systems, 255–258, *257, 258*
  wiring, 101, *102*

T

tagout, 83, *83*
tanks, 464, *465*
technical society, 42–43, *43*
telephone outlet symbols, *257*
terminal numbers, *337*
terminal strips
　for hardwiring, *389*, *399*, 399–404, *400*
　　control variations, 404
　　dual compressor application, *401*, 401–403
　　pump application, 403–404
thermal overload, 83
thermistors, 90–91, *91*
three-conveyor ladder diagrams, 154, *154*
three-phase motors, *99*
three-position valves, *436*, 436–438, *437*
three-way directional control valves, *432*, 432–434, *433*, *434*
three-way switches, 50, *52*
throws, 85, *85*
thyristors, *93*, 93–94
time logs, 196, *197*
timers, 88, *88*
title blocks, 9, *10*
　revision information, 10, *11*
trade associations, 42, *42*
tradesworkers, 183–184, *184*
transformers, *97*, 97–94, 224–225, *225*, *226*, *388*
　voltage rating, *367*
transistors, *94*, 94–95
triacs, *93*, 94
two-way directional control valves, 429–431, *430*, *431*
two-way switches, 50

U

underground electrical distribution, 59, 61, *62*
Underwriters Laboratories, Inc. (UL), 45, *45*
unidirectional pumps, *420*
United States Military Standards (Mil Standards), 43, *43*
unshielded twisted pair (UTP) cables, *264*, 264–266, *266*

utility company power distribution, *354*
utility plans, 136, *136*

V

valve actuators, 428, *428*
valve positions, *426*, 426–427
valve types
　check, 438, *439*
　control, 463, *464*
　directional control, *426*, 426–428, *427*
　flow control, 438, *438*
　manual, 462–463, *463*
　normally open/closed, 428–438
　pressure control, *439*, 439–442, *440*
valve ways, 427, *427*
vaporizer, *466*
variable capacitors, 91, *92*
variable displacement motors, 424, *424*
variable displacement pumps, 420, *420*
variable resistors, 90
VDV systems, 255–269, *256*
　abbreviations, *259*
　CSI MasterFormat™ division, 255
　detail drawings, 266–269, *267*, *268*, *269*
　floor plans, 261–266, *262*, *263*, *265*
　legends, 258, *260*
　riser diagrams, 260, *261*
　symbols, 255–258, *257*, *258*
vessels, 464, *465*

W

water tower application, 325–328, *327*
ways, 427, *427*
wire reference numbers, 333–334, *335*
wiring, 101, *102*
wiring detail drawing, 313, *313*
wiring diagram, 156, *157*, *374*
　HVAC systems, 309–310, *310*, *311*
wiring methods, *387*, 387–391
wiring variations, 395–396
work orders, 195, *196*

USING THE *PRINTREADING FOR INSTALLING AND TROUBLESHOOTING ELECTRICAL SYSTEMS* CD-ROM

Before removing the CD-ROM from the protective sleeve, please note that the book cannot be returned for refund or credit if the CD-ROM sleeve seal is broken.

System Requirements

To use this Windows®-compatible CD-ROM, your computer must meet the following minimum system requirements:
- Microsoft® Windows Vista™, Windows XP®, Windows 2000®, or Windows NT® operating system
- Intel® Pentium® III (or equivalent) processor
- 256 MB of available RAM
- 90 MB of available hard-disk space
- 800 × 600 monitor resolution
- CD-ROM drive
- Sound output capability and speakers
- Microsoft® Internet Explorer 5.5, Firefox 1.0, or Netscape® 7.1 web browser and Internet connection required for Internet links

Opening Files

Insert the CD-ROM into the computer CD-ROM drive. Within a few seconds, the home screen will be displayed allowing access to all features of the CD-ROM. Information about the usage of the CD-ROM can be accessed by clicking on USING THIS CD-ROM. The Quick Quizzes®, Illustrated Glossary, Resource Library, Prints, CSI MasterFormat™, Virtual Motor Enclosure, Flash Cards, and ATPeResources.com can be accessed by clicking on the appropriate button on the home screen. Clicking on the American Tech web site button (www.go2atp.com) accesses information on related educational products. Unauthorized reproduction of the material on this CD-ROM is strictly prohibited.

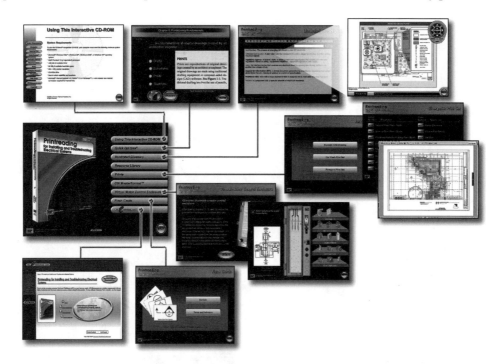

Intel and Pentium are registered trademarks of Intel Corporation or its subsidiaries in the United States and other countries. Microsoft, Windows Vista, Windows XP, Windows 2000, Windows NT, and Internet Explorer are either registered trademarks or trademarks of Microsoft Corporation in the United States and/or other countries. Adobe, Acrobat, and Reader are either registered trademarks or trademarks of Adobe Systems Incorporated in the United States and/or other countries. Netscape is a registered trademark of Netscape Communications Corporation in the United States and other countries. MasterFormat is a trademark of the Construction Specifications Institute. Quick Quiz and Quick Quizzes are registered trademarks of American Technical Publishers, Inc. All other trademarks are the properties of their respective owners.